SARS- and Other Coronaviruses

METHODS IN MOLECULAR BIOLOGY™

John M. Walker, SERIES EDITOR

METHODS IN MOLECULAR BIOLOGY™

SARS- and Other Coronaviruses

Laboratory Protocols

Edited by

Dave Cavanagh

Institute for Animal Health, Berkshire, UK

 Humana Press

Editor
Dave Cavanagh
Institute for Animal Health
Berkshire
UK
dave.cavanagh@bbsrc.ac.uk

Series Editor
John M. Walker
University of Hertfordshire
Hatfield, Hertfordshire
UK

ISBN: 978-1-58829-867-6   e-ISBN: 978-1-59745-181-9
DOI: 10.1007/978-1-59745-181-9

Library of Congress Control Number: 2008933583

Preface

The year 2003 was the year when the name "coronavirus" went around the world, somewhat further than the virus that sparked panic: severe acute respiratory syndrome coronavirus (SARS-CoV). It was spread rapidly by international, indeed transcontinental, travelers from its epicenter in China. The high mortality rate, around 10% among clinical cases, and the particularly high price paid by health care workers, spread fear globally. Public health facilities were stretched to the limit, and the effect on local economies was immense. Never before had a coronavirus made such an impact on the lives of the human population.

SARS enhanced interest in coronaviruses generally, including the hunt not only for SARS-CoV-like viruses in wild animals, but also for unrelated coronaviruses in wild and semi-domesticated animals. Prior to SARS we knew of only a dozen or so coronavirus species but it was always likely that there were many more. Coronavirologists knew, from decades of experience, that coronaviruses could be devastating, causing mortality, especially among young, and high economic loss among domestic animals. Moreover, they knew that a given coronavirus species was not limited to replication in one host species. Thus the potential existed for the spread of novel coronaviruses from wild animals to domestic animals and to man. It is with this potent threat in mind that I have included no fewer than seven chapters dealing with the detection and discovery of coronaviruses by nucleic acid approaches, with antibody-based approaches being described in three other chapters.

Although one can detect a virus these days without having to grow it in the laboratory, there is always a need to do this at some point, whether it be for the development of diagnostics and vaccines or for the study of pathogenesis, pathogenicity, variation, and other virological properties. Therefore, I have included several chapters on virus isolation and propagation.

For some purposes, e.g., structural studies (Section 3) and vaccine development, it is necessary to rigorously purify a virus after propagating it, an art that fewer virologists practice these days. Hence, two chapters contain detailed procedures for this. For other structural, and functional, studies and for raising antibodies for subsequent analytical use, it is sufficient, and, in the case of nonstructural proteins, essential, to express viral proteins individually. Several chapters address this task, including one that deals with crystallization of nonstructural proteins.

Having antibodies specific to each viral protein is immensely useful for structure-function studies, but getting good antibodies is often problematic. Hence, I have included several chapters with protocols for raising antibodies against the viral proteins, including peptides as well as proteins expressed *in vitro*, and as part of whole virions.

The penultimate section of this book comprises five approaches to the manipulation of coronavirus genomes. These technical achievements have revolutionized the study of coronavirus replication and the development of vaccines against coronaviruses. SARS demonstrated that these approaches, developed with long-known coronaviruses, could be readily adapted to the molecular cloning, and subsequent manipulation, of a new coronavirus. These protocols are here to assist those who will rise to the challenge of new coronaviruses, and to provide hard-won practical advice for those working with current coronaviruses.

Finally, there are two chapters that describe how to investigate aspects of the cell surface receptors for coronaviruses. One of these deals with the sugar moieties that frequently play a role in virus attachment and can affect pathogenicity and tropism, but which tend to be overlooked once a protein has been identified as being a receptor. The other chapter describes how recombinant vesicular stomatitis virus can be used to study coronavirus receptors and tropism, safely, and offers an approach to vaccine development.

I would like to thank all the authors in this book for their devotion to the task. I am full of admiration for what they have achieved in their research and their willingness to describe their protocols in minute detail, including the sort of practical advice that is never included in primary publications.

Although the protocols focus on coronaviruses, it seems to me that all the chapters have something to offer *every* virologist. Indeed, I believe that, among us, we have produced a protocol book for virologists in general, as well as for those, present and future, who, through choice or circumstance, work with coronaviruses.

Dave Cavanagh
Compton

Contents

Contributors

BRIAN D. ADAIR, PhD • *The Scripps Research Institute, La Jolla, CA, USA*

FERNANDO ALMAZÁN, PhD • *Centro Nacional de Biotecnología, CSIC, Department of Molecular and Cell Biology, Cantoblanco, Madrid, Spain*

MARIA ARMESTO, BSC, MSC, DPHIL • *Division of Microbiology, Institute for Animal Health, Compton Laboratory, Newbury, Berkshire, UK*

RALPH S. BARIC, PhD • *University of North Carolina, Departments of Epidemiology and Microbiology and Immunology, Chapel Hill, NC, USA*

BEN BERKHOUT, PhD • *Laboratory of Experimental Virology, Department of Medical Microbiology, Center for Infection and Immunity Amsterdam (CINIMA), Academic Medical Center, University of Amsterdam, Amsterdam, The Netherlands*

PAUL BRITTON, BSC, PhD • *Division of Microbiology, Institute for Animal Health, Compton Laboratory, Newbury, Berkshire, UK*

MICHAEL J. BUCHMEIER, PhD • *The Scripps Research Institute, La Jolla, CA, USA*

ROSA CASAIS, BSC, PhD • *Laboratorio de Sanidad Animal, Serida, Spain*

DAVE CAVANAGH, BSC, PhD, DSC • *Institute for Animal Health, Compton Laboratory, Newbury, Berkshire, UK*

FRANCESCA ANNE CULVER, BSC, PhD • *Institute for Animal Health, Compton Laboratory, Newbury, Berkshire, UK*

ANNIKEN DAABACH, BSC • *Center for Structural and Cell Biology in Medicine, Institute of Biochemistry, University of Lübeck, Lübeck, Germany*

ERIC F. DONALDSON, BS • *University of North Carolina, Departments of Epidemiology and Microbiology and Immunology, Chapel Hill, NC, USA*

LUIS ENJUANES, PhD • *Centro Nacional de Biotecnología, CSIC, Department of Molecular and Cell Biology, Cantoblanco, Madrid, Spain*

KLARA KRISTIN ERIKSSON, PhD • *Research Department, Kantonal Hospital St. Gallen, St. Gallen, Switzerland*

SHUETSU FUKUSHI, PhD • *Department of Virology I, National Institute of Infectious Diseases, Musashimurayama, Tokyo, Japan*

CARMEN GALÁN, PhD • *Centro Nacional de Biotecnología, CSIC, Department of Molecular and Cell Biology, Cantoblanco, Madrid, Spain*

STEPHANIE GLASER, MSC • *Center for Structural and Cell Biology in Medicine, Institute of Biochemistry, University of Lübeck, Lübeck, Germany*

JAMES S. GUY, PhD • *North Carolina State University, College of Veterinary Medicine, Raleigh, NC, USA*

BERT JAN HAIJEMA, PhD • *Virology Division, Faculty of Veterinary Medicine, Utrecht University, The Netherlands*

CORNELIS A. M. DE HAAN, PhD • *Virology Division, Faculty of Veterinary Medicine, Utrecht University, The Netherlands*

MUSTAFA HASOKSUZ, PhD • *Food Animal Health Research Program, Department of Veterinary Preventive Medicine, Ohio Agricultural Research and Development Center, The Ohio State University, Wooster, OH, USA; Istanbul University, Faculty of Veterinary Medicine, Department of Virology, Istanbul, Turkey*

RUTH M. HENNION, BSc • *Institute for Animal Health, Compton, Newbury, Berkshire, UK*

GEORG HERRLER, PhD • *Institute for Virology, University of Veterinary Medicine Hannover, Hannover, Germany*

ROLF HILGENFELD, PhD • *Center for Structural and Cell Biology in Medicine, Institute of Biochemistry, University of Lübeck, Lübeck, Germany*

LIA VAN DER HOEK, PhD • *Laboratory of Experimental Virology, Department of Medical Microbiology, Center for Infection and Immunity Amsterdam (CINIMA), Academic Medical Center, University of Amsterdam, Amsterdam, The Netherlands*

HÉLÈNE JACOMY, PhD • *Laboratory of Neuroimmunovirology, INRS-Institut Armand-Frappier, Laval, Québec, Canada*

MAARTEN F. JEBBINK, BSc • *Laboratory of Experimental Virology, Department of Medical Microbiology, Center for Infection and Immunity Amsterdam (CINIMA), Academic Medical Center, University of Amsterdam, Amsterdam, The Netherlands*

CHRISTINE MONCEYRON JONASSEN, PhD • *Section for Virology and Serology, National Veterinary Institute, Oslo, Norway*

BRENDA V. JONES, BA, C.BIOL, M.I.BIOL. *Institute for Animal Health, Compton, Newbury, Berkshire, UK*

ELS KEYAERTS, PhD • *Laboratory of Clinical and Epidemiological Virology, Department of Microbiology and Immunology, Rega Institute for Medical Research, University of Leuven, Belgium*

FRANCINE LAMBERT, BSc • *Laboratory of Neuroimmunovirology, INRS-Institut Armand- Frappier, Laval, Québec, Canada*

SANDRA LI, PhD STUDENT • *Laboratory of Clinical and Epidemiological Virology, Department of Microbiology and Immunology, Rega Institute for Medical Research, University of Leuven, Belgium*

CAROLYN E. MACHAMER, PhD • *Department of Cell Biology, The Johns Hopkins University School of Medicine, Baltimore, MD, USA*

DIVINE MAKIA, BSc • *Research Department, Kantonal Hospital St. Gallen, St.Gallen, Switzerland*

GABRIEL MARCEAU, MSc • *Laboratory of Neuroimmunovirology, INRS-Institut Armand-Frappier, Laval, Québec, Canada*

PAUL S. MASTERS, PhD • *Research Scientist and Chief, Laboratory of Viral Disease, Wadsworth Center, New York State Department of Health, Albany, NY, USA; Associate Professor, Department of Biomedical Sciences, School of Public Health, State University of New York at Albany, NY USA*

YVONNE VAN DER MEER, MSc, PhD • *Molecular Virology Laboratory, Department of Medical Microbiology, Leiden University Medical Center, Leiden, The Netherlands*

ELIEN MOËS, PhD STUDENT • *Laboratory of Clinical and Epidemiological Virology, Department of Microbiology and Immunology, Rega Institute for Medical Research, University of Leuven, Belgium*

RALF MOLL, PhD • *Center for Structural and Cell Biology in Medicine, Institute of Biochemistry, University of Lübeck, Lübeck, Germany*

BENJAMIN W. NEUMAN, PhD • *The Scripps Research Institute, La Jolla, CA, USA*

KAZUO OHNISHI, PhD • *Department of Immunology, National Institute of Infectious Diseases, Tokyo, Japan*

J. S. MALIK PEIRIS, MBBS, FRCPATH, DPHIL (OXON), HKAM (PATH), FRS • *Department of Microbiology, University of Hong Kong, Queen Mary Hospital, Pokfulam, Hong Kong, SAR*

AMANDA R. PENDLETON, PhD • *Department of Cell Biology, The Johns Hopkins University School of Medicine, Baltimore, MD, USA*

YVONNE PIOTROWSKI, DIPL.-BIOCHEM. • *Center for Structural and Cell Biology in Medicine, Institute of Biochemistry, University of Lübeck, Lübeck, Germany*

RAJESH PONNUSAMY, MSc • *Center for Structural and Cell Biology in Medicine, Institute of Biochemistry, University of Lübeck, Lübeck, Germany*

LEO L. M. POON, DPHIL (OXON) • *Department of Microbiology, University of Hong Kong, Queen Mary Hospital, Pokfulam, Hong Kong, SAR*

KRZYSZTOF PYRC, PhD • *Laboratory of Experimental Virology, Department of Medical Microbiology, Center for Infection and Immunity Amsterdam (CINIMA), Academic Medical Center, University of Amsterdam, Amsterdam, The Netherlands*

PETER J. M. ROTTIER, PhD • *Virology Division, Faculty of Veterinary Medicine, Utrecht University, The Netherlands*

LINDA J. SAIF, MS, PhD • *Food Animal Health Research Program, Department of Veterinary Preventive Medicine, Ohio Agricultural Research and Development Center, The Ohio State University, Wooster, OH, USA*

CHRISTEL SCHWEGMANN-WESSELS, PhD • *Institute for Virology, University of Veterinary Medicine Hannover, Hannover, Germany*

AMY C. SIMS, PhD • *University of North Carolina, Departments of Epidemiology and Microbiology and Immunology, Chapel Hill, NC, USA*

ERIC J. SNIJDER, MSC, PhD • *Molecular Virology Laboratory, Department of Medical Microbiology, Leiden University Medical Center, Leiden, The Netherlands*

FUMIHIRO TAGUCHI, DVM, PhD • *Department of Virology III, National Institute of Infectious Diseases, Tokyo, Japan*

PIERRE J. TALBOT, PhD • *Laboratory of Neuroimmunovirology, INRS-Institut Armand-Frappier, Laval, Québec, Canada*

VOLKER THIEL, PhD • *Research Department, Kantonal Hospital St. Gallen, St.Gallen, Switzerland*

YASUKO TSUNETSUGU-YOKOTA, MD, PhD • *Department of Immunology, National Institute of Infectious Diseases, Tokyo, Japan*

MARC VAN RANST, MD, PhD • *Laboratory of Clinical and Epidemiological Virology, Department of Microbiology and Immunology, Rega Institute for Medical Research, University of Leuven, Belgium*

LEEN VIJGEN, PhD • *Laboratory of Clinical and Epidemiological Virology, Department of Microbiology and Immunology, Rega Institute for Medical Research, University of Leuven, Belgium*

ANASTASIA VLASOVA, PhD • *Food Animal Health Research Program, Department of Veterinary Preventive Medicine, Ohio Agricultural Research and Development Center, The Ohio State University, Wooster, OH, USA*

ALFRED L. M. WASSENAAR, BSC • *Molecular Virology Laboratory, Department of Medical Microbiology, Leiden University Medical Center, Leiden, The Netherlands*

RIE WATANABE, DVM, PhD • *Department of Virology III, National Institute of Infectious Diseases, Tokyo, Japan*

MARK YEAGER, MD, PhD • *The Scripps Research Institute, La Jolla, CA, USA*

JESSIKA C. ZEVENHOVEN-DOBBE, BSC • *Molecular Virology Laboratory, Department of Medical Microbiology, Leiden University Medical Center, Leiden, The Netherlands*

Extra Notes from the Editor

A Guide to the Sections of this Book

Section 1: Detection and Discovery of Coronaviruses

See the following chapters, especially, for the use or detection of antibodies to detect coronaviruses:

Hasoksuz et al. (Section 1) for descriptions of ELISAs for detection of antibody and of antigen.

Lambert et al. (Section 2), for detection of coronaviruses in cell culture, using indirect immunoperoxidase staining.

Ohnishi (Section 5), for virus detection by immunofluorescence, immunoblotting, and antigencapture ELISAs.

Section 2: Isolation, Growth, Titration and Purification of Coronaviruses

See the following chapters also:

Hasoksuz et al. (Section 1) for virus growth and plaque titration.

Neuman and Buchmeier (Section 3) for virus purification.

De Haan et al. (Section 6) for virus propagation and plaque titration.

Donaldson et al. (Section 6) for virus propagation and plaque titration.

Schwegmann-Wessels and Herrler (Section 7) for virus propagation and purification.

Section 4: Expression of Coronavirus Proteins, and Crystallization

See also the following chapters:

Zevenhoven-Dobbe et al. (Section 5) for expression of coronavirus proteins in *Escherichia coli.*

Pendleton and Machamer (Section 5) for expression of coronavirus proteins in *Escherichia coli.*

I

DETECTION AND DISCOVERY OF CORONAVIRUSES

1

A Pancoronavirus RT-PCR Assay for Detection of All Known Coronaviruses

Leen Vijgen, Elien Moës, Els Keyaerts, Sandra Li, and Marc Van Ranst

Abstract

The recent discoveries of novel human coronaviruses, including the coronavirus causing SARS, and the previously unrecognized human coronaviruses HCoV-NL63 and HCoV-HKU1, indicate that the family *Coronaviridae* harbors more members than was previously assumed. All human coronaviruses characterized at present are associated with respiratory illnesses, ranging from mild common colds to more severe lower respiratory tract infections. Since the etiology of a relatively large percentage of respiratory tract diseases remains unidentified, it is possible that for a certain number of these illnesses, a yet unknown viral causative agent may be found. Screening for the presence of novel coronaviruses requires the use of a method that can detect all coronaviruses known at present. In this chapter, we describe a pancoronavirus degenerate primer-based method that allows the detection of all known and possibly unknown coronaviruses by RT-PCR amplification and sequencing of a 251-bp fragment of the coronavirus polymerase gene.

Key words: coronavirus; pancoronavirus RT-PCR; degenerate primers; sequencing; polymerase gene

1. Introduction

At present, viral culture is the "gold standard" for laboratory diagnosis of respiratory infections. Since coronaviruses are very difficult to grow in cell culture, accurate and sensitive diagnoses are not feasible by this technique. To overcome the lack of sensitivity and to obtain rapid diagnostic results, more sensitive molecular methods for the detection of human coronaviruses (HCoVs) have been developed, including reverse-transcriptase polymerase chain reaction

From: *Methods in Molecular Biology, vol. 454: SARS- and Other Coronaviruses,*
Edited by: D. Cavanagh, DOI: 10.1007/978-1-59745-181-9_1, © Humana Press, New York, NY

(RT-PCR), nested RT-PCR, and recently real-time RT-PCR *(1)*. Nevertheless, HCoVs are not often diagnosed in clinical laboratories, although the SARS epidemic and the identification of HCoV-NL63 drew attention to the clinical relevance of HCoVs. Multiplex RT-PCRs for the detection of common respiratory viruses in clinical specimens have now been supplemented with primer pairs for amplification of HCoV-OC43, 229E, and NL63 *(2,3)*.

In order to allow the detection of all known coronaviruses in one assay, consensus RT-PCRs have been developed, based on an alignment of conserved genome regions of several coronaviruses *(4)*. However, the recent identification of HCoV-NL63 demonstrated that this novel coronavirus could not be amplified by the consensus RT-PCR described by Stephensen and colleagues, which might be explained by the presence of several mismatches in the primer sequences. We designed a novel pancoronavirus RT-PCR in which we modified the coronavirus consensus RT-PCR primers based on an alignment of the HCoV-NL63 sequence and the corresponding sequences of 13 other coronaviruses. Theoretically, this pancoronavirus RT-PCR should amplify a 251-bp fragment of the polymerase gene of all coronaviruses known at present, and we tested this experimentally for the five known human coronaviruses (HCoV-NL63, 229E, OC43, HKU1, and SARS-CoV) and three animal coronaviruses (PHEV, FIPV, and MHV). The sensitivity of the pancoronavirus RT-PCR was found to be lower than for a nondegenerate RT-PCR, which can be explained by the high level of degeneracy of the pancoronavirus primers. Nevertheless, this pancoronavirus assay is a useful tool for screening sample collections for the presence of all known and potentially yet unknown coronaviruses *(5)*.

2. Materials

2.1. Sample Handling and RNA Extraction

1. Viral transport medium (VTM) (BD Diagnostic systems, USA) or phosphate buffered solution (PBS). Store at room temperature.
2. Filter Minisart plus 0.45 μm (Sartorius, Vivascience AG).
3. 2-ml syringe without a needle.
4. Glass pearls (5 mm diameter) (Merck MDA, VWR, Belgium).
5. Qiamp viral RNA mini kit (Qiagen, The Netherlands). Store at room temperature, except for the AVL lysis buffer, which should be stored at 4°C after addition of carrier RNA.

2.2. Pancoronavirus One-Step RT-PCR

1. Primers: Forward primer: Cor-FW (5′-ACWCARHTVAAYYTNAARTAYGC-3′) and reverse primer: Cor-RV (5′-TCRCAYTTDGGRTARTCCCA-3′) (Eurogentec, Seraing, Belgium). The primers are prepared in a stock solution of 100 μM.

A working solution of 15 μM is made for use in the RT-PCR. Store at −20°C until use.

2. Qiagen One-Step RT-PCR kit containing One-Step RT-PCR enzyme mix (a combination of Omniscript and Sensiscript reverse transcriptase and HotStarTaq DNA polymerase), One-Step RT-PCR buffer 5X, and dNTP mix, 10 mM each (Qiagen, The Netherlands). Store all reagents at −20°C.

3. RNAse-free water (Sigma-Aldrich, Belgium). Store at room temperature.

4. GeneAmp PCR system 9700 thermal cycler (Applied Biosystems, Foster City, CA, USA).

5. 1X Tris-borate EDTA (TBE) buffer: dilute 10X TBE buffer (Invitrogen Life Technologies) 1:10 with MilliQ water (Millipore). Store at room temperature.

6. PAGE gels (6% acrylamide/N,N′-methylenebisacrylamide in 1X TBE buffer containing 0.064% (w/v) ammonium persulfate and 150 μl tetramethylethylenediamine). After preparation of a batch of ten gels store at 4°C.

7. Molecular Weight Marker VI (MWM VI) (Roche, Mannheim, Germany). Store at 4°C.

8. Imagemaster VDS (Pharmacia, Uppsala, Sweden).

2.3. Sequencing and Sequence Analysis

1. MSB Spin PCRapace purification kit (Invitek, Westburg, The Netherlands). Store at room temperature.

2. Saekem GTG agarose for the recovery of nucleic acids (Cambrex, Belgium).

3. 1X Tris acetate EDTA (TAE) buffer: dilute 10X TAE buffer 1:10 with MilliQ water (Millipore). Store at room temperature.

4. QIAquick gel extraction kit (Qiagen, The Netherlands).

5. Forward primer: Cor-FW (5′-ACWCARHTVAAYYTNAARTAYGC-3′) or reverse primer Cor-RV (5′-TCRCAYTTDGGRTARTCCCA-3′) (Eurogentec, Seraing, Belgium). The primers are prepared in a stock solution of 100 μM. A working solution of 5 μM is made for use in the cycle sequencing reaction. Store at −20°C until use.

6. ABI PRISM BigDye Termination Cycle Sequencing Ready Reaction kit (Applied Biosystems, CA, USA). Store at −20°C.

7. Biometra T3000 thermocycler (Biometra, Westburg, The Netherlands).

8. Ethanol absolute.

9. Sodium acetate 3 M pH 4.6. Store at 4°C.

10. Formamide. Store at −20°C.

11. ABI Prism 3100 Genetic Analyzer (Applied Biosystems, Foster City, CA, USA).

3. Methods
3.1. Sample Handling and RNA Extraction

1. Nasopharyngeal and throat swabs: place the tip of the swab in a tube containing 3 ml VTM or PBS. After squeezing the tip of the swap and vortexing, take 140 μl of the solution for RNA extraction. Store the rest of the sample at −80°C.

2. Bronchoalveolar lavages and nasopharyngeal aspirates: take 140 μl of the sample for RNA extraction without prior dilution in VTM or PBS. Store the rest of the sample at −80°C.

3. Sputum samples: dilute about 100 to 300 mg of sputum in 3 ml VTM or PBS. Add glass pearls and vortex intensively until the solution containing the sputum is completely liquefied. Filter the solution using a 2-ml syringe without a needle and a 0.45-μm Minisart Plus filter (*see* **Note 1**). Take 140 μl of the filtrate for RNA extraction. Store the rest of the sample at −80°C.

4. Perform the RNA extraction on 140 μl of sample according to the manufacturer's instructions using the Spin Protocol (Qiagen QIAamp viral RNA mini kit handbook) (*see* **Note 2**). Store the obtained RNA extract (60 μl) at −80°C.

3.2. Pancoronavirus One-Step RT-PCR

1. Screening a sample for the presence of coronaviruses is performed by amplifying a 251-bp fragment of the coronavirus polymerase gene using the following primer set: Cor-FW (5′-ACWCARHTVAAYYTNAARTAYGC-3′) and Cor-RV (5′-TCRCAYTTDGGRTA RTCCCA-3′) (*see* **Fig. 1**) (*see* **Note 3**). Prepare the following master mix for RT-PCR according to the number of samples to be tested: 10 μl 5X QIAGEN OneStep RT-PCR Buffer, 2 μl dNTP mix (final concentration of 400 μM of each dNTP), 1.8 μl QIAGEN OneStep RT-PCR Enzyme Mix, 4 μM of forward and reverse primer, and RNase-free water to 40 μl per reaction (*see* **Notes 4** and **5**).

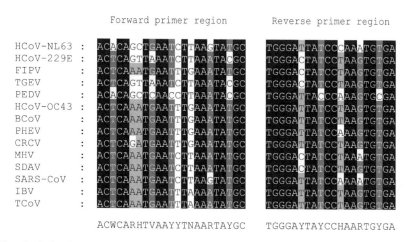

Fig. 1. Selection of primers for the novel pancoronavirus RT-PCR. Shown is the alignment of 14 coronaviral sequences of a conserved region of the polymerase gene. The forward (Cor-FW) and reverse (Cor-RV) primer sequences are shown at the bottom (Y=C/T, W=A/T, V=A/C/G, R=A/G, H=A/T/C, N=A/C/T/G).

2. Add 10 µl RNA-extract to the tube containing 40 µl of the master mix, leading to a final reaction volume of 50 µl (*see* **Notes 6** and **7**). The reaction is carried out with an initial reverse transcription step at 50°C for 30 min, followed by PCR activation at 95°C for 15 min, 50 cycles of amplification (30 sec at 94°C; 30 sec at 48°C; 1 min at 72°C), and a final extension step at 72°C for 10 min in a GeneAmp PCR system 9700 thermal cycler (Applied Biosystems, Foster City, CA, USA).

3. Load 9 µl of PCR product mixed with 1 µl loading dye on a 6% PAGE gel. Load 3 µl of MWM VI on each side of the gel (*see* **Note 8**). Run the gel at 200 V, 200 mA for 30 min. Visualize the bands by illuminating the gel with UV-light on an Imagemaster VDS system after staining the gel for 5 min with ethidium bromide (EtBr) (*see* **Note 9**).

4. Determination of the fragment size is performed by comparing the length of the bands to MWM VI. Samples from which a PCR product of approximately 251 bp is amplified are presumed to be coronavirus-positive and will be subject to further (sequence) analysis (*see* **Note 10**).

3.3. Sequencing and Sequence Analysis

1. If a single 251-bp band is visible after EtBr staining, purify the total amplification product by using the MSB Spin PCRapace purification kit according to the manufacturer's instructions (Invitek MSB Spin PCRapace purification kit handbook).

2. If more than one band is visible, including a 251-bp band, run the total amount of PCR product mixed with 5 µl loading dye on a 1.5% agarose gel (in 1X TAE buffer) on 100 V, 500 mA for 1 h. Stain for 10 min with EtBr, and cut each band from the gel with a sterile scalpel. Purify the fragments excised from the agarose gel using the QIAquick gel extraction kit according to the Spin Protocol in the QIAquick gel extraction kit handbook.

3. For sequencing, add 5 µl of purified amplification product to 4 µl of ABI PRISM BigDye Ready Reaction mix and 1 µl of a 5-µM solution of forward or reverse primer. The sequence reaction, which consists of 25 cycles of 30 sec at 96°C, 15 sec at 50°C, and 4 min at 60°C, is performed in a Biometra T3000 Thermocycler.

4. After cycle sequencing, precipitate the DNA products by adding 62.5 µl absolute ethanol, 3 µl sodium acetate, and 24.5 µl MilliQ water per sample. Mix the samples and leave them at room temperature for 15 min, followed by a centrifugation for 20 min at 13,000 rpm (*see* **Note 11**). Carefully aspirate the supernatant, wash the pellet by adding 150 µl 70% ethanol, and centrifuge the samples for 5 min at 13,000 rpm. Carefully aspirate the supernatant, and let the pellet dry for 15 min at room temperature. Dissolve the pellet in 15 µl of formamide and denature for 2 min at 95°C. The samples are then stored at −20°C before analysis on an ABI Prism 3100 Genetic Analyzer (Applied Biosystems, Foster City, CA, USA).

5. Chromatrogram sequencing files can be inspected with Chromas 2.2 (Technelysium, Helensvale, Australia), and contigs of the fragments sequenced by Cor-FW

and Cor-RV, respectively, can be prepared using SeqMan II (DNASTAR, Madison, WI). The contig sequence data can easily be arranged into a FASTA format by saving the consensus as a single text file in the Seqman program. The text file will automatically have the following format:

> >sequence title (hard return)
> GACGAGTAATTGC... (not containing hard returns)

Such a FASTA file can then be used for nucleotide similarity searches using the NCBI WWW-BLAST (basic local alignment search tool) server on the Gen-Bank DNA database release 155.0 *(6)*. The coronavirus type that is present in the sample can be determined by using this BLAST analysis, which provides coronavirus sequences that show the highest nucleotide similarity with the sequenced fragment (*see* **Note 12**). Furthermore, a FASTA file of the sequence data can be loaded into a multiple sequence alignment program such as CLUSTAL X version 1.82 *(7)*, and the resulting alignment can be manually edited in the

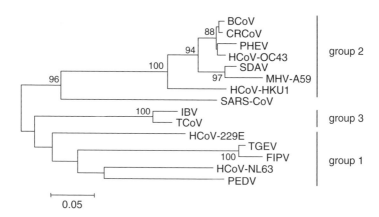

Fig. 2. Neighbor-joining phylogenetic tree of coronavirus partial polymerase gene sequence data (corresponding to the pancoronavirus fragment). The frequency of occurrence of particular bifurcations (percentage of 500 bootstrap replicate calculations) is indicated at the nodes. Bootstrap values over 75% are shown. The coronavirus sequences used here are available from GenBank under the following accession numbers: HCoV-NL63, AY567487; HCoV-229E, AF304460; PEDV, AF353511; transmissible gastroenteritis virus (TGEV), AF304460; feline infectious peritonitis virus (FIPV), AF124987; HCoV-OC43, AY391777; HCoV-HKU1, NC 006577; porcine hemagglutinating encephalomyelitis virus (PHEV), AF124988; bovine coronavirus (BCoV), AF391541; mouse hepatitis virus (MHV-A59), X51939; sialodacryoadenitis virus (SDAV), AF124990; canine respiratory coronavirus (CRCV), AY150273; SARS-CoV, AY313906; turkey coronavirus (TCoV), AF124991; and infectious bronchitis virus (IBV), Z30541.

GeneDoc Alignment editor *(8)*. Starting from a sequence alignment that compares nucleotides at homologous sites, i.e., sharing a common ancestor, a phylogenetic tree can be constructed using phylogenetic programs such as MEGA version 3.1 (**Fig. 2**) *(9)*. The construction of a phylogenetic tree based on sequence data of samples and coronavirus sequence data from GenBank will provide information regarding the genetic and evolutionary relatedness of a coronavirus detected in a sample to other known coronaviruses (*see* **Note 13**).

4. Notes

1. Sputum samples diluted in VTM or PBS should be filtered prior to analysis in order to remove large amounts of cellular debris. It can be difficult to completely liquefy the sputum sample, so the filter may become blocked. In this case, the only solution is to use another filter.

2. Especially during RNA extraction, the likelihood of cross-contamination is high. When performing the RNA extraction manually, one should change tips in between samples, as well as when adding wash buffer to the samples. Another precautionary measure that can be taken is to use tissue paper when opening the samples to avoid the formation of aerosols.

3. For modification of the coronavirus consensus RT-PCR primers *(4)*, an alignment of the polymerase gene sequences of 14 known coronaviruses was made (**Fig. 1**). Degenerate Cor-FW and Cor-RV primers were developed. The coordinates of Cor-FW and Cor-RV are 14017 and 14248, respectively, in the HCoV-NL63 complete genome sequence. The 14 coronavirus sequences used here are available from GenBank under the following accession numbers: HCoV-NL63, AY567487; HCoV-229E, AF304460; PEDV, AF353511; transmissible gastroenteritis virus (TGEV), AF304460; feline infectious peritonitis virus (FIPV), AF124987; HCoV-OC43, AY391777; porcine hemagglutinating encephalomyelitis virus (PHEV), AF124988; bovine coronavirus (BCoV), AF391541; mouse hepatitis virus (MHV-A59), X51939; sialodacryoadenitis virus (SDAV), AF124990; canine respiratory coronavirus (CRCV), AY150273; SARS-CoV, AY313906; turkey coronavirus (TCoV), AF124991; and infectious bronchitis virus (IBV), Z30541.

4. The sensitivity of the pancoronavirus RT-PCR assay was assessed by testing tenfold dilutions of HCoV-NL63 and HCoV-OC43 RNA. Primer concentrations were increased up to 4 μM per reaction to improve the sensitivity of the assay. While 50 copies of HCoV-OC43 RNA copies per microliter sample could be detected, the sensitivity for HCoV-NL63 was slightly lower, i.e., 5×10^3 RNA copies per microliter sample.

5. An important precautionary measure when performing RT-PCR is to avoid RNase contamination. Therefore, all materials should be RNase-free, and gloves and a gown should be worn even when preparing the (noninfectious) master mix. In addition, an RNase inhibitor (RNase OUT, Invitrogen Life Technologies; 20 units per reaction) can be added to the RT-PCR reaction to inhibit degradation

of the RNA sample owing to RNases. This measure should be taken especially when a positive control RNA sample no longer amplifies.

6. Include negative as well as positive controls in each RT-PCR. A negative control consists of a sample containing only the master mix without RNA (10 µl of water is added instead of RNA). As a positive control, RNA of a human coronavirus (e.g., HCoV-OC43) can be used.

7. When adding the RNA samples to the master mix, precaution must be taken to avoid RNA contamination. Such a precautionary measure is opening the tubes with tissue paper, thereby avoiding the formation of aerosols, as the latter could lead to cross-sample contamination.

8. It is advised to load the Molecular Weight Marker on both sides of the gel, as a gel can show the phenomenon of "smiling," which makes the determination of the fragment size difficult when the MWM is only present on one side of the gel.

9. Use two pairs of gloves when handling ethidium bromide, as this intercalating agent is highly mutagenic.

10. Amplification of a 251-bp fragment of the coronavirus polymerase gene was demonstrated for HCoV-NL63, HCoV-OC43, HCoV-229E, SARS-CoV, PHEV, FIPV, and MHV-A59 (**Fig. 3**) and for HCoV-HKU1 (data not shown).

11. The DNA pellet is often not visible. Therefore, during centrifugation the tubes are always put in the same orientation in the centrifuge, so that one knows at which side of the tube the pellet is located. The aspiration of supernatant should be performed carefully without touching the pellet.

12. The polymerase gene is a very conserved region in the coronavirus genome. Therefore, the percentage sequence similarity will be high among different coronavirus members belonging to the same group. Nevertheless, this sequence information allows a primary identification of the coronavirus type that is present in a sample.

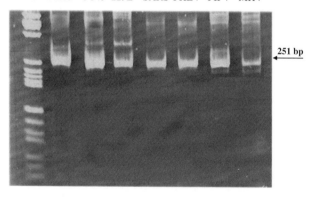

Fig. 3. Gel electrophoresis after pancoronavirus RT-PCR assay. The indicated band of 251 bp corresponds with the expected amplicon size. Molecular Weight Marker VI was used as a marker (Roche, Mannheim, Germany).

13. The genetic and evolutionary relatedness to known coronaviruses can be determined from the phylogenetic tree clustering pattern of the coronavirus sequence fragment, amplified from a sample. A phylogenetic tree is composed of nodes that are connected by branches. The external nodes represent the existing organisms or taxa, while the internal nodes represent the (hypothetical) ancestors of the existing taxa. The length of the horizontal branches is indicative of the evolutionary distance between the connected nodes, whereas the vertical branches are noninformative. Taxa that are present in the same cluster are monophyletic, which indicates that they show a (relatively) close genetic relatedness.

Acknowledgments

We would like to thank all the colleagues of the laboratory of Clinical and Epidemiological Virology, Department of Microbiology and Immunology, Rega Institute for Medical Research, University of Leuven, Belgium, and Dr. Lia van der Hoek and Prof. Dr. Ben Berkhout of the Department of Retrovirology, Academic Medical Center, University of Amsterdam, The Netherlands, for helpful comments and discussion. This work was supported by a postdoctoral fellowship (P.D.M.) of the Research Fund K.U. Leuven to Leen Vijgen.

References

1. van Elden, L. J., van Loon, A. M., van Alphen, F., Hendriksen, K. A., Hoepelman, A. I., van Kraaij, M. G., et al. (2004) Frequent detection of human coronaviruses in clinical specimens from patients with respiratory tract infection by use of a novel real-time reverse-transcriptase polymerase chain reaction. *J. Infect. Dis.* **189**, 652–657.
2. Bellau-Pujol, S., Vabret, A., Legrand, L., Dina, J., Gouarin, S., Petitjean-Lecherbonnier, J., et al. (2005) Development of three multiplex RT-PCR assays for the detection of 12 respiratory RNA viruses. *J. Virol. Meth.* **126**, 53–63.
3. Vabret, A., Mourez, T., Dina, J., van der Hoek, L., Gouarin, S., Petitjean, J., *et al.* (2005) Human coronavirus NL63, France. *Emerg. Infect. Dis.* **11**, 1225–1229.
4. Stephensen, C. B., Casebolt, D. B., and Gangopadhyay, N. N. (1999) Phylogenetic analysis of a highly conserved region of the polymerase gene from 11 coronaviruses and development of a consensus polymerase chain reaction assay. *Virus Res.* **60**, 181–189.
5. Moes, E., Vijgen, L., Keyaerts, E., Zlateva, K., Li, S., Maes, P., et al. (2005) A novel pancoronavirus RT-PCR assay: frequent detection of human coronavirus NL63 in children hospitalized with respiratory tract infections in Belgium. *BMC Infect. Dis.* **5**, 6.
6. Altschul, S. F., Gish, W., Miller, W., Myers, E. W., and Lipman, D. J. (1990) Basic local alignment search tool. *J. Mol. Biol.* **215**, 403–410.

7. Thompson, J. D., Gibson, T. J., Plewniak, F., Jeanmougin, F., and Higgins, D. G. (1997) The CLUSTAL X windows interface: flexible strategies for multiple sequence alignment aided by quality analysis tools. *Nucleic Acids Res.* **25**, 4876–4882.

8. Nicholas, K. B., Nicholas, H. B., and Deerfield, D. W. (2005) GeneDoc: analysis and visualization of genetic variation. *Embet News* **4**, 14.

9. Kumar, S., Tamura, K., and Nei, M. (2004) MEGA3: integrated software for molecular evolutionary genetics analysis and sequence alignment. *Brief Bioinform.* **5**, 150–163.

2

Detection of Group 1 Coronaviruses in Bats Using Universal Coronavirus Reverse Transcription Polymerase Chain Reactions

Leo L. M. Poon and J. S. Malik Peiris

Abstract

The zoonotic transmission of SARS coronavirus from animals to humans revealed the potential impact of coronaviruses on mankind. This incident also triggered several surveillance programs to hunt for novel coronaviruses in human and wildlife populations. Using classical RT-PCR assays that target a highly conserved sequence among coronaviruses, we identified the first coronaviruses in bats. These assays and the cloning and sequencing of the PCR products are described in this chapter. Using the same approach in our subsequent studies, we further detected several novel coronaviruses in bats. These findings highlighted the fact that bats are important reservoirs for coronaviruses.

Key words: bat coronaviruses; novel coronaviruses; molecular detection; RT-PCR

1. Introduction

Coronaviruses are a group of enveloped, positive-strand single stranded RNA viruses with a corona-like morphology. These viruses have genome sizes ranging from 28 to 32 kb, which makes them the biggest among the RNA viruses *(1)*. The genomes contain five major open reading frames (Orfs) that encode the replicase polyproteins (Orf 1a and Orf 1ab), spike (S), envelope (E), membrane (M), and nucleocapsid (N) proteins, in that order. Based on antigenic and genetic analyses, coronaviruses can be subdivided into three groups. Coronaviruses from groups 1 and 2 have been found to infect mammals. By contrast, avian species are known to be the natural hosts for group 3 viruses. Most coronaviruses

From: *Methods in Molecular Biology, vol. 454: SARS- and Other Coronaviruses*,
Edited by: D. Cavanagh, DOI: 10.1007/978-1-59745-181-9_2, © Humana Press, New York, NY

are pathogenic to their hosts *(2,3)*. Animal coronaviruses are known to cause respiratory, gastrointestinal, neurological, and hepatic diseases. By contrast, with the exception of SARS coronavirus, in the case of healthy individuals all previously known human coronaviruses were generally only associated with mild respiratory and gastrointestinal diseases.

The discovery of a novel coronavirus as the cause of SARS led to a resurgence in interest in these viruses and to the discovery of other novel coronaviruses in humans, namely NL-63 and HKU-1 *(4–6)*. The recognition of SARS-like coronaviruses in palm civets and other small mammals in live-game animal markets *(7)* prompted intensive surveillance for coronaviruses in wild animals. Following upon this interesting discovery, several novel coronaviruses were identified in the mammalian and avian species *(6,8–17)*. In particular, our initial studies first revealed that bats are important reservoirs for coronaviruses. By screening respiratory and fecal samples collected from a wide range of animals, we identified the first coronavirus in bats *(14)*. Subsequent studies by ourselves and others also indicated that there are number of group 1 and group 2 viruses circulating in bat species *(11,12,15,17)*. In particular, a SARS-like coronavirus was found in two of these studies independently *(11,12)*. These results suggest that bats might play a key role in the evolution of coronaviruses. Here we describe the procedures that were used in our virus surveillance studies.

2. Materials

2.1. Sampling

1. Viral transport medium: Dissolve 12.5 g of penicillin G sodium salt (Sigma-Aldrich), 50 mg of ofloxacin (Sigma-Aldrich), 0.1 g of nystatin (Sigma-Aldrich), 12.5 million units of polymycin B sulfate salt (Sigma-Aldrich), 250 mg gentamicin (Sigma-Aldrich), 1 g of sulfamethoxazole (Sigma-Aldrich), 0.2 g of sodium hydroxide (Merck), and 2.2 g of sodium hydrogen carbonate (Merck) in 1 liter of Medium 199 (Sigma-Aldrich). Adjust to pH 7.0–7.5 with sodium hydrogen carbonate solution and filter the solution using a 0.22-μm pore size filter. Store the filtered medium in aliquots in sterile screw cap tubes (1 ml/tube) (Axygen Scientific) at –20 °C (*see* **Note 1**).
2. Calcium alginate swabs with an ultrafine aluminum shaft (Fisher Scientific).
3. Portable ice bucket with reusable cooling packs.
4. Cotton work gloves.
5. Sampling net.

2.2. RNA Extraction

1. QIAamp virus RNA mini kit (Qiagen) (*see* **Note 2**).
2. Ethanol, 96–100%.
3. Autoclaved RNase-free water or its equivalent.

2.3. Reverse Transcription

1. SuperScript III reverse transcriptase, 200 U/μL (Invitrogen).
2. 5X first-strand buffer: 250 mM Tris-HCl (pH 8.3), 375 mM KCl, 15 mM MgCl$_2$ (Invitrogen).
3. 0.1 mM dithiothreitol (Invitrogen).
4. Random hexamers, 3 μg/μl (Invitrogen).
5. RNaseOUT recombinant ribonuclease inhibitor, 40 U/μl (Invitrogen).
6. Deoxynucleotide triphosphates (dNTP).
7. Autoclaved RNase-free water or equivalent.
8. Heating block or equivalent.

2.4. Polymerase Chain Reaction

1. AmpliTaq Gold DNA polymerase, 5 U/μl (Applied Biosystems).
2. 10X Gold PCR buffer (Applied Biosystems).
3. Deoxynucleotide triphosphates (dNTP).
4. 25 mM MgCl$_2$ solution (Applied Biosystems)
5. 10 μM PCR forward primer, 5′-GGTTGGGACTATCCTAAGTGTGA-3′
6. 10 μM PCR reverse primer, 5′- CCATCATCAGATAGAATCATCAT-3′
7. Themocycler (GeneAmp 9700, Applied Biosystems)

2.5. Gel Electrophoresis

1. 50X TAE buffer (Bio Rad).
2. Seakam LE agarose powder (Cambrex).
3. 6X gel loading buffer: 10 mM Tris-HCl (pH 7.6), 0.03% bromophenol blue, 0.03% xylene cyanol, 60% glycerol, and 60 mM EDTA.
4. 1 kb plus DNA ladder markers (Invitrogen).
5. Ethidium bromide, 10 mg/ml (*see* **Note 3**).
6. Agarose gel electrophoresis apparatus.
7. Power supply (PowerPac Basic, Bio-Rad).
8. Gel documentary machine or equivalent.

2.6. PCR Products Purification

1. QIAquick PCR purification kit (Qiagen).
2. Ethanol, 96–100%.

2.7. PCR Product Cloning

1. TOPO TA cloning kit with electrocompetent *E. coli* (Invitrogen).
2. Electroporation system (Gene Pulser Xcell, Bio-Rad).

3. Electroporation cuvettes (Gap width: 0.2 cm, Bio-Rad).
4. LB plate containing 50 μg/ml ampicillin.

2.8. Sequencing

1. BigDye Terminator v3.1 Cycle Sequencing kit (Applied Biosystems).
2. Spectrophotometer.
3. 3.2-μM sequencing primers (forward/reverse primers mentioned in Section 2.4)
4. Dye terminator removal columns (Genetix)
5. DNA analyzer (3700, Applied Biosystems) or equivalent.

3. Methods

The protocol described below was used to detect group 1 bat coronavirus in our previous studies *(14)*. The procedures and primer set for the assay can also detect other non-group-1 coronaviruses in other specimens. In our studies, this assay was also shown to be able to detect other common human coronaviruses in our hands (e.g., HKU1, NL63, OC43, and 229E).

As the primer set used in our studies would cross-react with a wide range of coronaviruses (**Fig. 1**), all the positive products from the PCR reactions should first be identified by DNA sequencing. Moreover, amplicons from our

```
                        Forward primer              Complementary sequence
                          sequence                    of reverse primer
                   GGTTGGGACTATCCTAAGTGTGA          ATGATGATTCTATCTGATGATGG
        Group 1    ──────────────────────►          ◄──────────────────────
        PEDV       5'-GGTTGGGATTACCCAAAGTGCGA---N₃₉₄---ATGATGATTCTTTCTGATGATGG-3'
        TGEV       5'-GGATGGGACTATCCTAAGTGTGA---N₃₉₄---ATGATGATTTTATCTGATGATGG-3'
        HCoV-229E  5'-GGATGGGACTATCCTAAGTGTGA---N₃₉₄---ATGATGATTCTTTCTGATGATAG-3'
        HCoV-NL63  5'-GGTTGGGATTATCCCAAATGTGA---N₃₉₄---ATGATGATTCTCTCTGATGACGG-3'

        Group 2
        BCoV       5'-GGTTGGGATTATCCTAAGTGTGA---N₃₉₄---ATGATGATTTTGAGTGATGATGG-3'
        MHV        5'-GGTTGGGACTATCCTAAATGTGA---N₃₉₄---ATGATGATTTTGAGTGATGATGG-3'
        SARS-CoV   5'-GGTTGGGATTATCCAAAATGTGA---N₃₉₄---ATGATGATTCTTTCTGATGATGC-3'
        HCoV-OC43  5'-GGTTGGGATTATCCTAAGTGTGA---N₃₉₄---ATGATGATTTTGAGTGATGATGG-3'

        Group 3
        IBV        5'-GGTTGGGATTATCCTAAGTGTGA---N₃₉₄---TTGATGATCTTGTCTGACGACGG-3'
```

Fig. 1. Sequence alignment of the target regions of representative coronaviruses. Forward and reverse primers were communicated through the World Health Organization's SARS etiology network by colleagues from the Centers for Disease Control and Prevention. The forward primer and complementary sequence of the reverse primer correspond to the nucleotide positions 15149 to 15171 and 15567-15589 of HCoV-OC43 (Genbank accession no: AY585229), respectively.

PCR assays were sometimes found to contain many single-nucleotide polymorphisms. To obtain more accurate sequencing data for our phylogenetic analysis, PCR products from these reactions were cloned into vectors (Section 3.8) and several clones from the same PCR reactions were chosen for sequencing analysis.

To enhance the detection rate of bat coronaviruses, we also developed a hemi-nested PCR assay to detect group 1 bat corornaviruses identified from our earlier study (Section 3.6) *(8)*. However, we did not evaluate the performance of this hemi-nested PCR assay to detect other coronaviruses.

3.1. Sample Collection

1. Trap the animal with a sampling net and handle it with appropriate protective equipment (e.g., cotton work gloves or leather gloves) (*see* **Notes 4** and **5**).
2. Examine the animal and record its physical parameters/features for species identification.
3. For collecting throat swab samples, insert swab into the animal's throat and leave it in place for a few seconds to absorb secretions. For collecting anal swab samples, insert swab into the anus and leave it in place for a few seconds (*see* **Note 6**).
4. Immediately place the swab into a sterile vial containing 1 ml of viral transport medium. Label each specimen properly.
5. Put the specimen vials on ice for transport to the laboratory. Do not freeze.
6. Vortex the specimen vial briefly in the laboratory. Transfer 500 µl of viral transport medium into a new sterile vial at –70°C for long-term storage. If the remaining half of the sample cannot be processed for RNA extraction (or virus isolation) on the same day, store it at 4°C until use (*see* **Note 7**).

3.2. RNA Extraction

1. For a new kit, perform the following procedures before specimen processing:

   a. Add 1 ml of buffer AVL to a tube of lyophilized carrier RNA (310 µg). Dissolve carrier RNA thoroughly. Transfer to the buffer AVL bottle and mix thoroughly. Store the buffer AVL at 4°C for up to 6 months.
   b. For every 19 ml of buffer AW1, add 25 ml of ethanol (96–100%). Mix well. Store the buffer AW1 at room temperature for up to 12 months.
   c. For every 13 ml of buffer AW2, add 30 ml of ethanol (96–100%). Mix well. Store the buffer AW1 at room temperature for up to 12 months (*see* **Note 2**).

2. Equilibrate all reagents to room temperature before use.
3. Pipette 140 µl of the sample into a 1.5-ml microcentrifuge tube.
4. Add 560 µl of prepared buffered AVL with carrier RNA to the microcentrifuge tube.
5. Briefly vortex the tubes for 15 sec and incubate at room temperature for 10 min.

6. Briefly centrifuge the microcentrifuge tube. Add 560 µl ethanol (96–100%) and mix by pulse vortexing for 15 sec.
7. Briefly centrifuge the microcentrifuge tube.
8. Transfer 630 µl of the solution from the tube to a QIAamp spin column placed in a 2-ml collection tube. Centrifuge at 6000 × g (8000 rpm) for 1 min. Place the spin column in a clean 2-ml collection tube. Discard the tube containing the filtrate.
9. Open the spin column and repeat step 8.
10. Add 500 µl buffer AW1. Centrifuge at 6000 × g (8000 rpm) for 1 min. Place the spin column in a clean 2-ml collection tube. Discard the tube containing the filtrate.
11. Add 500 µl buffer AW2. Centrifuge at 20,000 × g (14,000 rpm) for 3 min. Place the spin column in a clean 2-ml collection tube and centrifuge at 20,000 × g again for 1 min. Place the spin column in a clean 1.5-ml microcentrifuge tube. Discard the tube containing the filtrate.
12. Apply 60 µl buffer AVE equilibrated to room temperature directly on the membrane of the column. Close the cap and incubate at room temperature for 1 min.
13. Centrifuge at 6000 × g (8000 rpm) for 1 min. Collect the filtrate for cDNA synthesis. Store the RNA at –20°C or –70°C (*see* **Note 8**).

3.3. Reverse Transcription

1. Prepare a reverse transcription master mix sufficient for the designated number (*N*) of samples in a sterile 1.5-ml microcentrifuge tube as shown in **Table 1**.
2. Vortex and centrifuge the tube briefly. Keep the tube on ice.
3. Add 10 µl of master mix solution into separate 0.5-microcentrifuge tubes. Label the tubes accordingly and keep them on ice.
4. Add 10 µl of purified RNA sample into these tubes.
5. Vortex and centrifuge the tubes briefly.

Table 1
Composition of the Reverse Transcription Reaction

Reagent	Volume per reaction	Volume mix for N reactions	Final concentration
5X First-strand buffer	4 µl	4 × N µl	1×
0.1 mM DTT	2 µl	2 × N µl	0.01 mM
10 mM dNTP	1 µl	N µl	0.5 mM
Random primers (50 ng/µl)	1 µl	N µl	2.5 ng/µl
Reverse transcriptase (200 U/µl)	1 µl	N µl	200 U/reaction
Ribonuclease inhibitor	1 µl	N µl	40 U/reaction
Total volume of master mix	10 µl	10 × N µl	–

6. Stand the tubes at room temperature for 10 min and then incubate at 42°C for 50 min.
7. Inactivate the transcription reaction by incubating the tubes at 95°C for 5 min and then chill the samples on ice. Store the cDNA samples at –20°C (*see* **Note 9**).

3.4. PCR Assay

1. Prepare a PCR master mix sufficient for the designated number of samples in a sterile 0.5-ml microcentrifuge tube according to **Table 2**. Include at least 1 positive control and 1 negative control (water) for each run. Add additional controls (e.g., purified RNA from the studied samples) as necessary.
2. Vortex and centrifuge the tube briefly. Keep the tube on ice.
3. Aliquot 48 µl of the master mix into separate 0.5-ml microcentrifuge tubes and label the tubes accordingly.
4. Add 2 µl of cDNA generated from the reverse transcription reactions to the tubes. For the positive control, add 2 µl of coronavirus cDNA into the reaction. For the negative control, add 2 µl of autoclaved water.
5. Vortex and centrifuge the tubes briefly.
6. Run the PCR under the conditions shown in **Table 3**.
7. After the run, analyze the PCR products by gel electrophoresis. Before gel electrophoresis, the PCR reactions can be kept at –20°C for short-term storage.

3.5. Agarose Gel Electrophoresis

1. Place a gel-casting tray onto a gel-casting base. Level the base.
2. Prepare 1.5% agarose gel by weighing out 0.75 g of agarose powder. Add it to a 250-ml bottle containing 50 ml 1X TAE buffer. Microwave the bottle with a loosened cap until the gel starts to bubble and become transparent (*see* **Note 10**).

Table 2
Composition of the PCR Reaction

Reagent	Volume per reaction	Volume for N reactions	Final concentration
10X PCR buffer	5 µl	5 × N µl	1x
25 mM MgCl$_2$, 25 mM	5 µl	5 × N µl	2.5 mM
dNTP, 10 mM	1 µl	N µl	0.2 mM
Forward primers, 10 µM	1 µl	N µl	0.2 µM
Reverse primers, 10 µM	1 µl	N µl	0.2 µM
DNA polymerase (5U/µl)	0.2 µl	0.2 × N µl	1 U/reaction
Water	34.8 µl	34.8 × N µl	–
Total	48 µl	48 × N µl	–

Table 3
Conditions for the PCR Reaction

Step	Temperature	Time
1. Heat activation	94°C	10 min
2. Thermal cycling (45 cycles)		
Denaturing step	94°C	30 sec
Annealing step	48°C	30 sec
Extension	72°C	40 sec
3. Final extension	72°C	2 min
4. Soak	4°C	∞

 3. Cool the melted agarose to about 60°C and pour it into the gel-casting tray. Insert a comb into the tray.
 4. Allow the gel to solidify at room temperature.
 5. Remove the comb from the tray.
 6. Place the tray in the electrophoresis chamber with the wells on the cathode side.
 7. Fill the buffer chamber with 1X TAE buffer at a level that can cover the top of the gel.
 8. Mix 0.5 μl of the DNA markers with 1 μl of 6X gel loading dye and 4.5 μl of water on a parafilm sheet by repeated pipetting.
 9. Mix 5 μl of the PCR products with 1 μl of 6X gel loading dye on a parafilm sheet by pipetting up and down several times.
10. Apply the mixture to the corresponding well of the gel.
11. Close the lid of the electrophoresis apparatus and connect the electrical leads, anode to anode (red to red) and cathode to cathode (black to black).
12. Run the gel at 100 V for 30 min.
13. Turn off the power, remove the cover, and retrieve the gel.
14. Soak the gel in 1X TAE with 0.5 μg/ml ethidium bromide for 15 min. Wash the gel with water (*see* **Note 3**).
15. Place the gel on top of the transilluminator. Switch on the power of the gel documentation machine (*see* **Note 11**).
16. Adjust the position of the gel and record the results. The size of the expected product for coronaviruses is 440 bp (**Fig. 2**).

3.6. Hemi-Nested PCR for Group 1 Bat Coronaviruses (Optional)

 1. Mix 1 μl of the the PCR reaction from Section 3.4 with 9 μl of water. Use 2 μl of the diluted samples in the hemi-nested PCR reaction.
 2. Repeat the procedure as outlined in Section 3.4, replacing the reverse primer of the PCR reaction by the primer 5'ATCAGATAGAATCATCATAGAGA-3'.
 3. Repeat the steps in Section 3.5. The size of the expected PCR product is 435 bp.

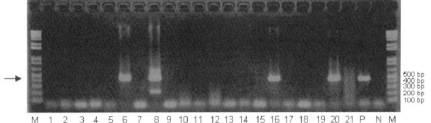

Fig. 2. PCR reaction specific for coronaviruses. The position of expected PCR products (440 nt) is highlighted by the arrow. Positive samples: lanes 5, 8, 16, and 20. Negative samples: lanes 1–5, 7, 9–15, 17—19, and 21. M, DNA markers; P: positive control; N: water control.

3.7. PCR Product Purification

1. When using a new PCR product purification kit, add ethanol (96–100%) to buffer PE according the product insert. Vortex the mixture briefly.
2. Mix 200 µl of buffer PB to 40 µl of PCR products from Section 3.4 or 3.6 in a clean 0.5-µl microcentrifuge tube.
3. Apply the mixture to a QIAquick column that is inserted in a 2-ml collection tube.
4. Centrifuge the tube at 20,000 × g (14,000 rpm) for 1 min. Discard the flow-through. Put the column back into the same collection tube.
5. Apply 0.75 ml of buffer PE to the column and repeat step 4.
6. Transfer the column in a sterile 1.5-ml microcentrifuge tube.
7. Apply 30 µl of buffer EB to the membrane of the column. Let it stand for 1 min.
8. Centrifuge the tube at 20,000 × g (14,000 rpm) for 1 min.
9. Discard the column. The PCR products collected in the 1.5-ml microcentrifuge tube are ready for sequencing work (Section 3.9). Alternatively, these samples could be stored at –20 °C until use (*see* **Notes 12** and **13**).

3.8. PCR Cloning (Optional)

1. On the day of the work, spread 40 µl of 40 mg/ml X-gal on selective agar plates, each of which contains 50 µg/ml ampicillin. Incubate the plates at 37°C until use.
2. Dilute the salt solution provided in the cloning kit by fourfold.
3. Thaw a vial of electrocompetent *E. coli* on ice.
4. Mix 4 µl of purified PCR products from Section 3.7 with 2 µl of fourfold diluted salt solution in a sterile 0.5-ml microcentrifuge tube by pipetting up and down several times.
5. Mix 1 µl TOPO vector into the reaction with a gentle pipetting.
6. Incubate for 5 min at room temperature.

7. Transfer 2 μl of reaction product into the thawed competent cells. Mix the contents by gently tapping the tube.
8. Carefully transfer the competent cells to an ice-chilled cuvette. Try to avoid forming bubbles in the cuvette.
9. Insert the cuvette into the electroporation chamber. Transfect the cells under these parameters: 2.5 kV, 200 Ω, and 25 μF.
10. Add 250 μl SOC medium provided from the cloning kit to the transfected cells.
11. Tranfect the solution into a 1.5-ml microcentrifuge tube and incubate it at 37°C with shaking.
12. Spread 100 μl of the incubated solution on a prewarmed selective plate. Incubate the plate at 37°C overnight in an inverted position.
13. Select 5 to 10 white colonies for PCR screening. Use sterile pipette tips to touch the selected colonies. Preserve the selected colonies by touching the pipette tips on a patch LB plate. Label the colonies accordingly. Incubate the plate at 37°C in an inverted position. Keep the plate at 4°C after an overnight incubation.
14. Transfer the rest of the contents attached to the pipette tips into 0.5-μl microcentrifuge tubes containing 20 μl of 10 mM Tris-HCl, pH 7.4. Label the tubes accordingly. Vortex the tubes briefly.
15. Heat the tubes at 95°C for 10 min. Then chill them on ice and centrifuge briefly.
16. Depending on the source of the PCR products, repeat the PCR protocol as described in Section 3.4 or 3.6. Use 2 μl of heat-treated supernatants as the DNA template in these reactions.
17. Analyze the size of the inserts of these clones according to the procedures described in Section 3.5. Only those with the expected product sizes (i.e., 440 bp for protocol 3.4; 435 bp for protocol 3.6) will be selected for PCR sequencing. Purify the positive samples as described in Section 3.7 before DNA sequencing.
18. Pick the desired clones off the patch plate and put them into 1 ml of LB containing 50 μg/ml of ampicillin. Grow the culture for 16–20 h. Mix 0.85 ml of the culture with 0.15 ml of sterile glycerol. Keep the transformed cells at –80°C for long-term storage.

3.9. PCR Sequencing

1. Measure the concentration of the purified amplicons using a spectrophotometer. The concentration of DNA in $\mu g/\mu l$ = (OD_{260} × volume of diluted sample)/ (20 × volume of sample used for dilution).
2. Prepare the sequencing reaction in 96-well plate format. Mix 5–10 ng of the purified products with 1 μl of diluted forward *or* reverse PCR primers (3.2 μM) used in Section 3.4 or 3.6. Adjust the final volume of the reaction to 12 μl with sterile water. Add 8 μl of Terminator Ready Reaction mix. Centrifuge the tube briefly. Keep the tube on ice.
3. Run the sequencing reactions under the conditions described in **Table 4**.

Table 4
Conditions for Sequencing Reactions

Step	Temperature	Time
1. Heat denaturing	94°C	1 min
2. Thermal cycling (25 cycles)		
Denaturing step	94°C	30 sec
Annealing step	50°C	15 sec
Extension	60°C	4 min
3. Soak	4°C	∞

4. Store the reaction products at 4°C until ready for purification.
5. Briefly vortex the dye removal columns to resuspend the gel.
6. Snap off the tip, remove the cap, and insert it into a clean microcentrifuge tube from the purification kit.
7. Centrifuge the tube at $1000 \times g$ (3300 rpm) for 2 min.
8. Discard the flow-through.
9. Apply 200 µl of water to each column. Centrifuge the tube at $1000 \times g$ (3300 rpm) for 2 min. Transfer the columns into clean microcentrifuge tubes.
10. Carefully apply the sequencing products (~20 µl) to the gel matrix in the columns.
11. Centrifuge the tube at $1000 \times g$ (3300 rpm) for 2 min. Discard the columns and store the tubes containing the purified sequencing products at –20°C.
12. Transfer 10 µl of the purified products to a 96-well sample loading plate. Seal the plate with a heat seal and load it into the DNA analyzer for analysis.
13. Run the DNA analyzer.
14. Analyze the deduced sequences (*see* **Notes 14** and **15**).

4. Notes

1. Viral transport medium contains a high concentration of antibiotic to inhibit bacterial growth.
2. Buffer AVL containing carried RNA might form white precipitates when it is stored at 4°C. The precipitates can be dissolved in the buffer by heating the bottle in a water bath. Cool the buffer to room temperature before use.
3. Ethidium bromide is a known mutagen and may be carcinogenic. Handle solutions of ethidium bromide with gloves.
4. For reasons of safety, staffs who are required to handle animals should have the appropriate inoculations (e.g., rabies vaccine for handling bats). Appropriate protective equipment should be worn when handling animals.
5. Bats are protected species in Hong Kong, and a permit from the relevant governmental office is required to perform the work.

6. For some small bat species, the finest swabs used in our studies are still too big to fit into the anal opening. In this situation, we use fecal instead of anal samples. To minimize possible contamination, these bats are kept individually in clean cotton bags for a few minutes. Fecal pellets collected in the cotton bags are sampled for further analysis.

7. In our study, we also tried to isolate viruses from our samples. As some viruses might be sensitive to repeat freeze and thaw cycles, we normally aliquot the original samples into two separate tubes. One tube is kept at –80°C for long-term storage and the other is kept at 4°C until it is ready for RNA extraction and virus isolation.

8. Contents in fecal samples and calcium alginate in swabs are known to inhibit RT-PCR assays. One should validate the protocols and test the reagents with specimens spiked with positive samples in advance. RNA purified from our recommended RNA extraction kit work well for us.

9. General procedures to prevent PCR cross-contamination should be strictly followed. Aerosol-resistant filtered pipette tips could minimize possible carryover of amplicons. Separate pipettes and areas are used for sample processing, PCR, and post-PCR analysis. It is essential to include multiple positive and negative controls in the PCR reactions when a large number of samples are tested at the same time.

10. Agarose solutions can be superheated in a microwave oven. Do not handle the bottle immediately after microwaving. Always wear heat-resistant gloves when handling melted agarose.

11. UV light can cause severe skin and eye damage. Wear safety glasses and close the photography hood before turning on the UV transilluminator.

12. Purified PCR products from Section 3.7 can be directly subjected to DNA sequencing. As quiescent species were found in our studied bat samples, we observed several single nucleotide polymorphisms in the products. To obtain more accurate sequencing data for our phylogenetic analysis, PCR products from these reactions were cloned into vectors (Section 3.8) and several clones from the same PCR reaction were randomly selected for further analysis.

13. On some occasions, nonspecific PCR products might be observed (lane 8, **Fig. 1**) and the desired PCR products have to be recovered by direct excisions from the agarose gel. As UV can damage DNA, it is important to use low-energy UV light to visualize and locate the desired product in this scenario. Try to excise the product as quickly as possible. We normally purify the excised products using a QIAquick gel extraction kit (Qiagen). The purified products can be subjected to cloning or sequencing as noted.

14. The products should be sequenced in both directions.

15. Deduced sequences are normally analyzed on both DNA and amino acid levels. Common sequence databases, such as Genbank (http://www.ncbi.nlm.nih.gov/Genbank/) can be used for the analysis. Common analytic freeware such as BioEdit (http://www.mbio.ncsu.edu/BioEdit/bioedit.html) and MEGA (http://www.megasoftware.net/) are commonly used analytical programs and are available on the Internet.

Acknowledgments

Funding for this research was provided by a Public Health Research Grant from the National Institute of Allergy and Infectious Diseases, USA, the Research Grant Council of Hong Kong (HKU 7343/04 M to LLMP), and the European Research Project SARS-DTV (Contract No: SP22-CT-2004).

References

1. Gorbalenya, A. E., Enjuanes, L., Ziebuhr, J., and Snijder, E. J. (2006) Nidovirales: evolving the largest RNA virus genome. *Virus. Res.* **117**, 17–37.
2. Saif, L. J. (2004) Animal coronavirus vaccines: lessons for SARS. *Dev. Biol. (Basel)* **119**, 129–140.
3. Weiss, S. R., and Navas-Martin, S. (2005) Coronavirus pathogenesis and the emerging pathogen severe acute respiratory syndrome coronavirus. *Microbiol. Mol. Biol. Rev.* **69**, 635–664.
4. Fouchier, R. A., Hartwig, N. G., Bestebroer, T. M., Niemeyer, B., de Jong, J. C., Simon, J. H., and Osterhaus, A. D. (2004) A previously undescribed coronavirus associated with respiratory disease in humans. *Proc. Natl. Acad. Sci. USA* **101**, 6212–6216.
5. van der Hoek, L., Pyrc, K., Jebbink, M. F., Vermeulen-Oost, W., Berkhout, R. J., Wolthers, K. C., Wertheim-van Dillen, P. M., Kaandorp, J., Spaargaren, J., and Berkhout, B. (2004) Identification of a new human coronavirus. *Nature Med.* **10**, 368–373.
6. Woo, P. C., Lau, S. K., Chu, C. M., Chan, K. H., Tsoi, H. W., Huang, Y., Wong, B. H., Poon, R. W., Cai, J. J., Luk, W. K., Poon, L. L., Wong, S. S., Guan, Y., Peiris, J. S., and Yuen, K. Y. (2005) Characterization and complete genome sequence of a novel coronavirus, coronavirus HKU1, from patients with pneumonia. *J. Virol.* **79**, 884–895.
7. Guan, Y., Zheng, B. J., He, Y. Q., Liu, X. L., Zhuang, Z. X., Cheung, C. L., Luo, S. W., Li, P. H., Zhang, L. J., Guan, Y. J., Butt, K. M., Wong, K. L., Chan, K. W., Lim, W., Shortridge, K. F., Yuen, K. Y., Peiris, J. S., and Poon, L. L. (2003) Isolation and characterization of viruses related to the SARS coronavirus from animals in southern China. *Science* **302**, 276–278.
8. Chu, D. K., Poon, L. L., Chan, K. H., Chen, H., Guan, Y., Yuen, K. Y., and Peiris, J. S. (2006) Coronaviruses in bent winged bats (Miniopterus spp.). *J. Gen. Virol.* **87**, 2461–2466.
9. East, M. L., Moestl, K., Benetka, V., Pitra, C., Honer, O. P., Wachter, B., and Hofer, H. (2004) Coronavirus infection of spotted hyenas in the Serengeti ecosystem. *Vet Microbiol* **102**, 1–9.
10. Jonassen, C. M., Kofstad, T., Larsen, I. L., Lovland, A., Handeland, K., Follestad, A., and Lillehaug, A. (2005) Molecular identification and characterization of novel coronaviruses infecting graylag geese (*Anser anser*), feral pigeons (*Columbia livia*) and mallards (*Anas platyrhynchos*). *J. Gen. Virol.* **86**, 1597–1607.

11. Lau, S. K., Woo, P. C., Li, K. S., Huang, Y., Tsoi, H. W., Wong, B. H., Wong, S. S., Leung, S. Y., Chan, K. H., and Yuen, K. Y. (2005) Severe acute respiratory syndrome coronavirus-like virus in Chinese horseshoe bats. *Proc. Natl. Acad. Sci. USA* **102**, 14040–14045.

12. Li, W., Shi, Z., Yu, M., Ren, W., Smith, C., Epstein, J. H., Wang, H., Crameri, G., Hu, Z., Zhang, H., Zhang, J., McEachern, J., Field, H., Daszak, P., Eaton, B. T., Zhang, S., and Wang, L. F. (2005) Bats are natural reservoirs of SARS-like coronaviruses. *Science* **310**, 676–679.

13. Pearks Wilkerson, A. J., Teeling, E. C., Troyer, J. L., Bar-Gal, G. K., Roelke, M., Marker, L., Pecon-Slattery, J., and O'Brien, S. J. (2004) Coronavirus outbreak in cheetahs: lessons for SARS. *Curr. Biol.* **14**, R227– R228.

14. Poon, L. L., Chu, D. K., Chan, K. H., Wong, O. K., Ellis, T. M., Leung, Y. H., Lau, S. K., Woo, P. C., Suen, K. Y., Yuen, K. Y., Guan, Y., and Peiris, J. S. (2005) Identification of a novel coronavirus in bats. *J. Virol.* **79**, 2001–2009.

15. Tang, X. C., Zhang, J. X., Zhang, S. Y., Wang, P., Fan, X. H., Li, L. F., Li, G., Dong, B. Q., Liu, W., Cheung, C. L., Xu, K. M., Song, W. J., Vijaykrishna, D., Poon, L. L., Peiris, J. S., Smith, G. J., Chen, H., and Guan, Y. (2006) Prevalence and genetic diversity of coronaviruses in bats from China. *J. Virol.* **80**, 7481–7490.

16. Wise, A. G., Kiupel, M., and Maes, R. K. (2006) Molecular characterization of a novel coronavirus associated with epizootic catarrhal enteritis (ECE) in ferrets. *Virology* **349**, 164–174.

17. Woo, P. C., Lau, S. K., Li, K. S., Poon, R. W., Wong, B. H., Tsoi, H. W., Yip, B. C., Huang, Y., Chan, K. H., and Yuen, K. Y. (2006) Molecular diversity of coronaviruses in bats. *Virology* **351**, 180–187.

3

Detection and Sequence Characterization of the 3′-End of Coronavirus Genomes Harboring the Highly Conserved RNA Motif s2m

Christine Monceyron Jonassen

Abstract

A remarkably conserved 43-nucleotide-long motif present at the 3′-end of the genomes of several members of the polyadenylated RNA virus families *Astroviridae*, *Coronaviridae*, and *Picornaviridae* can be used for the detection and sequence characterization of the viruses harboring it. The procedure makes use of a primer located in the most conserved core of s2m toward a generic anchored oligo(dT) primer in a semispecific PCR. This strategy allows the sequencing of some 50–100 nucleotides from the 3′-end of the virus genome, representing sufficient sequence information for initiation of further genomic characterization in a rapid amplification of cDNA ends (5′-RACE) and primer walking strategy.

Key words: coronavirus; s2m; PCR; sequencing; 5′-RACE; detection; diagnosis

1. Introduction

S2m is a conserved motif present in the genomes of several members of the RNA virus families *Astroviridae*, *Coronaviridae*, and *Picornaviridae* (*1*). These viruses have single-stranded positive-sense RNA genomes, 6.5 to 32 kb long, with a poly(A) tail at their 3′-ends. S2m is 43 nucleotides long and forms a highly conserved RNA structure (*2*). In these very different and rapidly evolving viruses, the remarkably conserved nature of s2m, in both RNA sequence and folding, suggests an essential, but as yet unknown function. This motif is located

From: *Methods in Molecular Biology, vol. 454: SARS- and Other Coronaviruses,*
Edited by: D. Cavanagh, DOI: 10.1007/978-1-59745-181-9_3, © Humana Press, New York, NY

50–150 nucleotides upstream of the 3' poly(A) tail of these virus genomes. Phylogenetic considerations and the genomic position of s2m suggest that ancestors of some of these viruses acquired s2m by horizontal gene transfer.

All of the known group 3 coronaviruses, including infectious bronchitis virus (IBV), turkey coronavirus, pheasant coronavirus, and the recently characterized coronaviruses infecting different wild bird species *(3)*, as well as the group 2 coronavirus associated with the severe acute respiratory syndrome (SARS) *(4,5)* have been found to have s2m.

As the core of s2m is highly conserved in nucleotide sequence, it is an ideal target for identification of the viruses that contain it, and, together with a coronavirus replicase sequence, it was one of the probes that first identified the etiology of SARS as a coronavirus *(4,6)*. In this chapter the use of s2m as a handle in amplification strategies for obtaining sequence information of novel coronavirus genomes is described.

2. Materials

2.1. RT-PCR

1. The reagents for reverse transcription (RT) include: Superscript III reverse transcriptase (200 U/μl) and 5X first-strand buffer (250 mM Tris-HCl, 375 mM KCl, 15 mM MgCl$_2$), DTT (100 mM), dNTP mix (10 mM each), and RNaseOUT recombinant RNase inhibitor (40 U/μL) (all reagents from Invitrogen).
2. The primers used in the initial RT and polymerase chain reaction (PCR) are: anchored oligo(dT)$_{20}$, 5'-(T)$_{20}$VN, (2.5 μl/μg) (Invitrogen) or anchored oligo(dT)$_{18}$, 5'-(T)$_{18}$VN, as reverse primers, and s2m-core: 5' CCG AGT A(C/G)G ATC GAG GG as the sense primer.
3. Reagents for PCR: dNTP mix (10 mM), RNase and DNase-free water (both from Invitrogen), and HotStar Taq DNA polymerase (5 U/μl) and 10X buffer (containing 15 mM MgCl$_2$) (Qiagen).
4. Thermocyclers (MJ Research) are used both for RT and PCR.

2.2. Sequencing and Analysis

1. Agarose (Applied Biosystems) for visualization of PCR products.
2. Sequencing primers are s2m-coreseq: 5'-GAG TA(C/G) GAT CGA GGG TAC, AV12: 5'-(T)$_{18}$GC for sequencing the initial RT-PCR products (3.1) or coronavirus gene-specific primers for sequencing the 5'-RACE PCR products (3.3).
3. Oligo software version 6.68 is used to check for primer hairpin structures and primer dimers and to calculate annealing temperatures for PCR and cycle sequencing.

4. BigDye Terminator Cycle Sequencing kit version 1.1 or version 3.1 (Applied Biosystems) is used for sequencing of short (from initial PCR) or long (5′-RACE PCR) PCR products, respectively.
5. Cycle sequencing reaction is performed on an MJ Research thermocycler.
6. 3130xl Genetic Analyser for sequencing (Applied Biosystems).
7. Sequence analysis programs: Sequencher version 4.1.4 and BioEdit version 7.0.1.

2.3. 5′-Race

1. All reagents for RT are described in Section 2.1.1.
2. For RNase treatment of newly synthesized cDNA, a master mix of two-thirds part RNase H (2 U/μl) (Invitrogen) and one-third part RNase T1 (100–150 U/μl) (Roche) is made. The mix is made fresh, and stored at 4°C until use.
3. cDNA purification: QIAquick PCR purification kit (Qiagen).
4. cDNA yields are measured using a NanoDrop Spectrophotometer (Saveen & Werner).
5. Tailing reagents: Terminal deoxynucleotidyl transferase (TdT) (15 U/μl), 5X TdT buffer (0.5 M potassium cacodylate, 10 mM $CoCl_2$, 1 mM DTT), and dCTP (2 mM) (all from Invitrogen)
6. Sense primer in the 5′-RACE PCR: 5′-RACE abridged anchor primer (5′-GGC CAC GCG TCG ACT AGT ACG GGI IGG GII GGG IIG) (Invitrogen).
7. PCR enzyme used in the long-range 5′-RACE PCR: BD Advantage 2 Taq polymerase with 10X BD Advantage 2 PCR buffer (400 mM Tricine-KOH, 150 mM KOAc, 35 mM $Mg(OAc)_2$, 37.5 μg/μl BSA, 0.05 % Tween 20, and 0.05 % Nonidet-P40) (Clontech).

3. Methods

Determining the 3′-end sequences of unknown RNA can be cumbersome, and the present method makes use of s2m in some of the coronavirus genomes as a handle for determination of the sequences between s2m and the 3′ poly(A) tail of these genomes. The method can be used on RNA extracted from clinical samples (e.g., tracheal or cloacal swabs) that have been confirmed to be positive for coronavirus, e.g., by RT-PCR and sequencing, using pancoronavirus primers *(7)* (*see* **Note 1**).

As s2m is located only about 100 nucleotides upstream of the 3′ poly(A) tail when it is present in the genomes of coronaviruses, only limited sequence information can be acertained from the initial amplification product obtained using a primer located within s2m toward a generic primer for polyadenylated RNA. This limited sequence information is, however, sufficient to design virus-specific primers that can be used in a 5′-RACE strategy that, together with primer walking, allow sequencing toward the 5′-end of the virus genomes *(8)*.

3.1. Initial Amplification Using an s2m-Based Primer toward a Generic Primer of Polyadenylated Genomes in Coronavirus Positive Samples

1. For RT, prepare the following mix (per sample): 4 µl 5X first-strand buffer, 1 µl 100 mM DTT, 1 µl dNTP (10 mM), 1 µl anchored oligo(dT) primer (*see* **Note 2**), 1 µl RNaseOUT (40 U/µl), and 1 µl Superscript III reverse transcriptase. The mix should be made fresh, and stored at 4°C until use.

2. Add 11 µl RNA extract to the 9-µl RT mix, with care taken to avoid air bubbles, as these can oxidize the reverse transcriptase. The RT is performed at 50°C for 30 min, followed by an enzyme inactivation step at 70°C for 15 min, and cooling to 4°C thereafter.

3. PCR is performed using the s2m-core primer toward a generic anchored oligo(dT) primer. Prepare the following PCR mix (per sample): 35.7 µl RNase and DNase-free water, 5 µl 10X PCR buffer, 1 µl dNTP (10 mM), 1 µl sense primer s2m-core (25 µM), 2 µl reverse primer anchored oligo(dT)$_{18}$ (25 µM) and 0.3 µl HotStar Taq DNA polymerase. The mix is made fresh, and stored at 4°C until use.

4. A 5-µl cDNA sample is added to 45 µl PCR mix. The following PCR program is performed: 95°C 15 min for activation of the polymerase, followed by 40 cycles (94°C, 40 sec; 55°C, 20 sec; 72°C, 40 sec); 72°C, 5 min; and cooling at 8°C

3.2. Sequencing and Analysis of the Initial PCR Products

1. 15 µl of the PCR products are visualized on a 2% agarose gel. If s2m is part of the coronavirus genome, one should obtain specific amplification of the 3′-end (*see* **Note 3**). The PCR products are purified directly prior to sequencing reaction, using the QIAquick PCR purification kit according to the manufacturer's instructions, with a final elution volume of 30 µl.

2. The cycle sequencing reaction is performed using the BigDye Terminator version 1.1, especially suited for short PCR products. Add 4 µl BigDye terminator mix, 1.3 µl primer s2m-coreseq (2.5 µM) (*see* **Note 4**) to 4.7 µl purified PCR product, and perform the cycle sequencing reaction as follows: 96°C for 1 min, followed by 25 cycles (96°C, 10 sec; 55°C, 5 sec; 60°C, 4 min), and cooling to 4°C thereafter.

3. The cycle sequencing products are purified from unincorporated labeled dideoxynucleotides and separated by capillary electrophoresis on a 3130xl Genetic Analyser for sequence determination, according to Applied Biosystems' instructions.

4. The sequence obtained gives information on the 3′-end of the coronavirus genome. As s2m is located about 100–150 nucleotides upstream of the poly(A) tail of the coronaviruses harboring it, 50–100 nucleotides can be determined that way, and two new primers are designed, which can be partly overlapping, for use in PCR and sequencing in the 5′-RACE strategy (*see* **Note 5**).

3.3. 5′-RACE for Sequencing and Primer Walking Strategy

1. A new RT is performed, with care taken not to vortex samples, mixing by gentle pipetting to avoid breakage of the genomic RNA molecule and achieve long-range

cDNA synthesis, to get long sequence stretches in the 5′-RACE procedure (*see* **Note 6**). A 10-μl RNA extract is added to 1 μl anchored oligo(dT)$_{20}$ primer (2.5 μg/μl), and incubated for 5 min at 65°C, to get rid of RNA secondary structures. The sample is then rapidly chilled on ice, for 1 min. The sample is then spun down, and 4 μl 5X first-strand buffer, 1 μl 10 mM dNTP, and 1 μl 100 mM DTT are added and mixed gently by pipetting up and down. The samples are then incubated for 1 min at 55°C, and 1 μl RNaseOUT (40 U/μl) and 2 μl Superscript III (200 U/μl) are added to each sample, and the samples are rapidly put back into the thermocycler used for RT. The RT is performed at 55°C, for 80 min, followed by enzyme inactivation at 70°C and cooling at 4°C thereafter (*see* **Note 7**).

2. The samples from reverse transcription are centrifuged for 20 sec at 1538 × *g* to collect all droplets, and incubated at 37°C in a thermocycler. A 1-μl RNase mix is then added to each sample (*see* **Note 8**) and mixed gently by pipetting up and down, and the samples are put back into the thermocycler for further incubation for 30 min at 37°C. When the incubation is finished, the samples are spun down and cooled.

3. The newly synthesized cDNA is purified using the QIAquick PCR purification kit according to the manufacturer's instructions, with the following modifications: all mixing steps are performed gently, and the sample is eluted in 30 μl RNase/DNase-free water prewarmed at 65°C. The cDNA amount is measured spectrophotometrically on a NanoDrop.

4. Cytosine residues are added to the 3'-ends of purified cDNA, in order to make a homopolymeric (C) tail to serve as a handle for semispecific amplification with a specially designed anchor primer. The tailing is performed as follows: 10 μl of the purified cDNA is added to 6.5 μl RNase/DNase-free water, 5 μl 5X tailing buffer, and 2.5 μl 2 mM dCTP. In order to avoid secondary structure in the cDNA that might impair the tailing, a short denaturation step is performed at 94°C for 3 min. The sample is then immediately chilled on ice and spun down. One μl TdT (15 U/μl) is then added, and the tailing reaction is performed at 37°C for 10 min, followed by an enzyme inactivation step at 65°C for 10 min. The sample is then spun down and cooled on ice.

5. The 5′-RACE PCR is performed as follows (*see* **Note 9**): 5 μl tailed sample is added to 45 μl PCR mix consisting of 34.5 μl RNase/DNase-free water, 5 μl 10X BD Advantage 2 PCR buffer, 1 μl dNTP (10 mM), 1 μl coronavirus-specific reverse primer (25 μM), 2.5 μl abridged anchor primer (10 μM), and 1 μl BD Advantage polymerase mix. The cycling conditions include an initial enzyme-activation step at 95°C for 1 min, followed by 35 cycles of 94°C for 30 sec, 55°C for 20 sec, and 68°C for 7 min (*see* **Note 10**). The cycling is followed by a step at 68°C for 7 min and cooling to 4°C thereafter.

6. Prior to sequencing reaction, the PCR product, appearing as a weak smear on a 1% agarose gel electrophoresis, is purified directly using the QIAquick PCR purification kit according to the manufacturer's instructions, with a final elution volume of 30 μl (*see* **Note 11**). Cycle sequencing is performed using BigDye Terminator version 3.1, for long PCR product sequencing, with the following reagents: 4 μl BigDye Terminator mix, 1.3 μl coronavirus-specific reverse primer 2 (2.5 μM), and 4.7 μl PCR product. The cycle sequencing program consists of an initial step

at 96°C for 1 min, followed by 25 cycles of 96°C for 10 sec, 50°C for 5 sec, and 60°C for 4 min, followed by cooling at 4°C.

7. The cycle sequencing products are purified from unincorporated labeled dideoxy-nucleotides and separated by capillary electrophoresis on a 3130xl Genetic Analyser for sequence determination, according to Applied Biosystems' instructions (*see* **Note 12**).

4. Notes

1. In addition to coronaviruses, this procedure has been successfully applied to amplify and sequence novel astrovirus and picornavirus genomes, and potentially all polyadenylated viral genomes harboring s2m can be identified and character-ized with this procedure. However, when used on biological samples that do not contain s2m-harboring viruses, nonspecific amplification of host cell ribosomal RNA is obtained on some occasions.

2. We have generally used 1 μl anchored oligo(dT)$_{18}$ (25 μ*M*) in the RT, but 1 μl Invitrogens anchored oligo(dT)$_{20}$ (2.5 μg/μl, corresponding to ca. 350 μM), or Invitrogens standard oligo(dT)$_{20}$ (50 μM), has also been used successfully as a reverse primer.

3. The obtained PCR products often give a smear on visualization on agarose gel, owing to serial mispriming of the anchored oligo(dT) primer on the 3' poly(A) tail, leading to long-range accumulation of As in the PCR products that can be observed upon sequencing. However, if amplification is successful, all PCR products, regardless of the number of terminal As, have the same start at s2m, as the s2m-core primer is highly specific, and the PCR product smear can be sequenced using s2m-core or s2m-coreseq primer.

4. As several of the viruses harboring s2m do terminate the genome with GC(A)n, AV12 (2.5 μM) can often be used for sequencing the opposite strand. If low signal intensity is obtained for the sequences, the following modifications can be made in the cycle sequencing program: lowering the annealing temperature to 50°C and/or adding 5–10 cycles in the cycle sequencing reaction.

5. Once a specific reverse primer can be designed as described in Section 3.2, an alternative strategy to 5'-RACE/primer walking is to amplify directly a long PCR product using upstream primers in conserved family-specific motifs, e.g., in the polymerase gene. This will be achieved more easily with viruses that have small genomes, such as astrovirus and picornavirus, especially when the polymerase is encoded toward the 3'- end of the genome, rather than with coronaviruses that have a large genome. In our laboratory, we were able to get sequence information by using the sense 2Bm primer designed by Stephensen et al. in 1999 *(7)* toward the coronavirus specific primer at the 3'-end of the coronavirus genome. How-ever, the PCR product yield was low, and only short sequence stretches could be obtained in this way.

6. The RT in the 5'-RACE procedure is laborious, as no mix containing all the reagents is prepared in advance, but, rather, primer and enzymes are added

individually to each sample. In addition, the 5′-RACE procedure described in the Invitrogen handbook suggests that the RNase treatment of the sample be performed on freshly synthesized cDNA, that has not been frozen. This is why the initial RT step described in Section 3.1, performed in order to check novel coronavirus genomes for the presence of s2m, makes use of a simpler RT protocol.

7. An increased amount of the Superscript III reverse transcriptase (400 U per reaction instead of the normally 200 U) is used for long-range cDNA synthesis in the 5′-RACE strategy. In addition, this enzyme allows the reaction to be performed at a temperature as high as 55°C, which minimizes secondary structures in RNA and is RNase H negative, which increases the yield of full-length cDNAs.

8. It is important to degrade the template RNA from the newly synthesized cDNA, to avoid hybrid renaturation prior to the tailing step. The RNase mix consists of both RNase H to hydrolyze RNA bound to cDNA, and RNase T1 to hydrolyze free RNA molecules.

9. The 5′-RACE PCR is semispecific, using an anchor primer as sense primer, which amplifies the poly(C)-tailed template, and a newly designed coronavirus-specific reverse primer. Even if care has been taken throughout the whole procedure to avoid synthesis of short cDNAs from template RNA, there might be some shorter cDNA molecules that will be more easily amplified in PCRs. It is therefore important to use a PCR enzyme with a proofreading activity, in order to get PCR products from the longer cDNA molecules as well.

10. The long elongation step is chosen to allow amplification of longer cDNAs.

11. As cDNA molecules of different lengths are generated during RT, the 5′-RACE PCR products will be of different lengths, so the 5′-RACE abridged anchor primer cannot be used in cycle sequencing. As all the 5′-RACE PCR products have aligned 3′-ends, the coronavirus-specific 5′-RACE reverse PCR primer or, preferably for increased specificity, a primer slightly upstream of this PCR primer is used in cycle sequencing.

12. Using this procedure, one should obtain at least 500 nucleotides, which would allow the design of new primers both in sense direction for sequence verification and reverse direction for acquisition of new sequence data toward the 5′-end of the genome in a primer walking strategy. The new primers can be used directly in step 5 of the 5′-RACE procedure, if all the steps have been optimized for long-range cDNA synthesis. However, usually no more than 2000–3000 nucleotides are efficiently sequenced from the initial 5′-RACE-RT, and the whole 5′-RACE procedure has to be repeated with a primer for cDNA synthesis as upstream as possible.

References

1. Jonassen, C. M., Jonassen, T. O., and Grinde, B. (1998) A common RNA motif in the 3' end of the genomes of astroviruses, avian infectious bronchitis virus and an equine rhinovirus *J. Gen. Virol.* **79**, 715–718.

2. Robertson, M. P., Igel, H., Baertsch, R., Haussler, D., Ares, M., Jr., and Scott, W. G. (2005) The structure of a rigorously conserved RNA element within the SARS virus genome *PLoS Biol.* **3**, E5.
3. Jonassen, C. M., Kofstad, T., Larsen, I. L., Løvland, A., Handeland, K., Follestad, A. and Lillehaug, A. (2005) Molecular identification and characterization of novel coronaviruses infecting graylag geese (*Anser anser*), feral pigeons (*Columbia livia*) and mallards (*Anas platyrhynchos*) *J. Gen. Virol.* **86**, 1597–1607.
4. Ksiazek, T. G., Erdman, D., Goldsmith, C. S., Zaki, S. R., Peret, T., Emery, S., Tong, S., Urbani, C., Comer, J. A., Lim, W., Rollin, P. E., Dowell, S. F., Ling, A. E., Humphrey, C. D., Shieh, W. J., Guarner, J., Paddock, C. D., Rota, P., Fields, B., DeRisi, J., Yang, J. Y., Cox, N., Hughes, J. M., Le Duc, J. W., Bellini, W. J., and Anderson, L. J. (2003) A novel coronavirus associated with severe acute respiratory syndrome *N. Engl. J. Med.* **348**, 1953–1966.
5. Marra, M. A., Jones, S. J., Astell, C. R., Holt, R. A., Brooks-Wilson, A., Butterfield, Y. S., Khattra, J., Asano, J. K., Barber, S. A., Chan, S. Y., Cloutier, A., Coughlin, S. M., Freeman, D., Girn, N., Griffith, O. L., Leach, S. R., Mayo, M., McDonald, H., Montgomery, S. B., Pandoh, P. K., Petrescu, A. S., Robertson, A. G., Schein, J. E., Siddiqui, A., Smailus, D. E., Stott, J. M., Yang, G. S., Plummer, F., Andonov, A., Artsob, H., Bastien, N., Bernard, K., Booth, T. F., Bowness, D., Czub, M., Drebot, M., Fernando, L., Flick, R., Garbutt, M., Gray, M., Grolla,A., Jones, S., Feldmann, H., Meyers, A., Kabani, A., Li,Y., Normand, S., Stroher, U., Tipples, G. A., Tyler, S., Vogrig, R., Ward, D., Watson, B., Brunham, R. C., Krajden, M., Petric, M., Skowronski, D. M., Upton, C., and Roper, R. L. (2003) The genome sequence of the SARS-associated coronavirus *Science* **300**, 1399–1404.
6. Wang, D., Urisman, A., Liu, Y. T., Springer, M., Ksiazek, T. G., Erdman, D. D., Mardis, E. R., Hickenbotham, M., Magrini, V., Eldred, J., Latreille, J. P., Wilson, R. K., Ganem, D., and Derisi, J. L. (2003) Viral discovery and sequence recovery using DNA microarrays *PLoS Biol.* **1**, E2.
7. Stephensen, C. B., Casebolt, D. B., and Gangopadhyay, N. N. (1999) Phylogenetic analysis of a highly conserved region of the polymerase gene from 11 coronaviruses and development of a consensus polymerase chain reaction assay *Virus Res.* **60**, 181–189.
8. Jonassen, C. M., Jonassen, T. O., Sveen, T. M., and Grinde, B. (2003) Complete genomic sequences of astroviruses from sheep and turkey: comparison with related viruses *Virus Res.* **91**, 195–201.

4

RT-PCR Detection of Avian Coronaviruses of Galliform Birds (Chicken, Turkey, Pheasant) and in a Parrot

Francesca Anne Culver, Paul Britton, and Dave Cavanagh

Abstract

Of the many primer combinations that we have investigated for the detection of avian coronaviruses, two have worked better than any of the others: they worked with the largest number of strains/samples of a given coronavirus and the most species of avian coronavirus, and they also produced the most sensitive detection tests. The primer combinations were: oligonucleotide pair 2Bp/4Bm, which is in a region of gene 1 that is moderately conserved among all species of coronavirus (*1*); and UTR11-/UTR41+, which are in a highly conserved part of the 3′ untranslated region of avian coronaviruses related to infectious bronchitis virus (*2*). The gene 1 primer pair enabled the detection of a new coronavirus in a green-cheeked Amazon parrot (*Amazon viridigenalis Cassin*). In this chapter we describe the use of these oligonucleotides in a one-step (single-tube) RT-PCR, and describe the procedure that we used to extract RNA from turkey feces.

Key words: pan-coronavirus PCRs; coronavirus detection; coronavirus discovery; avian coronaviruses; galliform birds

1. Introduction

In recent years sequence analysis has shown that turkey coronavirus (TCoV) and pheasant coronavirus (PhCoV) have a gene order that is the same as that of the longer-studied infectious bronchitis virus (IBV) of the domestic fowl (chicken): (5′-replicase-S-3-M-5-N-3′ *(2–5)*, placing them all in coronavirus group 3 *(6)*. Protein sequence identities are frequently greater than 80% among

From: *Methods in Molecular Biology, vol. 454: SARS- and Other Coronaviruses,*
Edited by: D. Cavanagh, DOI: 10.1007/978-1-59745-181-9_4, © Humana Press, New York, NY

these CoV species, except for the spike (S) protein of TCoV, which has only about 30% identity with that of IBV and PhCoV, and only about 17% with the S1 subunit of the S protein *(5)*. Recently, a coronavirus in quail has been shown to have an S1 sequence with about 80% identity with that of TCoV *(7)*. In gene 1, which encodes RNA replication functions, there are sequences with moderately high nucleotide identity among all known coronaviruses, which has led to the design of oligonucleotide combinations for the detection of all known coronaviruses—pancoronavirus primers (see, e.g., *(1)* and other chapters in this book). Herein we demonstrate the utility of the Stephensen primers for the detection of a coronavirus in parrots *(8)*.

The terminal region of the 3′UTR of group 3 coronaviruses is very highly conserved, including in coronaviruses from pigeon, duck, and goose *(9)*. In our field studies of TCoV, we used a 3′ UTR primer pair very successfully to detect TCoV in feces and gut contents [*(10)* and unpublished observations]. In this chapter we describe the conditions for use of these primers in conjunction with RNA extracted from feces.

2. Materials

2.1. RNA Extraction from Feces

1. We used the QIAamp DNA stool mini kit (Qiagen) for RNA extraction from feces, which has a special step designed to remove inhibitors that are present in feces (*see* **Notes 1 and 2**).
2. Absolute ethanol.

2.2. One-Step Reverse-Transcriptase Polymerase Chain Reaction (RT-PCR)

1. We performed a one step-RT-PCR using Ready-to-go RT-PCR beads (Amersham Biosciences), essentially as described by the manufacturer (*See* **Note 9**).
2. Nuclease-free water.
3. Oligonucleotide primers dissolved in nuclease-free water to a concentration of either 10 or 100 pmol/μl. The composition of the oligonucleotides is shown in **Table 1**.
4. Mineral oil.
5. Thermocycler.

2.3. Detection of PCR Products

1. Agarose.
2. Distilled water.
3. Ethidium bromide.
4. 10X Tris borate EDTA buffer (TBE): 1 M Tris, 0.9 M boric acid, 10 mM ethylenediaminetetracetic acid.

Table 1
Details of the Oligonucleotide Primers Described in this Chapter (*see Note 12*)

Primer name	Sequence	Length	Sense	IBV 5' position[a]	IBV 3' position[a]	Tm (°C)	Region amplified	Avian CoV amplified[b]	Reference
2 Bp	act ca(ag) (at)t(ag) aat (ct) t(agct) aaa ta(ct) gc	23	+	13953	13976	52	gene 1	I, T, Ph, Pa	(1)
4 Bm	tca ca(ct) tt(at) gga ta(ag) tcc ca	20	–	14184	14204	53	gene 1	I, T, Ph, Pa	(1)
UTR41+	atg tct atc gcc agg gaa atg tc	23	+	27342	27364	60	3' UTR	I, T, Ph	(2)
UTR11–	gct cta act cta tac cct a	22	–	27586	27607	56	3' UTR	I, T, Ph	(2)

[a] Relative to the genome of the Beaudette strain of IBV (Accession number AJ311317).
[b] The two primer combinations were tested using RNA extracted from IBV (I), TCoV, (T), pheasant coronavirus (PhCoV), and parrot coronavirus (PaCoV).

3. Methods

3.1. RNA Extraction from Feces

For RNA extraction from feces we used the QIAamp DNA stool mini kit (Qiagen), which has a step designed to remove inhibitors that are present in feces (*See* **Notes 1 and 2**). RNA extraction comprises three main steps: 0.2 g of each stool sample is lysed; a reagent in the kit called InhibitEX then absorbs any impurities (by removing PCR inhibitors present in feces—*see* **Note 3**); finally the RNA is purified on spin columns. The RNA is bound to a silica gel membrane, impurities are washed away, and then pure RNA is eluted from the spin column in nuclease-free water. We followed the manufacturer's instructions:

1. Perform RNA extraction from fresh or frozen material (*see* **Note 5**).
2. Use a pea-sized amount of feces, which is approximately 0.2 g, fresh or frozen (*see* **Note 5**).
3. If using fresh, i.e., not frozen, feces, homogenize the feces using a sterile spatula before dispensing into a microfuge tube (*see* **Note 6**). Feces should be kept on ice until buffer is added.
4. Add 1.4 ml of ASL buffer to each fecal sample. Vortex until thoroughly homogenized (30 sec to 1 min).
5. Heat fecal suspensions in a 70°C water bath for 5 min then vortex each sample for 15 sec.
6. Centrifuge each sample for 1 min to pellet any particulate matter.
7. Pipette 1.2 ml of supernatant into a new microfuge tube and discard the pellet (*see* **Note 8**).
8. Add one Inhibitex tablet to each sample and vortex for 1 min. Incubate the suspension for 1 min (*see* **Note 7**). Incubate at room temperature for 1 min before centrifuging the sample to pellet the Inhibitex tablet to which the inhibitors will be bound.
9. Pipette all the supernatant into a new 1.5-ml microfuge tube. Centrifuge the supernatant for 3 min.
10. Add 15 μl Proteinase K (provided in the kit) into a new 1.5-ml microfuge tube. To this add 200 μl of the supernatant and 200 μl AL buffer and mix vigorously. Then incubate for 10 min at 70°C.
11. To this, add 200 μl of 100% ethanol and mix thoroughly. Add the contents of this microfuge tube (steps 9, 10, and 11) to a QIAamp spin tube.
12. Centrifuge for 1 min and place the column into a new collection tube. To the spin column, add 500 μl of AW1 buffer. Centrifuge for 1 min.
13. Place spin column into a new collection tube. To the spin column, add 500 μl of AW2 buffer. Centrifuge for 3 min.
14. Place spin column into a new collection tube. Centrifuge for 1 min. Place spin column into a new microfuge tube. Pipette 200 μl of AE buffer onto the spin column (do not use water). Incubate for 1 to 5 min at room temperature before eluting RNA from the spin column by centrifugation for 1 min.

15. Upon elution, store the RNA on ice for as short a period of time as possible until it can be transferred to $-20°C$. It may be stored for several months at this temperature and may be thawed on ice a few times for subsequent RT-PCR.

3.2. One-Step Reverse-Transcriptase Polymerase Chain Reaction (RT-PCR)

We used a one-step RT-PCR, i.e., RT and PCR reagents all together in one microfuge tube throughout the procedure, using Ready-to-go RT-PCR beads (Amersham Biosciences) (*see* **Note 9**). Perform al RT-PCR steps on ice. It is advisable to use filter tips for pipettes when measuring volumes (*see* **Note 4**). For this one-step RT-PCR reaction:

1. Add nuclease-free water to a 0.5-ml microfuge tube containing beads (comprising, most importantly, lyophilized *Taq* DNA polymerase, dNTPs, murine Maloney leukemia virus reverse transcriptase, and a ribonuclease inhibitor).
2. Incubate microfuge tube containing nuclease-free water and bead on ice for 5 min, ensuring that the beads are resuspended in the water. The amount of water added depends on the volumes of RNA and primers; the final volume, after addition of oligonucleotide primers (see next step) should be 50 μl.
3. To this add 2 μl forward primer (1 pmol/μl) and 2 μl reverse primer (1 pmol/μl) and 1 to 10 μl of RNA (*see* **Note 10**). Reminder: upon addition of the primers and template, total reaction volume should be 50 μl.
4. Add a 50 μl mineral oil overlay to the microfuge tube while it is still on ice before placing the tubes into PCR thermocycler (*see* **Note 11**). The RT-PCR is performed in a thermocycler with a heated lid using the following program: $42°C$ for 30 min and $95°C$ for 5 min for the reverse transcriptase step. First-strand cDNA generated with these beads is used as a direct template for PCR. The amplification step is denaturation at $95°C$ for 1 min, primer annealing at the desired temperature for 1 min, and primer extension at $72°C$ for 3 min. These cycles are repeated 35 times, with a final step of $72°C$ for 10 min. Use $50°C$ for the gene 1 oligonucleotide combination 2Bp/4Bm, and $48°C$ for 3′ UTR oligonucleotide combination UTR11-/UTR41+. Visualize PCR products on a 1% agarose gel containing ethidium bromide (1 μg).

4. Notes

1. We used the Qiagen QIAamp stool extraction kit, which has been employed by others for extraction of RNA for pathogen detection, including the detection of SARS–CoV in human feces (*11*) and by ourselves for the detection of turkey coronavirus from turkey feces (*10*).

2. When RNA extraction or RT-PCR is being carried out in a laboratory on a routine basis it is important to put safeguards in place to prevent the detection of false positives or false negatives, which may occur owing to several sources of cross-contamination. For example, it is advisable to have separate areas for the handling of PCR reagents and areas in which the RT-PCR is set up. If possible each step of the process should be carried out in a separate room, e.g., one area for the storage of reaction components, one area for the reaction to be set up, another area where the reaction tube will be opened after PCR has been carried out, and another for the corresponding sequencing reactions.

3. The surfaces of the bench at which RNA extraction or RT-PCR is carried out should be prepared by cleaning with an RNAse inhibitor spray and 70% ethanol solution.

4. Filter tips should be used for all pipettes. If available, separate sets of pipettes should be used for RNA extraction and RT-PCR. The design of the experiment should incorporate positive and negative controls.

5. Freeze feces for storage. For subsequent sampling, simply scrape frozen feces off the surface; there is no need to thaw the feces.

6. We found that sterile ice-cream sticks (wooden birch sticks) were useful for this purpose, as they are cheap and disposable.

7. Feces contain materials that may adversely affect enzyme function or nucleic acid integrity. Although avian feces have a different composition from that of mammals, the QIAamp DNA stool kit of Qiagen was certainly effective when used with turkey and chicken feces.

8. When opening and closing microfuge tubes during RNA extraction, be careful of any aerosol or droplets spraying and contaminating work surfaces or other samples.

9. This is a one-step RT-PCR reaction to reduce the amount of handling required. In our laboratory it has been used to amplify RT-PCR products of between 200 bp and 2.5 kb in length.

10. We found that 2 µl of RNA was usually sufficient. However, if a very small amount of virus is suspected within fecal samples, up to 10 µl of RNA extracted from feces can be added without deleterious effect.

11. In our experience, better amplification of RNA products is obtained when mineral oil is added to the reaction, regardless of whether or not a thermocycler, with or without, a hot lid is used.

12. Even the conserved, 3′-terminal part of the 3′ UTR of group 3 avian coronaviruses exhibits some variation. It is likely that oligonucleotides UTR11– and UTR41+ will require modification to increase their universality for group 3 coronaviruses. For example, we modified our initial versions of these primers [UTR1– and UTR4+ (*12*)] to the current UTR11– and UTR41+ as more IBV sequences became available. We did this by moving the position of the oligonucleotides very slightly to avoid variation at the extreme 3′- end of the oligonucleotides. The data of Jonassen et al. (*9*) for group 3 coronaviruses in pigeon, mallard duck, and greylag goose suggests that our UTR11– and UTR41+ primers should be

modified slightly to increase the likelihood of them working with coronaviruses in a greater number of species. Modification might well include degeneracy, especially with respect to the 3′-most nucleotide.

Acknowledgments

We are grateful to the British Poultry Council, Merial Animal Health, Intervet UK, the Department of the Environment, Food and Rural Affairs, and the Biotechnology and Biological Sciences Research Council for support that contributed to this paper.

References

1. Stephensen, C. B., Casebolt, D. B., and Gangopadhyay, N. N. (1999) Phylogenetic analysis of a highly conserved region of the polymerase gene from 11 coronavirus and development of a consensus polymerase chain reaction assay. *Virus Res.* **60**, 181–189.
2. Cavanagh, D., Mawditt, K., Sharma, M., Drury, S. E., Ainsworth, H. L., Britton, P., and Gough, R. E. (2001) Detection of a coronavirus from turkey poults in Europe genetically related to infectious bronchitis virus of chickens. *Avian Pathol.* **30**, 365–378.
3. Breslin, J. J., Smith, L. G., Fuller, F. J., and Guy, J. S. (1999) Sequence analysis of the matrix/nucleocapsid gene region of turkey coronavirus. *Intervirology* 42, 22–29.
4. Cavanagh, D., Mawditt, K., Welchman, D. de B., Britton, P., and R. E. Gough (2002) Coronaviruses from pheasants (*Phasianus colchicus*) are genetically closely related to coronaviruses of domestic fowl (infectious bronchitis virus) and turkeys. *Avian Pathol.* **31**, 81–93.
5. Lin, T. L., Loa, C. C., and Wu, C. C. (2004) Complete sequences of 3' end coding region for structural protein genes of turkey coronavirus. *Virus Res.* **106**, 61–70.
6. Cavanagh, D. (2005) Coronaviruses in poultry and other birds. *Avian Pathol.* **34**, 439–448.
7. Circella, E., Camarda, A., Martella, V., Bruni, G., Lavazza, A., and Buonavoglia, C. (2007) Coronavirus associated with an enteric syndrome in a quail farm. *Avian Pathol.* **36**, 251–258.
8. Gough, R. E., Drury, S. E., Culver, F., Britton, P., and Cavanagh, D. (2006) Isolation of a coronavirus from a green-cheeked Amazon parrot (*Amazon viridigenalis* Cassin). *Avian Pathol.* **35**, 122–126.
9. Jonassen, C. M., Kofstad, T., Larsen I-L., Lovland, A., Handeland, K., Follestad, A., and Lillehaug, A. (2005) Molecular identification and characterization of novel coronaviruses infecting greylag geese (*Anser anser*), feral pigeons (*Columba livia*) and mallards (*Anas platyrhynchos*). *J. Gen. Virol.* **86**, 1597–1607.

10. Culver, F., Dziva, F., Cavanagh, D., and Stevens, M.P. (2006) Poult enteritis and mortality syndrome in turkeys in the UK. *Vet. Rec.* **159**, 209–210.

11. Thiel, V., Ivanov, K.A., Putics, A., Hertzig, T., Schelle, B., Bayer, S., Weißbrich, B., Snijder, E. J., Rabenau, H., Doerr, H.W., Gorbalenya, A. E., and Ziebuhr, J. (2003) Mechanisms and enzymes involved in SARS coronavirus genome expression. *J. Gen. Virol.* **84**, 2305–2315.

12. Adzhar, A., Shaw, K., Britton, P., and Cavanagh, D. (1996) Universal oligonucleotides for the detection of infectious bronchitis virus by the polymerase chain reaction. *Avian Pathol.* **25**, 817–836.

5

Detection of Group 2a Coronaviruses with Emphasis on Bovine and Wild Ruminant Strains

Virus Isolation and Detection of Antibody, Antigen, and Nucleic Acid

Mustafa Hasoksuz, Anastasia Vlasova, and Linda J. Saif

Abstract

Group 2a of the *Coronaviridae* family contains human and animal pathogens that include mouse hepatitis virus, rat coronavirus, human respiratory coronaviruses OC43 and the recently identified HKU1 strain, a newly recognized canine respiratory coronavirus, porcine hemagglutinating encephalomyelitis virus, equine coronavirus, bovine coronavirus (BCoV), and wild-ruminant coronaviruses. The presence of a hemagglutinin-esterase (HE) surface glycoprotein in addition to the viral spike protein is a distinguishing characteristic of most group 2a coronaviruses. BCoV is ubiquitous in cattle worldwide and is an economically significant cause of calf diarrhea, winter dysentery of adult cattle, and respiratory disease in calves and feedlot cattle. We have developed and optimized laboratory diagnostic techniques, including virus isolation in HRT-18 cell cultures, antibody and antigen ELISA, and RT-PCR, for rapid, sensitive, and reliable diagnosis of BCoV and related wild ruminant coronaviruses.

Key words: coronavirus; group 2a; bovine coronavirus (BCoV); wild-ruminant coronavirus; diagnostic tests; antibody detection; antigen detection; RT-PCR; polymerase chain reaction

1. Introduction

Coronaviruses cause a broad spectrum of diseases in domestic and wild animals, poultry, and rodents, ranging from mild to severe enteric, respiratory, and systemic disease, as well as the common cold or pneumonia in humans *(1–3)*.

From: *Methods in Molecular Biology, vol. 454: SARS- and Other Coronaviruses,*
Edited by: D. Cavanagh, DOI: 10.1007/978-1-59745-181-9_5, © Humana Press, New York, NY

Recently SARS-CoV emerged, likely from a wildlife reservoir, as a new CoV (group 2b) causing severe respiratory disease in humans *(4–10)*. Bats are a suspect reservoir for SARS-like CoVs with civet cats possibly playing a role as an intermediate host *(6,8–13)*.

The widespread prevalence of infections caused by group 2 coronaviruses, their extensive host range, the various disease manifestations, a high frequency of genomic recombination events, and the potential for interspecies transmission (BCoV, SARS-CoV) are characteristics that require continuous monitoring and improvement of diagnostic tests for these CoVs *(8,9)*. A summary of infections and standardized diagnostic tests available for group 2a CoVs is shown in **Table 1**.

Our laboratory has focused extensively on the study of bovine coronavirus (BCoV) and bovine-like CoVs for many years. Bovine coronavirus causes acute diarrhea in neonatal calves and winter dysentery in adult cattle causing large economic losses owing to diarrheal morbidity and mortality costs and decreased milk production *(2,27–31)*. The enteropathogenic coronaviruses were initially identified and isolated from neonatal calves with severe diarrhea *(31)*. Subsequently, coronaviruses were revealed to be a cause of winter dysentery in adult cattle *(2,32–35)* as well as respiratory disease (shipping fever pneumonia) *(29,36–40)*, and antigenic, biologic, and genetic variation among these strains was demonstrated *(28,37,41–43)*. Repeated upper respiratory BCoV infections occur frequently in calves, and subclinically infected animals may be reservoirs for BCoV *(28,30)*.

Recently, bovine-like CoVs were recognized as important enteric pathogens in captive wild ruminants from the United States, including Sambar deer (*Cervus unicolor*), white-tailed deer (*Odocoileus virginianus*), waterbuck (*Kobus ellipsiprymnus*) *(44)*, elk (*Cervus elephus*) *(45)*, and giraffe (*Giraffa camelopardalis*) *(45a)*. Coronaviruses isolated from these species were antigenically indistinguishable from BCoV *(44)*. Furthermore, some wild ruminants such as caribou (*Rangifer tarandus*), sitatunga (*Tragelaphus spekei*), musk oxen (*Ovibus moschatus*), white-tailed deer (*Odocoileus virginianus*), and mule deer (*Odocoileus hemionus*) were shown to have antibodies to BCoV *(44–47)*.

Detailed methods for cell culture propagation of the enteric BCoVs were described by Saif et al. in 1988 *(48)*. General procedures for the isolation of winter dysentery *(49)* and respiratory strains of BCoV in HRT-18 cell cultures (using cloned HRT-18 cells from the L. J. Saif laboratory) *(37)*, antibody and antigen ELISA, and RT-PCR have been optimized and standardized in our laboratory *(2,28,50,51)* and are described in detail in the following sections. For the antigen (Ag) ELISA tests, monoclonal antibodies (MAb) were produced and characterized by our laboratory and standardized for routine use in ELISA, immunoblotting, and immunohistochemistry *(41,50–52)*.

Table 1
Group 2a CoVs, Their Clinical Manifestations and Diagnostic Approaches

Virus	Disease	Test samples	Diagnostic tests	References
HuCoV OC-43 (Human coronavirus)	Respiratory infection	Respiratory secretions	RT-PCR, ELISA	(1,17,55)
HuCoV HKU1 (Human coronavirus)	Respiratory infection	Respiratory secretions	RT-PCR, ELISA	(3,18)
BCoV (Bovine coronavirus)	Enteric and respiratory infection	Feces and respiratory secretions, serum for serolgy	RT-PCR, ELISA, IFA, IEM	(2,27,28,38, 50,51,52)
MHV (Mouse hepatitis virus)	Respiratory and enteric infections, hepatitis, splenolysis, immune dysfunction, acute encephalitis, and chronic demyelinating disease of the brain and spinal cord	Liver, spleen, lung, intestines, feces, brain homogenates	RT-PCR, ELISA, IFA, infant mouse bioassay, MAP (mouse antibody production) test	(14–16,24, 26,56)
SDAV (Rat sialodacryo-adenitis virus) PRC (Parker's rat coronavirus)	Sialodacryoadenitis and respiratory illness	Salivary glands, nasal turbinates, trachea, lung, respiratory secretions, serum for serology	RT-PCR, IFA, ELISA	(23–26)
PHEV (Porcine hemagglutinating encephalomyelitis virus)	Two clinical forms: the vomiting and wasting disease (VWD) and the encephalitic forms (confined almost exclusively to pigs <4 weeks old)	Brain stem cell homogenates, throat swabs, serum for serology	RT-PCR, ELISA, HI, serum neutralization plaque reduction	(19)
ECoV (Equine coronavirus)	Enteric infection	Feces	RT-PCR	(22)
CRCoV (Canine respiratory coronavirus)	Respiratory infection	Respiratory secretions, serum for serology	RT-PCR, ELISA	(20,21)

2. Materials

2.1. Virus Isolation and Plaque Induction

1. Advanced Minimal Eagle's Medium (Advanced MEM) (Gibco, Invitrogen Corporation, Grand Island, NY, USA)
2. Trypsin-EDTA 1X (ethylendiamine tetraacetic acid) (1 mM) (Gibco).
3. Pancreatin solution (Sigma-Aldrich, St. Lois, MO, USA).
4. Fetal bovine serum, FBS (HyClone, Lawrenceville, GA, USA) (*see* **Note 1**).
5. L-Glutamine (200 mM).
6. Antibiotic-antimycotic 100X liq. solution (Gibco).
7. Neutral red.
8. DEAE-Dextran (diethylaminoethyl-Dextran) (Sigma-Aldrich).
9. Diluent #5 (Dil. #5): Minimal essential medium, MEM (Gibco) supplemented with 1% of antibiotic-antimycotic solution (Gibco) and 1% NaHCO₃; final pH should be 7.2.

2.2. Antigen-Capture ELISA for Detection of Viral Antigens

1. Coating buffer—0.1 M carbonate bicarbonate buffer: 1.59 g Na_2CO_3, 2.92 $NaHCO_3$, distilled water to 1000 ml. Adjust pH to 9.6 with HCl.
2. 96-well microtiter plates (Nunc, Rochester, NY, USA).
3. Coating MAbs: BCoV MAbs (developed against HE, S and N proteins of BCoV) *(50)*.
4. Phosphate buffered saline (PBS) 10X stock: 1.37 M NaCl, 27 mM KCl, 100 mM Na_2HPO_4, 18 mM KH_2PO_4 (adjust pH to 7.2 with HCl).
5. Wash buffer: PBS pH 7.2 containing 0.05% Tween-20.
6. Blocking buffer: 5% (w/v) nonfat dry milk (NFDM) in PBS.
7. Primary and secondary Abs-dilution buffer: PBS pH 7.2 containing 0.05% Tween-20.
8. Primary Ab: Guinea pig antiserum to BCoV Mebus strain.
9. Secondary Ab: Goat anti-guinea pig IgG horseradish peroxidase (HRP) conjugate (KPL, Gaithersburg, MD, USA).
10. TMB (3,3′,5,5′-tetramethylbenzidine) microwell peroxidase substrate (KPL).
11. Stop solution: 1 M phosphoric acid (H_3PO_4).

2.3. Antibody-Capture ELISA (Serological Test) for Detection of Virus-Specific Antibodies

1. Coating buffer: 0.1 M carbonate bicarbonate buffer (as described in Section 2.2).
2. 96-well microtiter plates (Nunc).
3. Coating MAbs: BCoV MAbs (developed against HE, S and N proteins of BCoV).

4. Phosphate buffered saline (PBS), as described in Section 2.2.
5. Wash buffer: PBS pH 7.2 containing 0.05% Tween-20.
6. Blocking buffer: 5% (w/v) NFDM in PBS.
7. Dilution buffer: PBS pH 7.2 containing 0.05% Tween-20.
8. Goat anti-bovine IgG HRP conjugate (KPL).
9. TMB microwell peroxidase substrate (KPL).
10. Stop solution: 1 M phosphoric acid (H_3PO_4).

2.4. RT-PCR Detection of Coronavirus RNA

1. TRIzol LS reagent (Invitrogen, Carlsbad, CA, USA)
2. 0.2 ml PCR tubes w/flat cap (Phenix, Hayward, CA, USA).
3. 10X thermophillic DNA polymerase buffer (Promega, Madison, WI, USA).
4. Magnesium chloride 25 mM (Promega) (*see* **Note 5**).
5. 0.2 M dNTPs mix (Promega).
6. Forward and reverse primers (see **Table 2**).
7. AMV Reverse Transcriptase (Promega).
8. RNasin (Promega).
9. Taq DNA polymerase (Promega).
10. Milli-Q diethylpyrocarbonate(DEPC)-treated water. Add DEPC (*see* **Note 6**) to Milli-Q water to a final concentration 0.1%, incubate for 2 h at 37°C with occasional shaking and autoclave (to dissociate DEPC).
11. 6X blue/range loading dye (Promega).
12. Ethidium bromide solution with concentration of 0.5 µg/ml.
13. Tris-acetate-EDTA (TAE) buffer: 2 M Tris-base, 2 M acetic acid, 0.05 M EDTA (pH 8.0).
14. Agarose I, biotechnology grade (Amresco, Solon, OH, USA).

3. Methods

3.1. Virus Isolation and Plaque Induction

3.1.1. Preparation of Fecal and Nasal Samples

Fecal and nasal samples are used for virus isolation, antigen capture ELISA, and RNA extraction for RT-PCR. Feces from animals (domestic cattle or wild ruminants), with or without clinical signs, should be collected in sterile fecal cups, put on ice, and transported. Then, 0.5 g of fecal sample is diluted in 4 ml of Dil. #5, vortexed, and centrifuged at 2000 × g for 30 min. The supernatant is aspirated and stored at −70°C until use. Samples should be filtered (using 0.22-nm filters) before inoculation onto the HRT-18 cell culture monolayers.

Paired sterile polyester-tipped swabs are used to collect nasal secretions from each nostril of domestic cattle or wild ruminants, put on ice and transported.

Table 2
Primer Sequences for Group 2a and BCoV or Bovine-like CoVs

Primers	Targeted region	Primer sequence 5′-3′	Product size	References
Pancorona-specific[a]	RdRp (RNA-dependent RNA polymerase) gene	Forward (IN-2 deg) GGGDTGGGAYTAY-CCHAARTGYGA	452 bp	(4)
For one-step RT-PCR		Reverse (IN-4 deg) TARCAVACAACISY-RTCRTCA		
Group 2-specific	Nucleoprotein gene	Forward (Gr2F) GAAGGCTCDGGAAR-GTCTG	298–304 bp[b]	Vlasova and Saif, unpublished
For one-step RT-PCR		Reverse (Gr2R) CCTCTYTYTHCCAAAA-CACTG		
BCoV-specific-1	Nucleoprotein gene	Forward (NOF) GCAATCCAGTAGTA-GAGCGT	729 bp	(28)
For RT-PCR		Reverse (NOR) CTTAGTGGCATCCTT-GCCAA		
BCoV-specific-2	Nucleoprotein gene	Forward (NF) GCCGATCAGTCCGACC-AATG	406 bp	(28)
For nested-PCR		Reverse (NR) AGAATGTCAGCCGGGG-TAG		

[a]The universal primers are modified from Ksiazek et al. (4) (see **Note 8**) for RNA sample screening in our lab.
[b]Band size varies from 298 bp for MHV and SDAV-Rat coronaviruses to 304 bp for others, depending on exact nucleoprotein gene sequence.

They are placed in 4 ml of Dil. #5, vortexed, and centrifuged at $2000 \times g$ for 30 min. Then the supernatant is aspirated and stored at $-70°C$ until use. Samples should be filtered (using 0.22-nm filters) before inoculation onto the HRT-18 cell culture monolayers.

3.1.2. Virus Isolation in HRT-18 Cell Cultures *(48)*

1. Monolayers of HRT-18 cell cultures, 3 to 5 days after seeding *(44,48,49)* into six-well plates are washed twice and incubated with Dil. #5 (3 ml per well) for 3 h at 37°C in a 5% CO_2 atmosphere.
2. Dil. #5 is aspirated from the wells and the cells are inoculated (in duplicate wells) with the fecal or nasal supernatants (200 μl per well), which are BCoV-positive by ELISA or RT-PCR. The supernatants are adsorbed for 1 h, during which time the plates are hand-rocked every 15 min. Then 3 ml of MEM containing pancreatin (5 μg/ml) is added to each well (*see* **Note 2**). The inoculated cells are incubated for 3 to 4 days at 37°C in a 5% CO_2 atmosphere. Cultures are examined daily for evidence of cytopathic effects (CPE).
3. The CPE, characterized by enlarged, rounded, and detached cells are usually observed approximately 72 h after inoculation, after five to seven initial blind passages (**Fig. 1**).
4. Immunofluorescence, Ag-ELISA, and RT-PCR tests can be used to confirm virus presence.

3.1.3. Plaque Induction in HRT-18 Cell Cultures

1. Monolayers of HRT-18 cell cultures, 3 to 5 days after seeding into six-well plates are washed twice and incubated with Dil. #5 (3 ml per well) for 3 h at 37°C in a 5% CO_2 atmosphere *(48)*.
2. Dil. #5 is aspirated from the wells and the cells are then inoculated (in duplicate wells) with CPE positive virus dilutions of 10^{-2} to 10^{-7} (200 μl per well).

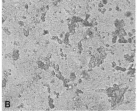

Fig. 1. (A) Mock-inoculated HRT-18 cell culture; (B) CPE in the HRT-18 cell culture inoculated 72 h previously with bovine-like coronavirus (isolated from giraffe feces).

3. After virus adsorption, plates are overlaid with MEM containing 1.6% noble agar plus 0.1% neutral red, 0.1% pancreatin, and 1% diethylaminoethyl dextran. Plates are inverted and incubated at 37°C in a 5% CO_2 atmosphere for 3 to 5 days.
4. Usually plaques appear within 3–5 days and diameters are approximately 0.8 to 1.5 mm.

3.2. Antigen-Capture ELISA for Detection of Viral Antigens

1. All wells of rows A, B, E, and F in 96-well flat bottom microtiter plates are coated with 100 μl of pooled BCoV MAbs (MAbs produced as mouse ascites against HE, S, and N proteins of BCoV) diluted in coating buffer as the capture antibody *(50)*. All wells of C, D, G, and H rows of plates are coated with 100 μl of BCoV negative mouse ascites (Sp2/0). Plates are incubated at 4°C overnight and rinsed four times with wash buffer.
2. The wells are blocked with 5% NFDM (200 μl per well) and incubated at room temperature for 2 h. The blocking solution is aspirated and the wells are washed four times with wash buffer.
3. The prepared fecal or nasal samples and positive and negative control samples (*see* **Note 3**) are added to four appropriate wells (two wells with positive and two wells with negative coating for each sample) (100 μl per well). For example: positive control sample to 1A, 1B, 1C, and 1D; negative control sample to 2A, 2B, 2C, and 2D; sample #1 to 3A, 3B, 3C, and 3D, etc. Plates are incubated at room temperature for 1 h in the dark and then washed four times with wash buffer.
4. Diluted guinea pig antiserum to BCoV (1:200–1:400) is added (100 μl per well) as the primary antibody and the plates are incubated at room temperature for 1 h (for nasal samples antiserum should be used twice as concentrated as for fecal samples: 1:200). Then plates are washed four times with wash buffer.
5. Goat anti-guinea pig IgG HRP conjugate (KPL) diluted 1:8000 is added (100 μl per well). The plates are then incubated at room temperature for 1 h and washed four times with wash buffer.
6. Finally, TMB microwell peroxidase substrate (KPL) is added (100 μl per well) and after 15 min incubation at room temperature, 1 M phosphoric acid (H_3PO_4) (50 μl per well) is applied as a stopping solution.
7. The cutoff value is calculated as the mean absorbance (450-nm wavelength) of the negative coating wells plus 3 standard deviations. Samples with an absorbance value higher than the cutoff value are considered positive.

3.3. Antibody-Capture ELISA (Serological Test) for Detection of Virus-Specific Antibodies in Bovine or Wild-Ruminant Serum (51)

1. Blood is collected and allowed to clot. Then serum is aspirated, heat inactivated (56°C, 30 min), and stored at –20°C until use.

2. All wells of rows A, B, E, and F in a 96-well flat bottom microtiter plate are coated with 100 μl of BCoV MAbs (developed against HE, S, and N proteins of BCoV) diluted in coating buffer as the capture antibody *(50,51)*. All wells of C, D, G, and H rows of the plate are coated with 100 μl of BCoV negative ascites (Sp2/0). Plates are incubated at 4°C overnight and rinsed four times with wash buffer.

3. The wells are then blocked with 5% NFDM (200 μl per well) and incubated at room temperature for 2 h. The blocking solution is discarded and the wells are washed four times with wash buffer.

4. Supernatants of HRT-18 cell cultures inoculated with BCoV Mebus strain are added to each well (100 μl per well). Plates are incubated at room temperature for 2 h in the dark and then washed four times with wash buffer.

5. The serum samples and positive and negative control serum samples (*see* **Note 1**; Section 2.3.3) are added to four appropriate wells (two wells with positive and two wells with negative coating per sample) (100 μl per well); e.g.: positive control sample to 1A, 1B, 1C, and 1D; negative control sample to 2A, 2B, 2C, and 2D; sample #1 to 3A, 3B, 3C, and 3D. Plates are incubated at 37°C for 1 h.

6. The plates are rinsed four times with wash buffer and goat anti-bovine IgG peroxidase conjugate (KPL) (1:4000) is added (or peroxidase conjugated Protein A, if ruminant sera fail to react with anti-bovine IgG), 100 μl per well. After incubation at 37°C for 1 h plates are washed four times with wash buffer.

7. Finally, TMB peroxidase substrate (KPL) is added (100 μl per well) and after 15 min incubation at room temperature, 1 M phosphoric acid (H_3PO_4) (50 μl per well) is applied as a stop solution.

8. The cutoff value is calculated as the mean absorbance (450-nm wavelength) from the negative coating wells plus 3 standard deviations. Samples with absorbance value higher than the cutoff value are considered positive.

3.4. RT-PCR Detection of Coronavirus RNA (ref. 28 and Vlasova and Saif, Unpublished)

3.4.1. RNA Extraction

TRIzol LS reagent is used for viral RNA extraction from fecal and nasal samples following the manufacturer's protocol (*see* **Note 7**). Briefly:

1. A 250-μl sample is mixed with 750 μl TRIzol LS in an Eppendorf tube, vortexed, and incubated for 5 min at room temperature.

2. Then 200 μl chloroform is added to each tube, vortexed for a short time, and incubated for 10 min at room temperature.

3. After centrifugation, at 12,000 × *g* for 15 min at 4°C, the supernatant (400 μl) is transferred to new Eppendorf tubes and an equal volume of 100% isopropyl alcohol is added to each tube. Tubes are vortexed and incubated for 10 min at room temperature.

4. Then tubes are centrifuged at 12,000 × *g* for 10 min at 4°C, the supernatant is discarded, and 800 μl of 75% EtOH is added to each tube.
5. Samples are vortexed and centrifuged at 7500 × *g* for 5 min at 4°C.
6. Finally, the supernatants are removed gently, the pellet is dried using a DNA Speed Vac dry machine (model DNA110, Savant INC, NY) low spin for 10 min and 40 μl RNase-free (DEPC-treated) water is added for elution.
7. After incubation at 55°–60°C for 10 min, RNA samples should be stored at −20°C or applied directly for one-step or nested RT-PCR.

3.4.2. One-Step RT-PCR

1. Each primer pair listed in Table 2 can be used for detection of the BCoV or bovine-like (wild-ruminant) CoV genomes. A pancorona-specific primer pair modified from Ksiazek et al. (*4*), targeted to the conserved region of the RNA-dependent RNA polymerase (RdRp), is capable of universal detection of known coronaviruses (*see* **Note 8**). Group 2a-specific primers were designed based on the consensus sequence of the nucleoprotein gene for group 2a coronaviruses (*see* **Note 9**). The BCoV-specific primers are targeted to the nucleoprotein and can also amplify a specific fragment for BCoV and the related wild-ruminant CoVs, but not other group 2 coronaviruses (*28*).
2. Reaction conditions are optimized using RT-PCR reagents commercially available from Promega, but they are easily adapted to enable the use of alternative reagents.
3. The typical PCR master mix should contain the following reagents in the 0.2-ml PCR tubes (RNAse/DNAse and Pyrogene safe): 10X thermophillic DNA polymerase buffer (2.5 μl); magnesium chloride 2.5 mM; 0.2 M dNTPs mix; forward and reverse primers 50 pmol each; AMV reverse transcriptase 2.5 U; RNasin 10 U; Taq DNA polymerase 1.25 U; adjust volume with Milli-Q DEPC-treated water to 23 μl. 2-μl RNAs extracted from the samples is added to each tube. RNA extracted from mock-inoculated HRT-18 cell culture supernatants or negative fecal or nasal samples and water should be used as negative controls and BCoV-infected HRT-18 cell culture supernatants or BCoV-positive fecal or nasal samples as positive controls.
4. The following RT-PCR cycles should be used: 42°C for 60 min, 94°C for 5 min; 35 cycles at 94°C for 0.5 min, 50°C/56°C (*see* **Note 10**) for 0.5 min, 72°C for 1 min, 72°C for 7 min, and hold at 4°C.

3.4.3. Nested PCR (*28*)

The nested-PCR protocol is as described for the RT-PCR procedure above (Section 3.4.2), but with minor modifications. For the nested-PCR master mix, 5 μl of undiluted RT-PCR product (cDNA) is used as a template instead of the RNA sample. The PCR master mix should not contain AMV reverse transcriptase and RNasin. For precautions against cross-contamination (*see* **Note 11**).

3.4.4. Electrophoresis

1. 10 μl of PCR product is mixed with 2μl of 6X Blue/Orange loading dye (Promega).
2. Load premixed samples in a 2% agarose gel in 1X TAE buffer containing ethidium bromide at a final concentration of 0.5 μg/ml. Add DNA molecular weight ladder also premixed with loading buffer.
3. Run the gel at 125 V for 25 min.
4. Visualize results using UV transilluminator (312-nm wavelength).

4. Conclusions

In summary, multiple diagnostic tests have been developed and standardized for the routine diagnosis of BCoV and the closely related ruminant Group 2a CoVs. These include antigen or antibody detection by ELISA, virus isolation in cell culture (HRT-18 cells), and, often the most sensitive method, detection of viral RNA by RT-PCR. These methods have provided evidence for the presence of CoVs closely related genetically or antigenically to BCoV, not only from wild ruminants, but also from dogs *(21)* and humans *(53)*. These findings and experimental transmission studies *(8,44,54)* confirm a broad host range for bovine-like group 2a CoVs and support a need to monitor their presence in other host species, including humans, by applying the diagnostic assays outlined in this chapter and the sequencing of new isolates.

5. Notes

1. FBS used for the cell cultures should be inactivated by heating at 56°C for 30 min and should be validated as bovine viral diarrhea virus-free.
2. After absorption of samples (1 h), washing the cells with Dil. #5 can help to avoid cytotoxic effects.
3. The positive control sample is from supernatant of HRT-18 cell cultures inoculated with BCoV Mebus strain and the negative control sample is from supernatants of mock-inoculated HRT-18 cell cultures. Alternatively, diluted feces from a BCoV- or mock-inoculated gnotobiotic calf can be used as positive and negative controls, respectively *(28,37,44)*.
4. Positive control serum sample is hyperimmune serum from a gnotobiotic calf inoculated with BCoV-440 and negative control serum sample is serum from a mock-inoculated gnotobiotic calf *(28,37,44)*.
5. All solutions should be prepared in water that has a resistivity of 18 MΩ-cm, unless stated otherwise.
6. Diethyl pyrocarbonate (DEPC) is a carcinogen, so a fume hood should be used when working with this reagent.
7. Some fecal samples from conventional animals can contain PCR inhibitors, especially if they were frozen. If such a problem occurs, dilution of extracted

RNA (1:10–1:20) can be helpful; if inhibitors persist, commercial kits (Epicentre, QIAGEN) optimized for DNA extraction from fecal samples should be used.

8. The conserved region of the RdRp was initially chosen as a target to allow universal detection of all CoVs. The original universal primer pair targeted to this region from Ksiazek et al. *(4)* was not optimal in RT-PCR with some CoV samples from wild ruminants. Thus, based on the consensus sequence (for all available RdRp sequences of coronaviruses from different groups) in all polymorphic positions, degenerative nucleotides were introduced. After these primers were tested with positive samples (validated in other assays), they demonstrated consistently positive results.

9. The coronavirus nucleoprotein gene was chosen for group-specific primer development because it has been shown to be variable in amino acid and nucleotide composition among the viruses that comprise the three coronavirus antigenic groups, but highly conserved within these groups.

10 The optimal melting temperature for group 2a- specific primers and nested-PCR is 56°C; for pancorona- and BCoV-specific primers, the optimal melting temperature is 50°C.

11. If RT-PCR sensitivity is low, a nested PCR method can increase PCR product yield. However, this will also increase the risk of cross-contamination. Standard precautions to avoid cross-contamination should be implemented. These include: use of gloves with frequent changing; employing of aerosol resistant pipettors and pipette tips; preparing of PCR master mixes in UV cabinets that should never be used for work with any sample materials or PCR DNA products; regular decontamination of work surfaces and all equipment with 10% chlorine or another commercially available disinfectant; separate working areas for each step of the PCR analysis as well as separate storage areas for PCR reagents and contaminated samples *(47)*.

References

1. Birch, C. J., Clothier, H. J., Seccull, A., Tran, T., Catton, M. C., Lambert, S. B., and Druce, J. D. (2005) Human coronavirus OC43 causes influenza-like illness in residents and staff of aged-care facilities in Melbourne, Australia. *Epidemiol. Infect.* **133**, 273–277.
2. Saif, L. J., and Heckert, R. A. (1990) Enteropathogenic coronaviruses. In: Saif, L. J., and Theil, K. W. (ed.) *Viral Diarrheas of Man and Animals.* CRC Press, Boca Raton, FL, pp. 185–252.
3. Woo, P. C. Y., Lau, S. K. P., Chu, C.-M., Chan, K.-H., Tsoi, H.-W., Huang, Y., Wong, B. H. L., Poon, R. W. S., Cai, J. J., Luk, W.-K., Poon, L. L. M., Wong S. S. Y., Guan, Y., Peiris, J. S. M., and Yuen, K.-Y. (2005) Characterization and complete genome sequence of a novel coronavirus, coronavirus HKU1, from patients with pneumonia. *J. Virol.* **79**, 884–895.

4. Ksiazek, T. G., Erdman, D., Goldsmith, C. S., Zaki, S. R., Peret, T., Emery, S., Tong, S., Urbani, C., Comer, J. A., Lim, W., Rollin, P. E., Dowell, S. F., Ling, A. E., Humphrey, C. D., Shieh, W. J., Guarner, J., Paddock, C. D., Rota, P., Fields, B., DeRisi, J., Yang, J. Y., Cox, N., Hughes, J. M., LeDuc, J. W., Bellini, W. J., Anderson, L. J., and Group, S. W. (2003) A novel coronavirus associated with severe acute respiratory syndrome. *N. Eng. J. Med.* **348**, 1953–1966.
5. Kuiken, T., Fouchier, R. A., Schutten, M., Rimmelzwaan, G. F., van Amerongen, G., van Riel, D., Laman, J. D., de Jong, T., van Doornum, G., Lim, W., Ling, A. E., Chan, P. K., Tam, J. S., Zambon, M. C., Gopal, R., Drosten, C., van der Werf, S., Escriou, N., Manuguerra, J. C., Stohr, K., Peiris, J. S., and Osterhaus, A. D. (2003) Newly discovered coronavirus as the primary cause of severe acute respiratory syndrome. *Lancet* **362**, 263–270.
6. Li, W., Wong, S. K., Li, F., Kuhn, J. H., Huang, I. C., Choe, H., and Farzan, M. (2006) Animal origins of the severe acute respiratory syndrome coronavirus: insight from ACE2-S-protein interactions. *J. Virol.* **80**, 4211–4219.
7. Peiris, J. S., Guan, Y., and Yuen, K. Y. (2004) Severe acute respiratory syndrome. *Nature Med.* **10**, S88–S97.
8. Saif, L. J. (2005) Comparative biology of animal coronaviruses: lessons for SARS. In: Peiris, M., Anderson, L. J., Osterhous, A. D. M. E., Stohr, K., and Yuen, K.Y. (ed.)*Severe Acute Respiratory Syndrome.* Blackwell, Malden, MA, pp. 84–99.
9. Saif, L. J. (2004) Animal coronaviruses: Lessons for SARS. In: Knobler, S., Mahmous, A., Lemon, S., Mack, A., Sivitz, L. and Oberholtzer, K. (ed.) *Learning from SARS: Preparing for the Next Disease Outbreak.* The National Academies Press, Washington, DC, pp. 138–149.
10. Wang, M., Yan, M., Xu, H., Liang, W., Kan, B., Zheng, B., Chen, H., Zheng, H., Xu, Y., Zhang, E., Wang, H., Ye, J., Li, G., Li, M., Cui, Z., Liu, Y. F., Guo, R. T., Liu, X. N., Zhan, L. H., Zhou, D. H., Zhao, A., Hai, R., Yu, D., Guan, Y. and Xu, J. (2005) SARS-CoV infection in a restaurant from palm civet. *Emerg. Infect. Diseases* **11**, 1860–1865.
11. Guan, Y., Zheng, B. J., He, Y. Q., Liu, X. L., Zhuang, Z. X., Cheung, C. L., Luo, S. W., Li, P. H., Zhang, L. J., Guan, Y. J., Butt, K. M., Wong, K. L., Chan, K. W., Lim, W., Shortridge, K. F., Yuen, K. Y., Peiris, J. S., and Poon, L. L. (2003) Isolation and characterization of viruses related to the SARS coronavirus from animals in southern China. *Science* **302**, 276–278.
12. Lau, S. K., Woo, P. C., Li, K. S., Huang, Y., Tsoi, H. W., Wong, B. H., Wong, S. S., Leung, S. Y., Chan, K. H., and Yuen, K. Y. (2005) Severe acute respiratory syndrome coronavirus-like virus in Chinese horseshoe bats. *Proc. Natl. Acad. Sci. USA* **102**, 14040–14045.
13. Li, W., Shi, Z., Yu, M., Ren, W., Smith, C., Epstein, J. H., Wang, H., Crameri, G., Hu, Z., Zhang, H., Zhang, J., McEachern, J., Field, H., Daszak, P., Eaton, B. T., Zhang, S., and Wang, L. F. (2005) Bats are natural reservoirs of SARS-like coronaviruses. *Science* **310**, 676–679.
14. Haring, J., and Perlman, S. (2001) Mouse hepatitis virus. *Curr. Opin. Microbiol.* **4**, 462–466.

15. Matthews, A. E., Weiss, S. R., and Paterson, Y. (2002) Murine hepatitis virus—a model for virus-induced CNS demyelination. *J. Neurovirol.* **8**, 76–85.

16. Yount, B., Denison, M. R., Weiss, S. R. and Baric, R. S. (2002) Systematic assembly of a full-length infectious cDNA of mouse hepatitis virus strain A59. *J. Virol.* **76**, 11065–11078.

17. Vijgen, L., Keyaerts, E., Moes, E., Thoelen, I., Wollants, E., Lemey, P., Vandamme, A. M. and Ranst M. V. (2005) Complete genomic sequence of human coronavirus OC43: molecular clock analysis suggests a relatively recent zoonotic coronavirus transmission event. *J. Virol.* **79**, 1595–1604.

18. Vabret, A., D. J., Gouarin, S., Petitjean, J., Corbet, S., and Freymuth F. (2006) Detection of the new human coronavirus HKUI: a report of 6 cases. *Clin. Infect. Dis.* **42**, 634–639.

19. Pensaert, M. B., and Andries, K. (1986) Hemagglutinating encephalomyelitis virus. In: Leman, A. D., Straw, B. E., Glock, R. D., Mengeling, W. L., Penny, R. H., Scholl, E. (ed.) *Diseases of Swine.* Iowa State University Press, Ames, pp. 310–315.

20. Erles, K., and Brownlie, J. (2005) Investigation into the causes of canine infectious respiratory disease: antibody responses to canine respiratory coronavirus and canine herpesvirus in two kennelled dog populations. *Arch. Virol.* **150**, 1493–1504.

21. Erles, K., Toomey, C., Brooks, H. W., and Brownlie, J. (2003) Detection of a group 2 coronavirus in dogs with canine infectious respiratory disease. *Virology* **310**, 216–223.

22. Guy, J. S., Breslin, J. J., Breuhaus, B., Vivrette, S., and Smith, L. G. (2000) Characterization of a coronavirus isolated from a diarrheic foal. *J. Clin. Microbiol.* **38**, 4523–4526.

23. Barker, M. G., Percy, D. H., Hovland, D. J., and MacInnes, J. I. (1994) Preliminary characterization of the structural proteins of the coronaviruses, sialodacryoadenitis virus and Parker's rat coronavirus. *Can. J. Vet. Res.* **58**, 99–103.

24. Homberger, F. R., Smith, A. L., and Barthold, S. W. (1991) Detection of rodent coronaviruses in tissues and cell cultures by using polymerase chain reaction. *J. Clin. Microbiol.* **29**, 2789–2793.

25. Percy, D., Bond, S., and MacInnes, J. (1989) Replication of sialodacryoadenitis virus in mouse L-2 cells. *Arch. Virol.* **104**, 323–333.

26. Smith, A. L. (1983) An immunofluorescence test for detection of serum antibody to rodent coronaviruses. *Lab. Anim. Sci.* **33**, 157–160.

27. Cho, K. O., Halbur, P. G., Bruna, J. D., Sorden, S. D., Yoon, K. J., Janke, B. H., Chang, K. O., and Saif, L. J. (2000) Detection and isolation of coronavirus from feces of three herds of feedlot cattle during outbreaks of winter dysentery-like disease. *J. Am. Vet. Med. Assoc.* **217**, 1191–1194.

28. Cho, K. O., Hasoksuz, M., Nielsen, P. R., Chang, K. O., Lathrop, S., and Saif, L. J. (2001) Cross-protection studies between respiratory and calf diarrhea and winter dysentery coronavirus strains in calves and RT-PCR and nested PCR for their detection. *Arch. Virol.* **146**, 2401–2419.

29. Hasoksuz, M., Hoet, A. E., Loerch, S. C., Wittum, T. E., Nielsen, P. R., and Saif, L. J. (2002) Detection of respiratory and enteric shedding of bovine coronaviruses in cattle in an Ohio feedlot. *J. Vet. Diagn. Invest.* **14**, 308–313.

30. Heckert, R. A., Saif, L. J., Hoblet, K. H., and Agnes, A. G. (1990) A longitudinal study of bovine coronavirus enteric and respiratory infections in dairy calves in two herds in Ohio. *Vet. Microbiol.* **22**, 187–201.

31. Mebus, C. A., Stair, E. L., Rhodes, M. B., and Twiehaus, M. J. (1973) Neonatal calf diarrhea: propagation, attenuation, and characteristics of a coronavirus-like agent. *Am. J. Vet. Res.* **34**, 145–150.

32. Saif, L. J. (1990) A review of evidence implicating bovine coronavirus in the etiology of winter dysentery in cows: an enigma resolved? *Cornell Vet.* **80**, 303–311.

33. Traven, M., Naslund, K., Linde, N., Linde, B., Silvan, A., Fossum, C., Hedlund, K. O. and Larsson, B. (2001) Experimental reproduction of winter dysentery in lactating cows using BCV—comparison with BCV infection in milk-fed calves. *Vet. Microbiol.* **81**, 127–151.

34. Tsunemitsu, H., Smith, D. R., and Saif, L. J. (1999) Experimental inoculation of adult dairy cows with bovine coronavirus and detection of coronavirus in feces by RT-PCR. *Arch. Virol.* **144**, 167–175.

35. Smith, D. R., Fedorka-Cray, P. J., Mohan, R., Brock, K. V., Wittum, T. E., Morley, P. S., Hoblet, K. H., and Saif, L. J. (1998) Epidemiologic herd-level assessment of causative agents and risk factors for winter dysentery in dairy cattle. *Am. J. Vet. Res.* **59**, 994–1001.

36. Thomas, L. H., Gourlay, R. N., Stott, E. J., Howard, C. J. and Bridger, J. C. (1982) A search for new microorganisms in calf pneumonia by the inoculation of gnotobiotic calves. *Res. Vet. Sci.* **33**, 170–182.

37. Hasoksuz, M., Lathrop, S. L., Gadfield, K. L., and Saif, L. J. (1999) Isolation of bovine respiratory coronaviruses from feedlot cattle and comparison of their biological and antigenic properties with bovine enteric coronaviruses. *Am. J. Vet. Res.* **60**, 1227–1233.

38. Lathrop, S. L., Wittum, T. E., Brock, K. V., Loerch, S. C., Perino, L. J., Bingham, H. R., McCollum, F. T., and Saif, L. J. (2000) Association between infection of the respiratory tract attributable to bovine coronavirus and health and growth performance of cattle in feedlots. *Am. J. Vet. Res.* **61**, 1062–1066.

39. Storz, J., Stine, L., Liem, A., and Anderson G. A. (1996) Coronavirus isolation from nasal swab sample in cattle with signs of respiratory tract disease after shipping. *J. Am. Vet. Med. Assoc.* **208**, 1452–1455.

40. Lin, X. Q., O'Reilly, K. L., and Storz, J. (2002) Antibody responses of cattle with respiratory coronavirus infections during pathogenesis of shipping fever pneumonia are lower with antigens of enteric strains than with those of a respiratory strain. *Clin. Diagn. Lab. Immunol.* **9**, 1010–1013.

41. Hasoksuz, M., Lathrop, S., Al-dubaib, M. A., Lewis, P., and Saif, L. J. (1999) Antigenic variation among bovine enteric coronaviruses (BECV) and bovine respiratory coronaviruses (BRCV) detected using monoclonal antibodies. *Arch. Virol.* **144**, 2441–2447.

42. Hasoksuz, M., Sreevatsan, S., Cho, K. O., Hoet, A. E., and Saif, L. J. (2002) Molecular analysis of the S1 subunit of the spike glycoprotein of respiratory and enteric bovine coronavirus isolates. *Virus Res.* **84**, 101–109.

43. Tsunemitsu, H., and Saif, L. J. (1995) Antigenic and biological comparisons of bovine coronaviruses derived from neonatal calf diarrhea and winter dysentery of adult cattle. *Arch. Virol.* **140**, 1303–1311.

44. Tsunemitsu, H., El-Kanawati, Z. R., Smith, D. R., Reed H. H. and Saif, L. J. (1995) Isolation of coronaviruses antigenically indistinguishable from bovine coronavirus from wild ruminants with diarrhea. *J. Clin. Microbiol.* **33**, 3264–3269.

45. Majhdi, F., Minocha, H. C., and Kapil, S. (1997) Isolation and characterization of a coronavirus from Elk calves with diarrhea. *J. Clin. Microbiol.* **35**, 2937–2942.

45a. Hasoksuz, M., Alekseev, K., Vlasova, A., Zhang, X., Spiro, D., Halpin, R., Wang, S., Ghedin, E., and Saif, L. J. (2007) Biologic, antigenic, and full-length genomic characterization of a bovine-like coronavirus isolated from a giraffe. *J. Virol.* **81(10)**, 4981–4990.

46. Chasey, D., Reynolds, D. J., Bridger, J. C., Debney, T. G., and Scott, A. C. (1984) Identification of coronaviruses in exotic species of Bovidae. *Vet. Rec.* **115**, 602–603.

47. Elazhary, M. A., Frechette, J. L., Silim, A., and Roy, R. S. (1981) Serological evidence of some bovine viruses in the caribou (*Rangifer tarandus caribou*) in Quebec. *J. Wildl. Dis.* **17**, 609–612.

48. Saif, L. J., Heckert, R. A., Miller, K. L., and Tarek, M. M. (1988) Cell culture propagation of bovine coronaviruses. *J. Tiss. Cul. Meth.* **11**, 139–145.

49. Benfield, D. A., and Saif, L. J. (1990) Cell culture propagation of a coronavirus isolated from cows with winter dysentery. *J. Clin. Microbiol.* **28**, 1454–1457.

50. Smith, D. R., Tsunemitsu, H., Keckert, R. A., and Saif, L. J. (1996) Evaluation of two antigen-capture ELISAs using polyclonal or monoclonal antibodies for the detection of bovine coronavirus. *J. Vet. Diagn. Invest.* **8**, 99–105.

51. Smith, D. R., Nielsen, P. R., Gadfield, K. L., and Saif, L. J. (1998) Further validation of antibody-capture and antigen-capture enzyme-linked immunosorbent assays for determining exposure of cattle to bovine coronavirus. *Am. J. Vet. Res.* **59**, 956–960.

52. Heckert, R. A., Saif, L. J., and Myers, G. W. (1989) Development of protein A-gold immunoelectron microscopy for detection of bovine coronavirus in calves: comparison with ELISA and direct immunofluorescence of nasal epithelial cells. *Vet. Microbiol.* **19**, 217–231.

53. Zhang, X. M., Herbst, W., Kousoulas, K. G., and Storz, J. (1994) Biological and genetic characterization of a hemagglutinating coronavirus isolated from a diarrhoeic child. *J. Med. Virol.* **44**, 152–161.

54. Ismail, M. M., Cho, K. O., Hasoksuz, M., Saif, L. J., and Saif, Y. M. (2001) Antigenic and genomic relatedness of turkey-origin coronaviruses, bovine coronaviruses, and infectious bronchitis virus of chickens. *Avian Dis.* **45**, 978–984.

55. Vijgen, L., Keyaerts, E., Moes, E., Mais, P., Duson, G., and Van Ranst, M. (2005) Development of one-step, real-time, quantitative reverse transcriptase PCR assays for absolute quantitation of human coronaviruses OC43 and 229E. *J. Clin. Microbiol.* **43**, 5452–5456.
56. Peters, R. L., Collins, M. J., O'Beirne, A. J., Howton, P. A., Hourihan, S. L., and Thomas, S. F. (1979) Enzyme-linked immunosorbent assay for detection of antibodies to murine hepatitis virus. *J. Clin. Microbiol.* **10**, 595–597.

6

Detection of SARS Coronavirus in Humans and Animals by Conventional and Quantitative (Real Time) Reverse Transcription Polymerase Chain Reactions

J. S. Malik Peiris and Leo L. M. Poon

Abstract

Severe acute respiratory syndrome is a novel human disease caused by a coronavirus of animal origin. Soon after the discovery SARS-CoV, several molecular assays were described for the detection of this virus. Of these, conventional and quantitative RT-PCR approaches were the primary tools for SARS-CoV RNA detection. In this chapter we describe a two-step conventional RT-PCR and a one-step quantitative RT-PCR that were used routinely in our laboratories during the SARS outbreak.

Key words: SARS coronavirus; RT-PCR; Molecular detection; clinical diagnosis

1. Introduction

Severe acute respiratory syndrome (SARS) is the first novel infectious respiratory disease in this century. The first known case of SARS was retrospectively identified in Foshan City, Gungdong, China, in late 2002 *(1)*. After its introduction to Hong Kong in mid-February 2003, the virus spread across many countries within weeks. On 15 March 2003, the World Health Organization (WHO) issued a travel advisory and officially recognized this atypical pneumonia as "SARS." Three research groups independently reported a novel group 2 coronavirus (CoV) as the etiology for the disease in late March *(2–4)*. Subsequently enormous efforts were taken to contain the disease. On 5 July 2003, the WHO declared that the epidemic had been contained worldwide. In this outbreak, over

From: *Methods in Molecular Biology, vol. 454: SARS- and Other Coronaviruses,*
Edited by: D. Cavanagh, DOI: 10.1007/978-1-59745-181-9_6, © Humana Press, New York, NY

8000 SARS patients were reported and 10% of them died from the disease. Several lines of evidence indicated that the disease was a result of spillovers of the virus from animals to humans *(5)*. In particular, SARS-like viral isolates that are almost identical to human isolates were recovered from infected Himalayan palm civets (*Paguma larvata*) *(6)*. Further studies also indicated that the virus is a distant relative of bat CoVs *(7–10)*, suggesting that bats might be natural carriers of the precursor of SARS-CoV. However, the natural reservoir of the virus has yet to be defined.

As detection of the SARS-CoV RNA in clinical specimens enabled prompt identification of patients who were at the early stage of disease onset, the focus of early diagnosis was primarily on the development of conventional and quantitative reverse transcriptase-polymerase chain reaction ((RT-PCR) assays during the outbreak *(5)*. Sporadic human cases were reported after the SARS epidemic, most of these patients having acquired the infection from laboratories or from infected palm civets *(5)*, which is an indication that SARS might reemerge in the future. This prompted many groups to develop more rapid, sensitive, and highly specific laboratory tests for SARS preparedness. Non-PCR-based nucleic acid amplification assays, such as loop-mediated isothermal amplification (LAMP) *(11,12)*, rolling circle amplification (RAC) *(13)*, and nucleic acid sequence-based amplification (NASBA) *(14)* were also developed for the detection of SARS-CoV. However, owing to the limited availability of clinical specimens, most of these novel molecular assays could not be evaluated to any great extent.

2. Materials

2.1. RNA Extraction

1. QIAamp virus RNA mini kit (Qiagen).
2. Ethanol, 96–100%.
3. Autoclaved RNase-free water or its equivalent.
4. Clinical samples stored in 1–3 ml of viral transport medium. For 1 liter of viral transport medium, dissolve 2 g of sodium bicarbonate (Merck), 5 g of bovine serum albumin (Sigma-Aldrich), 200 μg of vancomycin (Sigma-Aldrich), 18 μg of amikacin (Sigma-Aldrich), and 160 U of nystatin (Sigma-Aldrich) in 1 liter of Earle's balanced salt solution (Sigma-Aldrich) and filter the solution using a 0.22-μm pore size filter (*See* **Note 1**).

2.2. Reverse Transcription

1. SuperScript II reverse transcriptase, 200 U/μl (Invitrogen).
2. 5X first-strand buffer: 250 mM Tris-HCl (pH 8.3), 375 mM KCl, 15 mM MgCl$_2$ (Invitrogen).

3. 0.1 mM dithiothreitol (Invitrogen).
4. Random hexamers, 150 ng/μl (Invitrogen).
5, RNaseOUT recombinant ribonuclease inhibitor, 40 U/μl (Invitrogen).
6. Deoxynucleotide triphosphates (dNTP) mix, 10 mM each.
7. Autoclaved RNase-free water or equivalent.
8. Heating block or equivalent.

2.3. Polymerase Chain Reaction

1. AmpliTaq Gold DNA polymerase, 5 U/μl (Applied Biosystems).
2. 10X Gold PCR buffer (Applied Biosystems).
3. Deoxynucleotide triphosphates (dNTP) mix, 10 mM each.
4. 25 mM $MgCl_2$ solution (Applied Biosystems).
5. 10 μM PCR forward primer, 5′-TACACACCTCAGCGTTG-3′.
6. 10 μM PCR reverse primer, 5′- CACGAACGTGACGAAT -3′.
7. Themocycler (GeneAmp 9700, Applied Biosystems) (*see* **Note 2**).

2.4. Gel Electrophoresis

1. 50X TAE buffer (Bio Rad).
2. Seakam LE agarose powder (Cambrex).
3. 6X gel loading buffer: 10 mM Tris-HCl (pH 7.6), 0.03% bromophenol blue, 0.03% xylene cyanol, 60% glycerol, and 60 mM EDTA.
4. 1 kb plus DNA ladder markers (Invitrogen).
5. Ethidium bromide, 10 mg/ml.
6. Agarose gel electrophoresis apparatus.
7. Power supply (PowerPac Basic, Bio-Rad).
8. Gel documentary machine or equivalent.

2.5. Quantitative RT-PCR

1. TaqMan EZ RT-PCR Core Reagents kits (Applied Biosystems).
2. 50 μM PCR forward primer, 5′-CAGAACGCTGTAGCTTCAAAAATCT-3′.
3. 50 μM PCR reverse primer, 5′ TCAGAACCCTGTGATGAATCAACAG-3′.
4. 10 μM probe, 5′-(FAM)TCTGCGTAGGCAATCC(NFQ)-3′ (FAM, 6-carboxy-fluorescein; NFQ, nonfluorescent quencher; Applied Biosystems).
5. Quantitative PCR machine (ABI Prism 7000 Sequence Detection System, Applied Biosystems)
6. PCR reaction plates (MicroAmp optical 96-well reaction plate, Applied Biosystems)
7. Optical adhesive covers (Applied Biosystems)

8. Benchtop centrifuge (Allegra X-15R, Beckman Coulter) with microplate carriers (SX4750μ, Beckman Coulter).

3. Methods

The protocols described below were routinely used for our clinical diagnosis of SARS during the outbreak *(5,15,16)*. The PCR assays were based on a short viral RNA sequence deduced from our initial studies *(5)*. Sections 3.2 to 3.4 describe a manual RT-PCR assay that allows testing clinical samples in laboratories with conventional PCR machines. Section 3.5 describes a one-step RT-PCR assay using a quantitative PCR platform. In our evaluation, the performance of the quantitative RT-PCR assay is better than the manual RT-PCR assays *(17)*. In addition, as viral load was found to be a good indicator for disease severity *(5)*, the quantitative results generated from the real-time RT-PCR might provide additional data for prognosis (*see* **Notes 3–6**).

3.1. RNA Extraction

1. For a new kit, perform the following procedures before specimen processing:
   a. Add 1 ml of buffer AVL to a tube of lyophilized carrier RNA (310 μg). Dissolve carrier RNA thoroughly. Transfer to the buffer AVL bottle and mix thoroughly. Store the buffer AVL at 4°C for up to 6 months.
   b. For every 19 ml of buffer AW1, add 25 ml of ethanol (96–100%). Mix it well. Store the buffer AW1 at room temperature for up to 12 months.
   c. For every 13 ml of buffer AW2, add 30 ml of ethanol (96–100%). Mix well. Store buffer AW1 at room temperature for up to 12 months (*see* **Note 7**).
2. Equilibrate all reagents to room temperature before use.
3. Transfer 140 μl of the sample into a 1.5-ml microcentrifuge tube.
4. Add 560 μl of prepared buffered AVL with carrier RNA to the microcentrifuge tube.
5. Briefly vortex the tubes for 15 sec and incubate at room temperature for 10 min.
6. Briefly centrifuge the microcentrifuge tube. Add 560 μl ethanol (96–100%) and mix by pulse-vortexing for 15 sec.
7. Briefly centrifuge the microcentrifuge tube.
8. Transfer 630 μl of the solution from the tube to a QIAamp spin column placed in a provided 2-ml collection tube. Centrifuge at 6000 × *g* (8000 rpm) for 1 min. Place the spin column in a clean 2-ml collection tube. Discard the tube containing the filtrate.
9. Open the spin column and repeat step 8.
10. Add 500 μl buffer AW1. Centrifuge at 6000 × *g* (8000 rpm) for 1 min. Place the spin column in a clean 2-ml collection tube. Discard the tube containing the filtrate.

11. Add 500 µl buffer AW2. Centrifuge at 20,000 × *g* (14,000 rpm) for 3 min. Place the spin column in a clean 2-ml collection tube and centrifuge at 20,000 × *g* for 1 min. Place the spin column in a clean 1.5-ml microcentrifuge tube. Discard the tube containing the filtrate.
12. Apply 50 µl buffer AVE equilibrated to room temperature directly on the membrane of the column. Close the cap and incubate at room temperature for 1 min.
13. Centrifuge at 6000 × *g* (8000 rpm) for 1 min. Collect the filtrate for cDNA synthesis. Store the RNA at –20°C or –70°C.

3.2. Reverse Transcription

1. Prepare a reverse transcription master mix sufficient for the designated number of samples in a sterile 1.5-ml microcentrifuge tube as shown in **Table 1**.
2. Vortex and centrifuge the tube briefly. Keep the tube on ice.
3. Add 10 µl of master mix solution into separate 0.5-ml microcentrifuge tubes. Label the tubes accordingly and keep them on ice.
4. Add 10 µl of purified RNA samples into these tubes.
5. Vortex and centrifuge the tubes briefly.
6. Stand the tubes at room temperature for 10 min and then incubate at 42°C for 50 min.
7. Inactivate the transcription reaction by incubating the tubes at 95°C for 5 min and then chill the samples on ice. Store the cDNA samples at –20°C (*see* **Note 8**).

3.3. PCR Assay

1. Prepare a PCR master mix sufficient for the designated number of samples in a sterile 1.5-ml microcentrifuge tube, according to **Table 2**. Include at least one positive control and one negative control (water) for each run. Add additional controls (e.g., purified RNA from the studied samples) as necessary.

Table 1
Components of Reverse Transcription Reaction

Reagent	Volume per reaction	Volume mix for N reactions	Final concentration
5X First strand buffer	4 µl	4 × N µl	1×
0.1 mM DTT	2 µl	2 × N µl	0.01 mM
10 mM dNTP	1 µl	N µl	0.5 mM
Random primers (150 ng/µl)	1 µl	N µl	7.5 ng/µl
Reverse transcriptase (200 U/µl)	1 µl	N µl	200 U/reaction
Ribonuclease inhibitor (optional)	1 µl	N µl	40 U/reaction
Total volume of master mix	10 µl	10 × N µl	–

Table 2
Components of the PCR

Reagent	Volume per reaction	Volume for N reactions[a]	Final concentration
10X PCR buffer	5 μl	5 × N μl	1×
MgCl$_2$, 25 mM	5 μl	5 × N μl	2.5 mM
dNTP, 10 mM	0.5 μl	0.5 × N μl	0.1 mM
Forward primers, 10 μM	1.25 μl	1.25 × N μl	0.25 μM
Reverse primers, 10 μM	1.25 μl	1.25 × N μl	0.25 μM
DNA polymerase (5U/μl)	0.25 μl	0.25 × N μl	1.25 U/reaction
Water	34.75 μl	34.75 × N μl	–
Total	48 μl	48 × N μl	–

[a]N = number of 1.5 ml tubes.

2. Vortex and centrifuge the tube briefly. Keep the tube on ice.
3. Aliquot 48 μl of the master mix into separate 0.5-ml microcentrifuge tubes and label the tubes accordingly.
4. Add 2 μl of cDNA generated from the reverse transcription reactions to these tubes. For the positive control, add 2 μl of SARS-CoV cDNA into the reaction. For the negative control, add 2 μl of autoclaved water.
5. Vortex and centrifuge the tubes briefly.
6. Run the PCR under the conditions shown in **Table 3**.
7. After the run, analyze the PCR products by gel electrophoresis. Alternatively, the products can be kept at –20°C for short-term storage.

3.4. Agarose Gel Electrophoresis

1. Place a gel-casting tray onto a gel-casting base. Level the base.

Table 3
Conditions for the Nonquantitative PCR

Step	Temperature	Time
1. Heat activation	94°C	8 min
2. Thermal cycling (40 cycles)		
Denaturing step	95°C	30 sec
Annealing step	50°C	40 sec
Extension	72°C	15 sec
3. Final extension	72°C	2 min
4. Soak	4°C	∞

2. Prepare 2% agarose gel by weighing out 1 g of agarose powder. Add it into a 250-ml bottle containing 50 ml 1X TAE buffer. Microwave bottle with a loosened cap until the gel starts to bubble and becomes transparent (*see* **Note 9**).
3. Cool the melted agarose to about 60°C and pour it into the gel-casting tray. Insert a comb into the tray.
4. Allow the gel to solidify at room temperature.
5. Remove the comb from the tray.
6. Place the tray into the electrophoresis chamber with the wells at the cathode side.
7. Fill the buffer chamber with 1X TAE buffer at a level that can cover the top of the gel.
8. Mix 0.5 μl of the DNA markers with 2 μl of 6X gel loading dye and 9.5 μl of water on a parafilm sheet by repeated pipetting.
9. Mix 10 μl of the PCR products with 2 μl of 6X gel loading dye on a parafilm sheet by pipetting up and down several times.
10. Apply the mixture to the corresponding well of the gel.
11. Close the lid of the electrophoresis apparatus and connect the electrical leads, anode to anode (red to red) and cathode to cathode (black to black).
12. Run the gel at 100 V for 30 min.
13. Turn off the power, remove the cover, and retrieve the gel.
14. Soak the gel in 1X TAE with 0.5 μg/ml ethidium bromide for 15 min. Wash the gel briefly with water (*see* **Note 10**).
15. Place the gel on top of the transilluminator. Switch on the power of the gel documentation machine (*see* **Note 11**).
16. Adjust the position of the gel and record the results. The size of the expected product for CoV is 182 bp (*see* **Note 12**).

3.5. Quantitative RT-PCR

1. Turn on the quantitative RT-PCR machine. Activate the Detection Manager from the supplied software and confirm that the reporter, quencher, and passive reference dyes are FAM, NFQ, and ROX, respectively. Set the cycle condition according to **Table 4**.
2. In the reaction plate template, input the necessary information for the corresponding samples (e.g., positive standard, negative control, or name of the clinical specimen). Include at least one set of tenfold serially diluted positive controls with known copy numbers of the target sequence (e.g., 10^6 to 10 copies/reaction) and three negative controls (water) in each run. For the positive controls, key in the copy numbers of the target sequence used in the corresponding reactions.
3. Prepare a PCR master mix sufficient for the designated number of samples in a sterile 2.5-ml screw cap tube according to **Table 5**. Add additional controls (e.g., purified RNA from the studied samples) as necessary.
4. Close the cup. Vortex and centrifuge the tube briefly.

Table 4
Conditions for the Quantitative PCR

Step	Temperature	Time
1. UNG treatment	50°C	2 min
2. Reverse transcription	60°C	40 min
3. Heat inactivation	95°C	5 min
4. Thermal cycling (50 cycles)		
Denaturing	95°C	15 sec
Annealing and extension	55°C	1 min

Table 5
Components of the Quantitative PCR

Reagent	Volume per reaction	Volume for N reactions	Final Concentration
Water	6.2 μl	6.2 × N μl	–
5X TaqMan EZ buffer	5 μl	5 × N μl	1×
Manganese acetate, 25 mM	3 μl	3 × N μl	3.0 mM
dATP, 10 mM	0.75 μl	0.75 × N μl	0.3 mM
dUTP, 10 mM	1.5 μl	1.5 × N μl	0.6 mM
dCTP, 10 mM	0.75 μl	0.75 × N μl	0.3 mM
dGTP, 10 mM	0.75 μl	0.75 × N μl	0.3 mM
Forward primers, 50 μM	0.4 μl	0.4 × N μl	0.8 μM
Reverse primers, 50 μM	0.4 μl	0.4 × N μl	0.8 μM
Probe, 10 μM	1 μl	1 × N μl	0.4 μM
rTth DNA polymerase (2.5U/μl)	1 μl	1 × N μl	2.5 U/reaction
AmpErase UNG (1 U/μl)	0.25 μl	0.25 × N μl	0.25 U/reaction
Total	21 μl	21 × N μl	–

5. Aliquot 21 μl of the master mix into the corresponding wells of the reaction plate.
6. Add 4 μl of the samples into the corresponding wells carefully (*see* **Note 13**).
7. Seal the reaction plate with an adhesive cover. Make sure each reaction well is sealed properly.
8. Briefly centrifuge the reaction plate.
9. Insert the plate to the quantitative PCR machine and perform the RT-PCR cycle.
10. After the reaction, examine the threshold cycles (Ct) and the amplification curves of the reactions. For a good experiment, the Ct values deduced from the standards should correlate well with the $\log_{10}$ copy numbers of the target sequence used in these reactions (**Fig. 1A**). Positive clinical samples will generate amplification

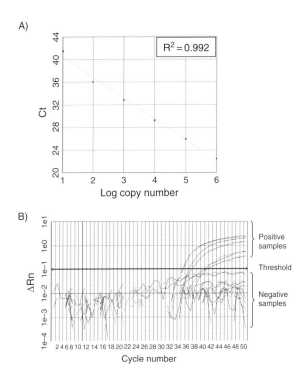

Fig. 1. Quantitative RT-PCR assay for SARS-CoV: (A) Standard curve for quantitative analysis of ORF 1b of SARS-CoV. The threshold cycle (Ct) is the number of PCR cycles required for the fluorescent intensity of the reaction to reach a predefined threshold. The Ct is inversely proportional to the logarithm of the starting concentration of the input target. The correlation coefficient (R^2) between these two parameters is shown. (B) An amplification plot of fluorescence intensity against the PCR cycle. The fluorescence signals for positive and negative samples are indicated. The X-axis denotes the cycle number of a quantitative PCR assay. The Y-axis denotes the fluorescence intensity.

signals above the threshold (**Fig. 1B**). By contrast, signals from the water controls and negative samples will be below the threshold line. Based on the Ct values from the reference standards, the amount of input target in the positive reactions will be calculated by the software automatically (*see* **Notes 14** and **15**).

4. Notes

1. Viral transport medium contains a high concentration of antibiotic to inhibit bacterial growth.
2. The primers and probe used in these assays are perfectly matched to the sequences deduced from SARS-CoV in humans and civets, including those isolated in 2004.

3. In our patient cohort, respiratory samples (e.g., nasopharyngeal aspirate, throat swab) collected from patients within the first week of disease onset have the highest positive rates for SARS-CoV. By contrast, fecal samples have the highest positive rate after the first week of onset. However, to increase the chance of identifying SARS patients in a nonepidemic period, we recommend testing multiple specimens available from suspected patients.

4. For respiratory samples isolated from early disease onset, the detect rates could be enhanced by increasing the initial extraction volume of the NPA sample from 140 to 560 μl *(18)*.

5. Personal protection equipment should be worn by the health care worker taking specimens from suspect or probable SARS patients (http://www.who.int/csr/sars/infectioncontrol/en/)

6. For extracting RNA from suspected infectious samples, the procedure must be handled in a Biosafety Level (BSL) 2 containment with BSL 3 work practices (http://www.who.int/csr/sars/biosafety2003_12_18/en/).

7. Buffer AVL containing carrier RNA might form white precipitates when it is stored at 4°C. The precipitates can be dissolved in the buffer by heating the bottle in a water bath. Cool the buffer to room temperature before use.

8. General procedures to prevent PCR cross contamination should be strictly followed. Aerosol-resistant filtered pipette tips can minimize possible carryovers of amplicons. Separate pipettes and areas are used for sample processing, PCR, and post-PCR analysis. It is essential to include multiple positive and negative controls in the PCR reactions when a large number of samples are tested at the same time.

9. Agarose solutions can be superheated in a microwave oven. Do not handle the bottle immediately after microwaving. Always wear heat-resistant gloves when handling melted agarose.

10. Ethidium bromide is a known mutagen and may be carcinogenic. Handle solutions of ethidium bromide with gloves.

11. UV light can cause severe skin and eye damage. Wear safety glasses and close the photography hood before turning on the UV transilluminator.

12. The conventional RT-PCR protocol is highly specific to SARS-CoV isolated from respiratory samples. However, we observed a few false-positive results from RNA isolated from stool *(15)*. To overcome this problem, all of our positive fecal samples were retested by the quantitative RT-PCR as described in Section 3.5 or a SYBR green-based RT-PCR assay *(19)* for confirmation.

13. When performing step 6 in Section 3.5, the RNA samples, including those positive standards, must be handled with extreme care. Cross-contamination might lead to false-positive or unreliable quantitative results.

14. The amplification curves of all positive samples in the quantitative RT-PCR assays must be examined individually. We occasionally find some clinical specimens yielding high backgrounds and the analytical program might misclassify these samples as positive.

15. To exclude the negative results owing to the poor recovery of RNA, the poor performance of RT-PCR reaction, the presence of PCR inhibitors, or human error, we subsequently modified our quantitative RT-PCR assays to a duplex assay. The revised test allows simultaneous detection of SARS-CoV and endogenous 18S rRNA derived from host cells *(20)*. The primers and probe for 18S rRNA are commercially available (TaqMan Ribosomal RNA Control Reagents, Applied Biosystems).

Acknowledgments

We acknowledge research funding from Public Health Research Grant from the National Institute of Allergy and Infectious Diseases, USA, The Research Grant Council of Hong Kong (HKU 7343/04 M to LLMP), European Research Project SARS-DTV (contract no: SP22-CT-2004).

References

1. Zhong, N. S., Zheng, B. J., Li, Y. M., et al. (2003) Epidemiology and cause of severe acute respiratory syndrome (SARS) in Guangdong, People's Republic of China, in February, 2003. *Lancet* **362**, 1353–1358.
2. Drosten, C., Gunther, S., Preiser, W., et al. (2003) Identification of a novel coronavirus in patients with severe acute respiratory syndrome. *N. Engl. J. Med.* **348**, 1967–1976.
3. Ksiazek, T. G., Erdman, D., Goldsmith, C. S., et al. (2003) A novel coronavirus associated with severe acute respiratory syndrome. *N. Engl. J. Med.* **348**, 1953–1966.
4. Peiris, J. S., Lai, S. T., Poon, L. L., et al. (2003) Coronavirus as a possible cause of severe acute respiratory syndrome. *Lancet* **361** 1319–1325.
5. Poon, L. L., Guan, Y., Nicholls, J. M., Yuen, K. Y., and Peiris, J. S. (2004) The aetiology, origins, and diagnosis of severe acute respiratory syndrome. *Lancet Infect. Dis.* **4**, 663–671.
6. Guan, Y., Zheng, B. J., He, Y. Q., et al. (2003) Isolation and characterization of viruses related to the SARS coronavirus from animals in southern China. *Science* **302**, 276–278.
7. Lau, S. K., Woo, P. C, Li, K. S., et al. (2005) Severe acute respiratory syndrome coronavirus-like virus in Chinese horseshoe bats. *Proc. Natl. Acad. Sc.i USA* **102** 14040–14045.
8. Li, W., Shi, Z., Yu M., et al. (2005) Bats are natural reservoirs of SARS-like coronaviruses. *Science* **310**, 676–679.
9. Ren, W., Li, W., Yu, M., et al. (2006) Full-length genome sequences of two SARS-like coronaviruses in horseshoe bats and genetic variation analysis. *J. Gen. Virol.* **87**, 3355–3359.
10. Tang, X. C., Zhang, J. X., Zhang, S. Y., et al. (2006) Prevalence and genetic diversity of coronaviruses in bats from China. *J. Virol.* **80**, 7481–7490.

11. Poon, L. L., Leung, C. S., Tashiro, M., et al. (2004) Rapid detection of the severe acute respiratory syndrome (SARS) coronavirus by a loop-mediated isothermal amplification assay. *Clin. Chem.* **50**, 1050–1052.

12. Hong, T. C., Mai, Q. L., Cuong, D.V., et al. (2004) Development and evaluation of a novel loop-mediated isothermal amplification method for rapid detection of severe acute respiratory syndrome coronavirus. *J. Clin. Microbiol.* **42**, 1956–1961.

13. Wang, B., Potter, S. J., Lin, Y., et al. (2005) Rapid and sensitive detection of severe acute respiratory syndrome coronavirus by rolling circle amplification. *J. Clin. Microbiol.* **43**, 2339–2344.

14. Keightley, M. C, Sillekens, P., Schippers, W., Rinaldo, C., and George, K. S. (2005) Real-time NASBA detection of SARS-associated coronavirus and comparison with real-time reverse transcription-PCR. *J. Med. Virol.* **77**, 602–608.

15. Chan, K. H., Poon, L. L., Cheng, V. C., et al. (2004) Detection of SARS coronavirus in patients with suspected SARS. *Emerg. Infect. Dis.* **10**, 294–299.

16. Yam, W. C., Chan, K. H., Poon, L. L., et al. (2003) Evaluation of reverse transcription-PCR assays for rapid diagnosis of severe acute respiratory syndrome associated with a novel coronavirus. *J. Clin. Microbiol.* **41**, 4521–4524.

17. Poon, L. L., Chan, K. H., Wong, O. K., et al. (2004) Detection of SARS coronavirus in patients with severe acute respiratory syndrome by conventional and real-time quantitative reverse transcription-PCR assays. *Clin. Chem.* **50**, 67–72.

18. Poon, L. L., Chan, K. H., Wong, O. K., et al. (2003) Early diagnosis of SARS coronavirus infection by real time RT-PCR. *J. Clin. Virol.* **28**, 233–238.

19. Poon, L. L., Wong, O. K., Chan, K. H., et al. (2003) Rapid diagnosis of a coronavirus associated with severe acute respiratory syndrome (SARS). *Clin. Chem.* **49**, 953–955.

20. Poon, L. L., Wong, B. W., Chan, K. H., et al. (2004) A one step quantitative RT-PCR for detection of SARS coronavirus with an internal control for PCR inhibitors. *J. Clin. Virol.* **30**, 214–217.

7

Detection of New Viruses by VIDISCA

Virus Discovery Based on cDNA-Amplified Fragment Length Polymorphism

Krzysztof Pyrc, Maarten F. Jebbink, Ben Berkhout, and Lia van der Hoek

Abstract

Virus discovery based on *c*DNA-*A*FLP (amplified fragment length polymorphism) (VIDISCA) is a novel approach that provides a fast and effective tool for amplification of unknown genomes, e.g., of human pathogenic viruses. The VIDISCA method is based on double restriction enzyme processing of a target sequence and ligation of oligonucleotide adaptors that subsequently serve as priming sites for amplification. As the method is based on the common presence of restriction sites, it results in the generation of reproducible, species-specific amplification patterns. The method allows amplification and identification of viral RNA/DNA, with a lower cutoff value of 10^5 copies/ml for DNA viruses and 10^6 copies/ml for the RNA viruses. Previously, we described the identification of a novel human coronavirus, HCoV-NL63, with the use of the VIDISCA method.

Key words: VIDISCA; virus discovery; detection; diagnosis; cDNA-AFLP; amplification; RT-PCR.

1. Introduction

To date, there is still a variety of human diseases of unknown etiology, including several chronic diseases such as amyotrophic lateral sclerosis (ALS) and multiple sclerosis (MS), but also acute infections such as Kawasaki disease and multiple respiratory diseases *(1,2)*. A viral origin has been suggested for many of these diseases, emphasizing the importance of a continuous search for new viruses. Identification of previously unrecognized viral agents in patient samples

From: *Methods in Molecular Biology, vol. 454: SARS- and Other Coronaviruses,*
Edited by: D. Cavanagh, DOI: 10.1007/978-1-59745-181-9_7, © Humana Press, New York, NY

is of great medical interest, but remains a major technical challenge. Identification of novel viral pathogens is difficult with the virus discovery tools known to date. Several problems are encountered when searching for new viruses. First, most of the unidentified viruses do not replicate *in vitro*, at least not in the cells that are commonly used in viral diagnostics. Second, the molecular biology techniques previously employed to identify unknown viruses have their specific drawbacks. Several techniques are in use for virus discovery, e.g., universal primer PCR, random priming based PCR, and representational difference analysis (RDA). Although every technique has proven to be useful for virus discovery in certain circumstances, they all have serious limitations and restrictions.

Universal PCR primers should amplify new members of an already known virus family, but this method has two major drawbacks. First, a choice for a specific virus family has to be made. This limits the possibility of identifying a member of an unsuspected family or the founding member of a totally new one. Second, the universal primers may simply not match the genome sequence of novel members of a virus family. This is illustrated by the lack of success of universal coronavirus primers that were designed before the new members—SARS-CoV, HCoV-NL63, and HCoV-HKU1—were identified. None of the studies that used such primers was able to detect a novel human coronavirus *(3,4)*. Obviously, such primers gradually improve once more family members are known.

Another technique uses nonspecific amplification of viral sequences in a random priming PCR at low annealing temperatures. However, most ingredients of this assay contain contaminating DNA. For instance, the enzymes used may contain trace amounts of DNA from the bacteria in which they are produced. This contaminating DNA is also amplified and it is therefore not possible to determine at an early stage whether amplification products represent a new virus or contaminating nucleic acids. This can be resolved only after excessive cloning and sequencing. Therefore, high throughput screening of many clinical samples is impractical. Moreover, this technique has only been successful with viruses that replicate *in vitro*, in which case cell culture supernatant was used as input for the assay *(5)*.

Representational difference analysis (RDA) is a subtractive hybridization technique that enriches for nucleic acid sequences that are present in one tissue but absent or present at lower concentration in an otherwise identical tissue sample. RDA utilizes PCR to generate sets of nucleic acids in a target and a (negative control) tester sample. After subtractive hybridization, there is selective amplification of target-enriched sequences. The method was developed for tissue material and not for nontissue samples such as serum/plasma or virus culture supernatants *(6)*. The fact that these liquid samples have low concentrations of DNA and RNA in the tester sample may restrain the selective amplification of an unknown viral target. A disadvantage of this technique is that it requires a negative control tissue from the same person from whom the diseased tissue was obtained.

We recently developed a general, simple, and easy to use new virus discovery method that allows large-scale screening for any RNA or DNA virus in samples such as serum/plasma or virus culture supernatant *(7)*. The method is based on the cDNA-AFLP technique *(8)* (*Virus dis*covery *c*DNA-*A*FLP: VIDISCA). The main feature of VIDISCA is that prior knowledge of the genome sequence is not required as the presence of restriction enzyme sites is sufficient to guarantee PCR amplification.

VIDISCA begins with a treatment to selectively enrich for viral nucleic acid, which includes a centrifugation step to remove residual cells and mitochondria (**Fig. 1**). In addition, a DNase treatment is used to remove interfering chromosomal DNA and mitochondrial DNA from degraded cells, whereas RNases in the sample will degrade RNA. During this step, the viral nucleic acid is specifically protected within the viral particle. Next, DNase/Rnases are inactivated and the viral nucleic acids are subsequently extracted from the particles, RNA is reverse transcribed into cDNA, and second-strand synthesis is performed to make dsDNA (from a viral RNA or DNA genome). The dsDNA is digested with

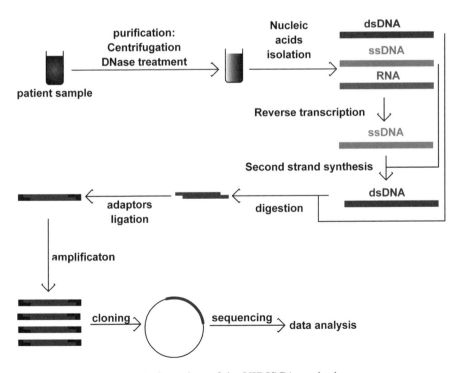

Fig. 1. Overview of the VIDISCA method.

frequently cutting restriction enzymes that are likely to be present in every viral target (*Hin*P1-I and *Mse*-I), and *Hin*P1-I- and *Mse*-I-anchors are ligated to the digested DNA. Essential to the method is that the restriction enzymes remain active during the ligase reaction, thus preventing concatamerization of digested fragments. The anchors themselves are not removed because they are designed in such a way that the restriction site is lost.

The target is subsequently PCR-amplified with primers that anneal to the anchor sequences, followed by a round of selective amplification with primers that are extended with one nucleotide (G, A, T, or C). Thus, 16 primer combinations are used and each sample is compared to a representative negative control (negative serum, plasma, or supernatant from an uninfected culture). The PCR fragments that are specific to the "infected" clinical sample can then be cloned and sequenced. Because amplification is based on the presence of restriction sites, the PCR is reproducible (in duplicate samples the same fragments are amplified) and these PCR products can be distinguished from background amplification. The assay is relatively high-throughput as multiple samples (about ten) can be tested per cycle of VIDISCA.

We were able to amplify viral nucleic acids from EDTA-plasma of a person with hepatitis B virus infection and a person with an acute parvovirus B19 infection. Using urine, we could detect adenoviral DNA and influenza B RNA in two patients. The technique can also detect HIV-1 and picornaviruses in cell culture. These results illustrate that the VIDISCA technique has the capacity to identify both RNA and DNA viruses directly from patient material or from cell cultures. In fact, it was the first experiment with a suspected virus culture that led to identification of a novel human coronavirus [HCoV-NL63 *(9)*]. Only three human coronaviruses were known at that time: HCoV-229E, HCoV-OC43, and SARS-CoV *(10,11)*, and HCoV-NL63 represents the fourth species. The rapid identification of this novel coronavirus demonstrates the power of our virus discovery tool, which can now be used to test large sample sets suspected of containing viral pathogens.

2. Materials

2.1. Pretreatment of the Sample

1. DNase I, RNase-free at concentration 2 U/μl (Ambion)
2. DNase buffer: 10X concentrated (Ambion)
3. Sterile HPLC pure water (Baker)

2.2. Nucleic Acid Isolation

1. L2 buffer: 0.1 M Tris-HCl pH 6.4. Prepare by mixing 12.1 g Tris, 9.4 ml of 32% HCl, and adjust with sterile water (Baker) to 1 liter *(12)*.

2. L2 solution: prepare by mixing 480 g guanidine thiocyanate (SIGMA) and 400 ml of L2 buffer *(12)*.

3. L6 solution: prepare by mixing 480 g guanidine thiocyanate (SIGMA), 88 ml of 0.2 M EDTA, 10.4 g Triton X-100 (Merck), and 400 ml of L2 buffer *(12)*.

4. Silica. Prepare 60 g of silicon dioxide (Sigma) in a 500-ml glass graduated cylinder and adjust the volume to 500 ml with sterile water (Baker). Resuspend the silica with vortexing and incubate at room temperature for 25 h. Remove 430 ml of top liquid. Adjust the volume with sterile water (Baker) to 500 ml and resuspend. Incubate at room temperature for 5 h and remove 440 ml of water. Resuspend the silica and stir adding 600 μl of 32% HCl. Aliquot and autoclave. Store at room temperature *(12)*.

5. 70% ethanol (Merck)

6. 100% acetone (Merck)

7. Sterile HPLC pure water (Baker)

2.3. Reverse Transcription, Digestion, Ligation, and PCR Amplification

1. Sequenase 2.0, T7 DNA polymerase at concentration 13 U/μl (Amersham Biosciences).

2. Random primers (hexamers; Amersham Biosciences). Working solution 1 μg/μl.

3. RNase H, 5 U/μl (Amersham Biosciences).

4. MMLV-RT (Moloney murine leukemia virus reverse transcriptase enzyme; 200 U/μl; Invitrogen)

5. CMB buffer (10X): 100 mM Tris-HCl pH 8.3, 500 mM KCl, 1% Triton-X100. Prepare by mixing 1 ml of 2 M Tris-HCl pH 8.3, 5 ml of 2 M KCl, 2 ml of 10% Triton X-100, and 12 ml of sterile water (Baker). Store at –20°C in 250 μl portions.

6. SEQII buffer (10X): 350 mM Tris-HCl pH 7.5, 250 mM NaCl, 175 mM $MgCl_2$. Prepare by mixing 2.25 ml sterile water (Baker), 3.5 ml of 1 M Tris-HCl pH 7.5, 2.5 ml of 1 M NaCl, 1.75 ml of 1 M $MgCl_2$. Store at –20°C in 100 μl portions.

7. Magnesium chloride (100 mM).

8. dNTP's (25 mM of each; Amersham Biosciences).

9. PCR buffer (10X): 100 mM Tris-HCl pH 8.3, 500 mM KCl, 100 mg BSA. Prepare by mixing 10 ml of 2 M Tris-HCl pH 8.3, 25 ml of 2 M KCl, 60 ml of sterile water (Baker), and 5 ml of BSA (Bovine serum albumin, 20 mg/ml; Roche). Store at –20°C in 250 μl portions.

10. First PCR primer set: *Hin*P1-I standard primer 5′-GAC GAT GAG TCC TGA CCG C-3′ and *Mse*-I standard primer 5′-CTC GTA GAC TGC GTA CCT AA-3′.

11. Nested PCR primer set: *Hin*P1-I-X Selective primers 5′-GAC GAT GAG TCC TGA CCG CA-3′; 5′-GAC GAT GAG TCC TGA CCG CT-3′; 5′-GAC GAT GAG TCC TGA CCG CC-3′; and 5′-GAC GAT GAG TCC TGA CCG CG-3′.

12. Nested PCR primer set: *Mse*-I-X Selective primers. *Mse*-I-A: 5′-CTC GTA GAC TGC GTA CCT AAA-3′; *Mse*-I-T: 5′-CTC GTA GAC TGC GTA CCT AAT-3′; *Mse*-I-C: 5′-CTC GTA GAC TGC GTA CCT AAC-3′; *Mse*-I-G: 5′-CTC GTA GAC TGC GTA CCT AAG-3′.

13. *Hin*P1-I anchors. Top strand: 5′-GAC GAT GAG TCC TGA C-3′; Bottom strand: 5′-CGG TCA GGA CTC AT- 3′. Oligonucleotides should be diluted to the 10 μM concentration.

14. *Mse*-I anchor: Top strand: 5′-CTC GTA GAC TGC GTA CC-3′; Bottom strand: 5′-TAG GTA CGC AGT C-3′. Oligonucleotides should be diluted to the 10 μM concentration.

15. *Mse*-I restriction enzyme, 10 U/μl (New England Biolabs). BSA and NEB-2 buffer are included.

16. *Hin*P1-I restriction enzyme. 10 U/μl (New England Biolabs).

17. Ligase, 5 U/μl (Invitrogen).

18. Ligase buffer (Invitrogen).

19. UltraPure™ Phenol:Chloroform:Isoamyl Alcohol (25:24:1, v/v) (Invitrogen).

20. 3 M sodium acetate (pH 5.2).

21. 100% ethanol.

22. 70% ethanol.

23. AmpliTaq® DNA Polymerase polymerase (5 U/μl; Applied Biosystems).

24. Sterile HPLC pure water (Baker).

2.4. Gel Electrophoresis and Gel Extraction

1. MetaPhor agarose (Cambrex).
2. Agarose MP (Roche).
3. Ethidium bromide (BioRad).
4. Tris-Borate-EDTA buffer (Sigma).
5. A 25-bp DNA ladder (Invitrogen).
6. Smart ladder DNA size marker (Eurogentec).
7. Sterile razor blades.
8. QIAquick gel extraction kit.
9. Sterile HPLC pure water (Baker).
10. Agarose gel loading buffer: 0.1% orange G, 30% glycerol in 0.5X TBE.

2.5. Cloning and Sequencing

1. TOP10 *E. coli* chemically competent bacteria (Invitrogen).
2. TOPO TA dual promoter or TOPO 2.1 TA cloning kit (Invitrogen).
3. Luria-Broth (LB, Gibco) agar plates supplemented with ampicilline.
4. BigDye terminator kit (Applied Biosystems).
5. M13 reverse primer and T7 primer (10 μM and 1 μM) (Eurogentec).

3. Methods

3.1. Sample Purification

1. Upon receipt samples are stored without thawing at –80°C to preserve the nucleic acids. On the day of the assay, the sample is thawed, vortexed, and 110 μl is immediately centrifuged at room temperature at 13,500 rpm (in a microfuge) for

10 min, in order to remove the cells, cell debris, and insoluble particles such as mucus. Every analyzed sample is tested in duplicate and in every experiment an appropriate negative control is included. The negative control can be a sample of the same type derived from a healthy person or virus-negative cell culture of the same cell type if the pathogen was cultured.

2. Immediately after centrifugation, 100 μl of sample is transferred into a fresh tube. Care should be taken that pelleted material is not transferred. If the sample is exceptionally full of cells/insoluble material, the primal volume may be increased as needed.

3. DNase treatment. DNase I solution is prepared in the nucleic acid free environment by mixing 15 μl of DNase I enzyme, 15 μl of DNase I buffer, and 20 μl of sterile water per 100 μl of the original sample. Subsequently, the DNase I solution is added to the sample material and incubated at 37°C for 45 min.

3.2. Nucleic Acid Isolation Using the Boom Method (see Note 1)

1. Immediately after the DNase I treatment, 900 μl of L6 solution is added to the sample to lyse the material (*see* **Notes 2–5**). The lysis is done at room temperature for 10 min. Sample should by thoroughly mixed by inverting and vortexing.

2. 40 μl of silica is added and the sample is incubated at room temperature with gentle shaking for 10 min.

3. Sample is centrifuged (13,200 rpm) for 10 sec to pellet the silica particles, and the L6 supernatant is discarded.

4. The pelleted silica is washed twice with 900 μl of L2. After addition of L2 solution, the sample is vortexed thoroughly until no pellets or large particles are visible and centrifuged for 10 sec at 13,200 rpm. Washing with L2 is necessary to remove all traces of Triton-X100 and EDTA that may inhibit the following enzymatic reactions.

5. The sample is washed twice with room temperature 70% ethanol and once with 100% acetone, in the same manner as described above for L2. Ethanol is added to wash out the guandine thiocyanate and residual traces of detergent and EDTA, whereas the acetone washing is needed primarily to speed up the drying process.

6. After the removal of the acetone, silica is dried for 5 min at 56°C with the lid open.

7. To elute bound nucleic acids, 50 μl of sterile water is added and the sample is vortexed until all silica particles are in suspension and incubated at 56°C for 10 min with shaking (500 rpm). After the elution, the sample is centrifuged for 2 min at 13,200 rpm. About 30 μl of the liquid fraction is transferred into a fresh tube. Samples should be stored at 80°C until needed.

3.3. Reverse Transcription and Second-Strand Synthesis

3.3.1. RT Reaction

The reverse transcription (RT) reaction mixture is assembled under nucleic acid and nuclease free conditions and consists of a two-step reaction.

1. Two RT solutions (I and II) are prepared according to the formula:

   a. RT solution I (10 μl per sample)

      i. 2.5 μl of random primers
      ii. 3.0 μl of 10X concentrated CMB buffer
      iii. 2.4 μl of MgCl$_2$
      iv. 2.1 μl of sterile water

   b. RT solution II (20 μl per sample)

      i. 2.0 μl of 10X concentrated CMB buffer
      ii. 1.0 μl of MMLV-RT enzyme
      iii. 0.8 μl of dNTPs
      iv. 16.2 μl of sterile water

2. The nucleic acids isolated by the Boom method are centrifuged (1 min, 13,200 rpm) in order to remove the residual silica particles.
3. After centrifugation, 20 μl of the supernatant is mixed with 10 μl of RT solution I and incubated at room temperature for 2 min in order to support primer-template annealing.
4. After 2 min of incubation, 20 μl of RT solution II is added and samples are incubated for 90 min at 37°C to allow efficient reverse transcription. The RT reaction is followed by 5 min 95°C to deactivate the enzyme.

3.3.2. Second-Strand Synthesis

1. After the RT reaction, the resulting single-stranded cDNA cannot be used as a template for restriction enzyme cleavage or adaptor ligation. Therefore, second-strand synthesis is performed using RNase H to digest any residual RNA and Sequenase enzyme to synthesize the second strand DNA.
2. Second-strand reaction mixture (100 μl per sample):

   a. 10 μl of 10 × concentrated SEQII buffer
   b. 2.0 μl of Sequenase 2.0
   c. 1.5 μl of RNase H
   d. 1.0 μl of dNTPs
   e. 85.5 μl of water

   The mixture is added to the 50.0 μl RT reaction product

3. Incubate for 90 min at 37°C.

3.3.3. Phenol/Chloroform Extraction

1. After the second-strand synthesis, the sample (150 μl) is mixed with an equal volume of UltraPure^{TM} phenol: chloroform: isoamyl alcohol and vortexed vigorously until the two phases are completely mixed.

2. The separation of phases is done by centrifugation (2 min, 13,200 rpm, room temperature) and the water phase containing ds-DNA is collected into a new tube. The usual volume recovery rate is about 93%.
3. Subsequently, the recovered ds-DNA is precipitated with ethanol. Water phase is mixed with 350 μl of 100% ethanol and 14 μl of 3 M sodium acetate and incubated for 14 h at –20°C.
4. The ds-DNA is pelleted by 25 min centrifugation (15,000 rpm) at 4°C, supernatant is discarded and the pellet is washed with 200 μl of freshly prepared 70% ethanol (centrifugation for 25 min at 15,000 rpm at 4°C).
5. The ethanol is discarded, and the pellet is air dried for 15 min at room temperature.
6. The pellet is dissolved in 30 μl of sterile water by incubation at room temperature.

3.4. Construction of the Adaptors

1. Adaptors are ds-DNA oligonucleotides with its 3′-ends designed to anneal to the cleavage site in the target DNA molecule. Introduction of the single mutation in the region recognized by the restriction enzyme prevents the cleavage of ligated adaptor-target DNA molecule (**Fig. 2**). Adaptors are homemade, using single-stranded oligonucleotides.
2. To prepare the functional adaptor, the bottom and top oligonucleotides are annealed by adding:

   a. 20.0 μl of top adaptor (*Mse*-I or *Hin*P1-I)
   b. 20.0 μl of bottom adaptor (*Mse*-I or *Hin*P1-I)
   c. 5.0 μl of ligase buffer
   d. 49.0 μl of sterile water

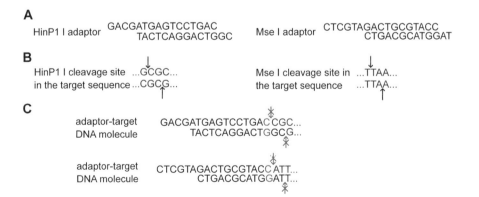

Fig. 2. Ligation of adaptors to the target sequence: (A) *Hin*P1-I and *Mse*-I adaptors. (B) the restriction sites specific for *Hin*P1-I and *Mse*-I enzymatic cleavage. (C) The product of ligation of the adaptors to the target sequence. Incorporation of a single nucleotide in the region of cleavage result in a nonfunctional cleavage site.

3. Premixes for each adaptor should be prepared independently.
4. Solution should be heated up to 65°C for 5 min and cooled down slowly to the room temperature.
5. Prepared adaptors are stored at –20°C.

3.5. Digestion of Target Sequence and Ligation of Adaptors

1. Digestion of the ds-cDNA is performed with restriction enzymes. In the protocol described here we included only digestion with *Hin*P1-I and *Mse*-I enzymes, but this combination may be altered *(13)*. The use of two different restriction enzymes is essential, as we observed that fragments that are cleaved on both sides by the same enzyme have less of a chance to be amplified by PCR. If one is planning to change the enzyme combination, care should be taken that the combination of restriction enzymes generates restriction fragments from virtually all templates.
2. For digestion of ds-cDNA material, the premix is prepared, containing per sample (10 µl per sample):

   a. 4.0 µl of NEB-2 buffer
   b. 4.0 µl of 1:10 diluted BSA
   c. 1.0 µl of *Hin*p1-I restriction enzyme
   d. 1.0 µl of *Mse*-I restriction enzyme

3. 10 µl of the digestion premix is added to 30 µl of purified ds-cDNA and incubated for 2 h at 37°C.
4. The ligation mix should be prepared before the digestion reaction is finished. The ligation premix contains per sample (15 µl per sample):

   a. 1.0 µl of *Hin*P1-I adaptor
   b. 1.0 µl of *Mse*-I adaptor
   c. 2.0 µl of 5X concentrated ligase buffer
   d. 1.0 µl of ligase
   e. 10.0 µl of sterile water

5. 15 µl of the ligation premix is added to the digested sample (40 µl). There is no inactivation step between the digestion and ligation, as restriction activity prevents generation of the concatameric forms of the target templates. Fragments that are properly ligated with adaptors will not be cleaved, because of the point mutation introduced (**Fig. 2**). The ligation should be performed for 2 h at 37°C.

3.6. PCR Reactions

The main part of the VIDISCA method is amplification of the genetic material without prior knowledge of the sequence. The preprocessed ds-cDNA with adaptors can be now amplified using the primers specific for the adaptors. During development of the method it was determined that a single PCR round does not provide sufficient specificity and sensitivity, so a second "nested" PCR is

Table 1
First PCR Thermocycling Profile

Time (min)	Temperature	Number of cycles
5	94°C	1 cycle
1	94°C	
1	55°C	20 cycles
2	72°C	
10	72°C	1 cycle
∞	4°C	1 cycle

included in the protocol. This PCR uses primers that are similar to the primers used in the first PCR, but with one nucleotide added to the 3′-end of the primers.

1. The first PCR reaction is done with the standard PCR thermocycling program (**Table 1**) and is optimized for 50 µl reaction.
2. 10 µl of ligated sample is mixed with 40 µl of PCR mix. The PCR mix is prepared as described below (the volumes are calculated per sample):

   a. 31.25 µl of sterile water
   b. 0.75 µl of $MgCl_2$
   c. 5.0 µl of 10X concentrated PCR buffer
   d. 0.5 µl of dNTPs
   e. 1.0 µl of *Hin*p1-I standard primer (10 µM)
   f. 1.0 µl of *Mse*-I standard primer (10 µM)
   g. 0.5 µl of AmpliTaq® DNA polymerase

3. The PCR reaction thermocycling is performed according to the scheme presented in **Table 1** (*see* **Note 6**). After successful thermocycling the sample can be store at –20°C until needed.
4. Second PCR—selective amplifications: The second, nested PCR reaction is necessary to provide high specificity and sensitivity. This selective PCR is performed with primers with sequence identical as the standard primers, but with an additional nucleotide on its 3′ part. This additional nucleotide is outside the adaptor sequence and thus belongs to the unknown material (**Fig. 2**). Use of an additional nucleotide allows separation of the reactions in 16 different primer combinations and enables better analysis of the sample. To have a selectivity that is required when one wants to amplify only those fragments with a 100% match, the thermocycling profile is designed to increase the specificity of reaction by using the starting annealing temperature of 65°C, which gradually decreases during first ten cycles to 56°C.
5. The PCR mix is prepared as described below (47.5 µl per sample). The *Hin*pI-X and *Mse*I-X primers denote primers with an additional 3′-nucleotide.

Table 2
Touch Down PCR profile

Time	Temperature	Number of cycles
5 min	94°C	1 cycle
60 sec	94°C	
60 sec	65–56°C[a]	10 cycles
90 sec	72°C	
30 sec	94°C	
30 sec	56°C	23 cycles
60 sec	72°C	
10 min	72°C	1 cycle
∞	4°C	1 cycle

[a] −1°C per cycle for each successive cycle.

    a. 40.3 μl of sterile water
    b. 0.75 μl of $MgCl_2$
    c. 5.0 μl of 10X concentrated PCR buffer
    d. 0.2 μl of dNTPs
    e. 0.5 μl of HinpI-X primer (10 μM)
    f. 0.5 μl *Mse*-I-X primer (10 μM)
    g. 0.25 μl AmpliTaq® DNA polymerase

6. 16 PCR premixes are prepared with different primer combinations (*Hinp*1-I - G,C,A,T; *Mse*-I - G,C,A,T) and 47.5 μl per sample of each premix is combined with 2.5 μl of the first PCR product. The second PCR thermocycling profile is presented in **Table 2** (*see* **Note 6**). The PCR product may be analyzed immediately or stored at −20°C until needed.

3.7. Gel Analysis of the PCR Product and Purification of the Amplified DNA

1. The second PCR product is analyzed on agarose gel. Most of generated fragments are less than 300 bp in size. Owing to the need for high-quality separation and small differences in fragment sizes, the MetaPhor agarose is being used (it allows differentiation among fragments varying 1 bp in size). Additionally, the MetaPhor agarose provides an easy setup and high-throughput processing for gel analysis and purification, compared to the polyacrylamide gels. The MetaPhor agarose gel is prepared as described below.
2. 150 ml of 0.5X concentrated TBE buffer is poured into an Erlenmeyer flask and stirred with a magnetic stirrer. 4 g of MetaPhor agarose is weighted and gently poured into the Erlenmeyer flask while mixing. Addition of all agarose powder at once will result in clumping of the agarose. The solution is stirred for another

10 min to soak the agarose grains and heated in the microwave for 60 sec with low power. After the primary heating, the agarose is stirred and heated in 30-sec cycles (low power) with extensive stirring in between. All the agarose is solubilized during a final heating step for 60 sec with medium power. The agarose is cooled down to ~65°C and 10 μl of ethidium bromide (10 mg/ml) is added. The agarose is poured into the electrophoresis tray and combs are inserted. It is crucial to remove all air bubbles from the gel (e.g., with a pipette tip). The agarose is solidified at room temperature and further incubated at 4°C for at least 20 min (the incubation at 4°C improves the gel resolution). The gel is positioned in the electrophoresis box filled with 0.5X TBE buffer

3. 15 μl of the second PCR product is mixed with 5 μl of the loading buffer and the samples are layered on the prepared agarose gel. 5 μl of the 25-bp ladder is used as a DNA size marker. Electrophoretic separation is performed at 150 V for about 1 h.

4. Immediately after the electrophoresis is completed, the gel is analyzed on the UV transilluminator. A picture is taken for analysis and the gel is stored at 4°C, wrapped in plastic (Saran Wrap). The picture of the gel is used to search for fragments that are present in the sample of interest and not in the control sample All the fragments that are present exclusively in the sample of interest are marked on the picture (**Fig. 3**). If the bands appear very faint on the gel, the PCR products can be concentrated by vacuum centrifuge and reanalyzed on a MetaPhor gel. After fragment selection, the gel is again positioned on the UV transilluminator and the selected bands are excised with sterile razors (about 100 mg per slice) and stored in coded 1.5-ml Eppendorf tubes at 4°C (*see* **Note 7**). After excision of all bands, a second picture of the gel should be taken to document the proper excision.

5. The DNA fragments from the gel are extracted with the QIAquick gel extraction kit following the manufacturer's protocol. The gel slices are solubilized in 600 μl of QG buffer and 100 μl of isopropanol is added. After extraction, resulting DNA is dissolved in 30 μl of EB buffer. Alternatively, any other gel extraction method may be used.

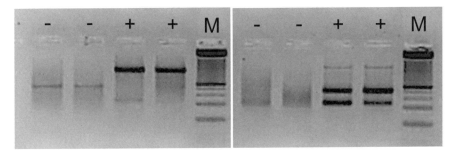

Fig. 3. Representative VIDISCA fragments on the MetaPhor agarose gel. Samples with '+' were supernatants of LLC-MK2 cells infected with HCoV-NL63 and samples '–' were supernatants of control LLC-MK2 cells.

3.8. Cloning, Selection of Plasmids, and Sequencing

1. Purified DNA is subsequently cloned into a vector and transformed into bacteria. The usual procedure is to use the TOPO cloning kit from Invitrogen (TOPO TA Cloning® Kit Dual Promoter). Cloning is done with 0.5 μl of vector, 0.5 μl of salt solution, and 2 μl of gel purified DNA. Chemically competent TOP10 *E. coli* (Invitrogen) bacteria are transformed using the TOPO reaction sample (10 μl of bacteria per reaction). The *E. coli* are plated on LB agar plates supplemented with ampicilline. The growth on the LB plates is carried on for 16 h at 37°C. Eight colonies per plate are collected with a pipette tip into 50 μl of the BHI medium supplemented with ampicilline on the 96-well PCR plate. The suspended bacteria are subjected directly to a colony-PCR procedure, described below.

2. Colony PCR. The PCR mix is prepared by mixing (45 μl per sample):

    a. 0.5 μl of M13 reverse primer (10 μM)
    b. 0.5 μl of T7 primer (10 μM)
    c. 5.0 μl of 10X concentrated PCR buffer
    d. 0.5 μl of dNTPs
    e. 0.75 μl of MgCl₂
    f. 0.2 μl of AmpliTaq® DNA Polymerase
    g. 37.55 μl of sterile water

3. 5 μl of suspended *E. coli* bacteria in BHI medium is added to the PCR mixture.

4. The thermocycling is performed as described in **Table 3**. After the PCR is completed, 10 μl of the PCR product is mixed with gel loading dye and analyzed on 0.8% agarose MP gel with a Smart ladder DNA size marker. A representative picture of such a gel is shown in **Fig. 4**.

5. The lanes that seem to contain the plasmid with proper insert are selected, and corresponding PCR products are subjected to sequencing reactions.

6. Sequencing reactions are performed on the colony-PCR product with the BigDye chemistry, using the M13 reverse and T7 primer, according to the manufacturers' instructions (Applied Biosystems).

Table 3
Colony PCR Thermocycling Scheme

Time(min)	Temperature	Number of cycles
5	95°C	1 cycle
1	95°C	
1	55°C	25 cycles
2	72°C	
10	72°C	1 cycle
∞	4°C	1 cycle

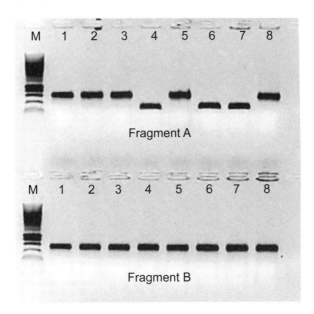

Fig. 4. Gel analysis of colony PCR of VIDISCA fragments cloned into the TOPO 2.1 vector.

3.9. Data Analysis

The sequence data obtained in the survey is analyzed with the BLAST server (http://www.ncbi.nlm.nih.gov/BLAST/). The raw sequence is edited to remove the sequence derived from the vector and the adaptors. This procedure can be done manually or using designated program, e.g., CodonCode (http://www.codoncode.com/). After the cleanup, the sequences are analyzed for their quality and only those that show a clear, single signal are exported in FASTA format for further analysis. Once imported to the BioEdit program (http://www.mbio.ncsu.edu/BioEdit/bioedit.html), the sequences are subjected to batch BLAST analysis with default settings. This batch analysis allows preselection of the sequences of interest, as mRNA and rRNA fragments are frequently found as background. All results that indicate the presence of a virus, or an unknown sequence should be selected and reanalyzed with the BLAST server (nblast) with the expectation number of 1000 against all databases. If the results are still not clear the following steps might be taken:

1. Analysis against translated database (tblastx)
2. Search the conserved domain database (rpsblast)
3. Analysis against virus database (nblast)

The sequences in tblastx and rpsblast that display similarity to viral sequences should be considered as possibly unknown pathogens. If the sequence is analyzed against a viral database, care should be taken with each hit, because virtually all fragments show some similarity to viral sequences. In that case, the pathogen might be considered identified only if the results from different fragments from one sample show similarity to the same virus family.

In all cases, it is essential to design a diagnostic primer set and retest the original material for the presence of the pathogen. It is only when the pathogen can be detected by the diagnostic (RT)-PCR in the original sample that efforts to sequence the entire genome can be undertaken.

4. Notes

1. For the nucleic acid isolation, any highly efficient method may be used. It is not advisable to use TRIzol isolation, as that is intended for isolation of nucleic acids from cells and tissues.
2. The L6 buffer lysis is sufficient to inactivate the virus. After a 10-min incubation it is safe to process the sample in a normal biochemistry laboratory.
3. The L6 and L2 buffer contain concentrated guanidine thiocyanate (GTC) and thus should be considered as highly toxic. Remember to store the GTC waste separately with addition of one-tenth volume of 1 N sodium hydroxide to prevent GTC degradation.
4. All RNA and cDNA handling before the first PCR should be performed in a nucleic-acid-free environment. The sequence independent amplification will result in overamplification of contaminating DNA.
5. The use of chlorine as a decontaminant should be limited as it may decrease the viability of reverse transcription enzyme.
6. If the thermocycling is performed in a PCR machine that does not include heating of the cover, two drops of paraffin oil should be layered on top of the PCR solution to prevent evaporation during the PCR reaction.
7. It is advised to use a fresh razor for each band during excision. The exposure of the gel to UV light should be limited, as such exposure results in DNA degradation.

References

1. Burgner, D., and Harnden, A. (2005) Kawasaki disease: what is the epidemiology telling us about the etiology? *Int. J. Infect. Dis.* **9**, 185–194.
2. Fujinami, R. S., von Herrath, M. G., Christen, U., and Whitton, J. L. (2006) Molecular mimicry, bystander activation, or viral persistence: infections and autoimmune disease. *Clin. Microbiol. Rev.* **19**, 80–94.
3. Stewart, J., Talbot, P., and Mounir, S. (1995) Detection of coronaviruses by the polymerase chain reaction. In: Becker, Y., and Darai, G. (eds.) *Diagnosis of Human Viruses by Polymerase Chain Reaction Technology*, Springer-Verlag, New York, pp. 316–327.

4. Stephensen, C. B., Casebolt, D. B., and Gangopadhyay, N. N. (1999) Phylogenetic analysis of a highly conserved region of the polymerase gene from 11 coronaviruses and development of a consensus polymerase chain reaction assay. *Virus Res.* **60**, 181–189.

5. Drosten, C., Gunther, S., Preiser, W., van der Werf, S., Brodt, H. R., Becker, S., Rabenau, H., Panning, M., Kolesnikova, L., Fouchier, R. A., Berger, A., Burguiere, A. M., Cinatl, J., Eickmann, M., Escriou, N., Grywna, K., Kramme, S., Manuguerra, J. C., Muller, S., Rickerts, V., Sturmer, M., Vieth, S., Klenk, H. D., Osterhaus, A. D., Schmitz, H., and Doerr, H. W. (2003) Identification of a novel coronavirus in patients with severe acute respiratory syndrome. *N. Engl. J. Med.* **348**, 1967–1976.

6. Chang, Y., Cesarman, E., Pessin, M. S., Lee, F., Culpepper, J., Knowles, D. M., and Moore, P. S. (1994) Identification of herpesvirus-like DNA sequences in AIDS-associated Kaposi's sarcoma. *Science* **266**, 1865–1869.

7. van der Hoek L., Pyrc, K., Jebbink, M. F., Vermeulen-Oost, W., Berkhout, R. J., Wolthers, K. C., Wertheim-van Dillen, P. M., Kaandorp, J., Spaargaren, J., and Berkhout, B. (2004) Identification of a new human coronavirus. *Nature Med.* **10**, 368–373.

8. Bachem, C. W., van der Hoeven, R. S., de Bruijn, S. M., Vreugdenhil, D., Zabeau, M., and Visser, R. G. (1996) Visualization of differential gene expression using a novel method of RNA fingerprinting based on AFLP: analysis of gene expression during potato tuber development. *Plant J.* **9**, 745–753.

9. van der Hoek L., Pyrc, K., Jebbink, M. F., Vermeulen-Oost, W., Berkhout, R. J., Wolthers, K. C., Wertheim-van Dillen, P. M., Kaandorp, J., Spaargaren, J., and Berkhout, B. (2004) Identification of a new human coronavirus. *Nature Med.* **10**, 368–373.

10. Holmes, K. V., and Lai, M. M. C. (1996) *Coronaviridae*: The viruses and their replication. In: Fields, B. N., Knipe, D. M., Howley, P. M., et al. (eds.) *Fields Virology*. Lippincott-Raven, Philadelphia, pp. 1075–1093.

11. Drosten, C., Gunther, S., Preiser, W., van der Werf, S., Brodt, H. R., Becker, S., Rabenau, H., Panning, M., Kolesnikova, L., Fouchier, R. A., Berger, A., Burguiere, A. M., Cinatl, J., Eickmann, M., Escriou, N., Grywna, K., Kramme, S., Manuguerra, J. C., Muller, S., Rickerts, V., Sturmer, M., Vieth, S., Klenk, H. D., Osterhaus, A. D., Schmitz, H., and Doerr, H. W. (2003) Identification of a novel coronavirus in patients with severe acute respiratory syndrome. *N. Engl. J. Med.* **348**, 1967–1976.

12. Boom, R., Sol, C. J., Salimans, M. M., Jansen, C. L., Wertheim-van Dillen, P. M., and van der Noordaa, J. (1990) Rapid and simple method for purification of nucleic acids. *J. Clin. Microbiol.* **28**, 495–503.

13. Bachem, C. W., van der Hoeven, R. S., de Bruijn, S. M., Vreugdenhil, D., Zabeau, M., and Visser, R. G. (1996) Visualization of differential gene expression using a novel method of RNA fingerprinting based on AFLP: analysis of gene expression during potato tuber development. *Plant J.* **9**, 745–753.

II

ISOLATION, GROWTH, TITRATION, AND PURIFICATION OF CORONAVIRUSES

8

Titration of Human Coronaviruses, HCoV-229E and HCoV-OC43, by an Indirect Immunoperoxidase Assay

Francine Lambert, Hélène Jacomy, Gabriel Marceau, and Pierre J. Talbot

Abstract

Calculation of infectious viral titers represents a basic and essential experimental approach for virologists. Classical plaque assays cannot be used for viruses that do not cause significant cytopathic effects, which is the case for strains 229E and OC43 of human coronavirus (HCoV). An alternative indirect immunoperoxidase assay (IPA) is herein described for the detection and titration of these viruses. Susceptible cells are inoculated with serial logarithmic dilutions of samples in a 96-well plate. After viral growth, viral detection by IPA yields the infectious virus titer, expressed as "Tissue Culture Infectious Dose" ($TCID_{50}$). This represents the dilution of a virus-containing sample at which half of a series of laboratory wells contain replicating virus. This technique is a reliable method for the titration of HCoV in biological samples (cells, tissues, or fluids).

Key words: human coronavirus; HCoV-229E; HCoV-OC43; cell and tissue samples; titration; immunoperoxidase assay; $TCID_{50}$

1. Introduction

HCoV were first isolated in the mid-1960s from patients with upper respiratory tract disease (1–3). As described by McIntosh (4), the identification of coronaviruses in clinical samples was a very difficult task until the 1970s, as HCoV induced subtle or nonexistent cytopathic effects, and many cells types were not susceptible to the virus. Commonly, HCoV-OC43 (ATCC: VR-759)

From: *Methods in Molecular Biology, vol. 454: SARS- and Other Coronaviruses,*
Edited by: D. Cavanagh, DOI: 10.1007/978-1-59745-181-9_8, © Humana Press, New York, NY

was grown and maintained on BS-C-1 (ATCC: CCL-26), RD (ATCC: CCL-136) or HRT-18 (also called HCT-8; ATCC: CCL-224) cell lines. On the other hand, HCoV-229E (ATCC: VR-740) was grown and maintained on WI-38 (ATCC: CCL-75), MRC-5 (ATCC: CCL-171) or L-132 (ATCC: CCL-5) cell lines. Interestingly, cells from the central nervous system (CNS) were reported to be highly susceptible to HCoV replication. The SK-N-SH (ATCC: HTB-11) neuroblastoma and H4 (ATCC: HTB-148) neuroglioma cell lines were highly susceptible to infection, as well as astrocytoma cell lines U-87 MG (ATCC: HTB-14), U-373 MG, and GL-15 *(5–8)*. Even though it was reported that HCoV-229E infectious titers could be determined by plaque assay on specific cell lines, such as MRC-5 or L-132 *(6)*, this proved not to be a reliable assay and, as for HCoV-OC43, an alternative assay was required: an indirect immunoperoxidase assay (IPA) on coronavirus-susceptible cells to quantify infectious virus in biological samples.

The immunoperoxidase assay is an enzymatic antigen detection technique that uses the enzyme horseradish peroxidase (HRP) to label antigen-antibody complexes. The principle of the IPA technique is that a specific antibody recognizes and binds to its specific antigen to yield an antibody-antigen complex. For detection of these complexes, the primary antibody is either directly labeled with HRP or remains unlabeled, with detection achieved by a labeled secondary antibody. If a secondary antibody is used, it must be generated against the immunoglobulins of the animal species in which the primary antibody was produced. The enzyme substrate, 3,3'-diaminobenzidine (DAB) is then converted by HRP to a precipitating brown end-product.

2. Materials

2.1. Preparation of Tissue Samples

1. Digital scales.
2. Ethanol 70% (v/v).
3. Sterile dissection kit.
4. Ice.
5. Sterile PBS.
6. Centrifuge.
7. Sterile tubes: 15 ml for big pieces of tissue or 5 ml for small pieces.

2.2. Preparation of Cell Samples

1. Sterile PBS.
2. Centrifuge.
3. Sterile tubes.
4. Cell medium.

2.3. Preparation of Susceptible Cells

1. L-132 cell line (human lung epithelium; ATCC: CCL5) for HCoV-229E (*see* **Note 1**).
2. HRT-18 cell line (human adenocarcinoma rectal (*9*)) for HCoV-OC43 (*see* **Note 1**).
3. Alpha-MEM (alpha minimum essential medium; Invitrogen).
4. Fetal bovine serum (FBS, Hyclone).
5. Trypsin at 0.25% (w/v) for the L-132 cell line or trypsin/EDTA at 0.05% (w/v) for the HRT-18 cell line.
6. 96-well plate flat bottom for tissue culture.
7. Multichannel pipette and sterile tips.
8. Incubator 37°C with 5% (v/v) CO_2.

2.4. Immunoperoxidase Assay

2.4.1. Infection of Cells with Samples to Test

1. Alpha MEM with 1% (v/v) FBS.
2. Sterile paper towels.
3. Sterile tips.
4. Incubator 33°C with 5% (v/v) CO_2.

2.4.2. Virus Detection

1. 100% methanol.
2. Hydrogen peroxide 30% (Sigma).
3. PBS.
4. Primary antibodies (*see* **Note 2**):

   a. HCoV-229E: An ascites fluid or culture supernatant from mouse MAb 5-11H.6 (*7, 10*), directed against the surface glycoprotein (S) of HCoV-229E, or equivalent polyclonal antiserum

   b. HCoV-OC43: An ascites fluid or culture supernatant from mouse MAb 1-10C.3 (*8*), directed against the surface glycoprotein (S) of HCoV-OC43, or equivalent polyclonal antiserum.

5. Secondary antibody: anti-mouse immunoglobulins conjugated to horseradish peroxidase (HRP; from KPL) at 1/500 dilution in PBS.
6. Multichannel pipette and tips.
7. Developing solution: 25–50 μg/100 ml of freshly made DAB (3,3'-diaminobenzidine tetrahydrochloride) solution in PBS. Filter this solution on Whatman paper then add 0.01% hydrogen peroxide (33 μl of H_2O_2 30%). This solution is light sensitive and toxic.
8. Nonsterile paper tissue.

9. Light microscope.
10. Incubator 37°C.

3. Methods

3.1. Preparation of Tissue Samples

1. Prepare sterile tubes for each organ and determine the weight of the empty tubes.
2. Quickly dissect the organs of interest in sterile conditions and keep them on ice. Rinse instruments in 70% (v/v) ethanol between each dissection.
3. Determine the weight of each biological sample.
4. Homogenize tissue samples to 10% (w/v) sterile PBS with a Polytron homogenizer in a laminar flow hood.
5. Centrifuge tubes at 4°C, for 20 min at 1000 × g.
6. Collect supernatants in new sterile tubes and process to immunoperoxidase detection, or immediately freeze at −80°C and store until assayed.

3.2. Preparation of Cell Samples

3.2.1. Adherent Cells

1. To determine extracellular viral production, collect supernatants in sterile tubes from infected cells (to be tested) at appropriate times following infection. For estimation of intracellular viral titers, remove medium, wash cells in warm (37°C) sterile PBS, then add the same volume or a known volume of medium and perform three cycles of freeze/thaw at −80°C to lyse cells and release viral particles in the medium.
2. Collect medium samples in sterile tubes, centrifuge for 5 min at 1000 × g.
3. Collect supernatants in new sterile tubes and process to IPA detection, or immediately freeze at −80°C and store until assayed.

3.2.2. Nonadherent Cells

1. Centrifuge nonadherent cells at 1000 × g for 5 min.
2. Collect supernatants in sterile tubes; they represent the extracellular portion of infectious virus.
3. Resuspend the cell pellet in the same volume (corresponding to the supernatant) or in a known volume of culture medium.
4. Perform three cycles of freeze/thaw at −80°C to lyse cells and release the viral particles in the medium; they represent the intracellular portion of the infectious virus.
5. Centrifuge for 5 min at 1000 × g.
6. Collect supernatants in new sterile tubes and process to IPA detection, or immediately freeze at −80°C and store until assayed.

3.3. Preparation of Susceptible Cells

To verify the productivity of HCoV infection in biological samples, an immunoperoxidase assay is performed on coronavirus-susceptible cells. Suspensions of susceptible cells are prepared from confluent monolayers by trypsin treatment (trypsin [0.25% (w/v)] for the L132 cell line or trypsin/EDTA [0.05% (w/v)] for the HRT-18 cell line). Trypsin is then inactivated by the addition of culture medium supplemented by 10% (v/v) with fetal bovine serum (FBS). Susceptible cells are plated in a 96-well plate for tissue culture using a multichannel pipette. Dispense 100 µl/well of cell suspension as follows.

- HCoV-229E: L-132 cells are seeded at 70,000 cells/ml if inoculation of samples is to be performed 2 days later or 50,000 cells/ml if inoculation of samples is to be performed 3 days later, in alpha-MEM supplemented with 10% (v/v) FBS. Incubate at 37°C in a humid atmosphere containing 5% (v/v) CO_2.
- HCoV-OC43: HRT-18 cells are plated at 70,000 cells/ml if inoculation of samples is to be performed 3 days later or 50,000 cells/ml if inoculation of samples is to be performed 4 days later, in alpha-MEM supplemented with 10% (v/v) FBS. Incubate at 37°C in a humid atmosphere containing 5% (v/v) CO_2.

3.4. Immunoperoxidase Assay

3.4.1. Infection of Susceptible Cells with Samples to Be Tested

1. When susceptible cells (L-132 for HCoV-229E or HRT-18 for HCoV-OC43) are at 70–80% confluence in the laminar flow hood, flick the medium of the 96-well plate. Remove residual medium by wrapping each plate in a sterile paper tissue and gently flicking it face down onto sterile paper.
2. Add 100 µl of alpha-MEM supplemented with 1% (v/v) FBS to each well.
3. Each sample should be tested in four adjacent columns in the same plate. Put 100 µl/well of aliquots to be tested in the first row.
4. Inoculate with serial logarithmic dilutions of infected samples: Add 11 µl/well of the same sample again in four wells in the second row of the 96-well plate. Mix with a multichannel pipette by pipetting up and down in the pipette tip three times and transfer 11 µl/well in the third row and so forth to the last row of the plate (**Fig. 1**). Change tips after every dilution and discard the remaining 11 µl.
5. Include a positive control in each experiment by inoculating four wells of reference virus.
6. Incubate in a humidified chamber at 33°C with 5% (v/v) CO_2.

   a. HCoV-229E: 5 days
   b. HCoV-OC43: 4 days

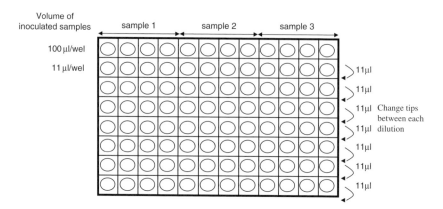

Fig. 1. Schematic representation of a 96-well plate illustrating inoculation and serial logarithmic dilutions of infected samples for titration.

3.4.2. Virus Detection

1. Remove medium completely by flicking the plates onto nonsterile paper towels in the laminar flow hood and delicately rinse the cells by gently filling the wells with PBS with a multichannel pipette. Flick the PBS and remove residual PBS by wrapping each plate in a paper tissue.
2. Fix cells with 100% methanol containing 0.3% (v/v) hydrogen peroxide for 15–30 min at room temperature.
3. Remove fixative by flicking the plates over a sink and remove the residual liquid by flicking it face down onto several paper towels lying on the bench top and then let it totally air-dry, face up for approximately 15–30 min.
4. Prepare an antibody solution specific to the virus at the appropriate dilution in PBS. Add 100 µl of specific viral-antibody to each well and incubate plates for 2 h at 37°C.
5. Completely remove the medium by flicking it over a sink. Rinse the plate by filling the wells with PBS. Flick the PBS into the sink and rinse with PBS twice more, flicking the PBS into the sink after each rinse. Remove residual PBS by flicking it face down onto nonsterile paper towels lying on the bench top.
6. Add 100 µl to each well of secondary antibody at 1/1000 dilution in PBS.
7. Incubate plate for 2 h at 37°C without CO_2 (*see* **Note 3**).
8. 30 min before the end of incubation, prepare a solution of DAB (detection reagent).
9. Rinse the plate three times in PBS as in step 5.
10. Add 100 µl of developing reagent, DAB solution in PBS, with 0.01% (v/v) hydrogen peroxide to each well and incubate 10–20 min at room temperature, or until the positive control is stained.

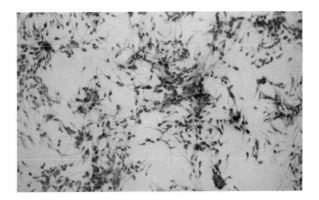

Fig. 2. Example of HRT-18 cells positive for HCoV-OC43. Monolayer of HRT-18 cells were inoculated with an infected biological sample for titration. Four days postinoculation, an indirect immunoperoxidase assay was performed and cells positive for viral antigens appear brown after DAB detection.

11. Stop the reaction with one wash with water as in step 5 and fill each well with 100 μl deionized water.
12. Read plates with a light microscope to quantify all wells presenting stained cell (**Fig. 2**)

3.4.3. Determination of Titers

Infectious virus titers (tissue culture infectious dose at 50%, $TCID_{50}$) are calculated by the Karber method (**11**):

$$DICT_{50} = D - [d(S - 0.5)]$$

where $D = -\log_{10}$ of the last dilution showing 100% (4 wells/4 wells) of virus positive wells; $d = -\log_{10}$ of dilution factor. Example: $-\log_{10} 10 = -1.0$; $S =$ number of all wells presenting virus, including those showing 100% of viral positive wells; this last one representing the unit and other dilutions being a fraction of this unit. For an example see **Fig. 3** (*also see* **Note 4**).

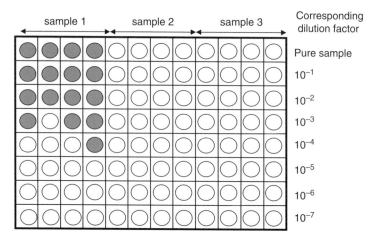

Fig. 3. Schematic representation of a 96-well plate after the immunoperoxidase assay: viral antigen positive wells (as soon at least two cells are stained brown) are illustrated in gray for the first sample.

Variables:

10^{-1} = 4/4 positive wells = 1.0

10^{-2} = 4/4 positive wells = 1.0 (*) (**)

10^{-3} = 3/4 positive wells = 0.75 (**)

10^{-4} = 1/4 positive wells = 0.25 (**)

10^{-5} = 0/4 positive wells = 0.0

(*) used for D and (**) used for S

Calculation:

$$
\begin{aligned}
\mathrm{DICT}_{50} &= D - [d\,(S - 0.5)] \\
&= (*) - [d\,((**) - 0.5)] \\
&= 2 - [\,-1\,((1 + 0.75 + 0.25) - 0.5)] \\
&= 2 - [\,-1(2 - 0.5)] \\
&= 2 - [-1(1.5)] \\
&= 2 - [-1.5] \\
&= 2 + 1.5 \\
&= 3.5 \log_{10}\ (\text{in } 100\ \mu\text{l as each well contain } 100\ \mu\text{l})
\end{aligned}
$$

The infectious titer of sample 1 is 4.5 $\log_{10}$/ml or $10^{4.5}$ TCID$_{50}$ /ml.

For organs or intracellular viral titers, keep in mind the dilution factor used. Normally tissue samples are diluted at 10% (w/v), so virus titers will be multiplied by 10. In example 1, the virus titer would be $10^{5.5}$ TCID$_{50}$ /ml.

4. Notes

1. This protocol can be adapted for many other cell culture systems. HCoV-229E could be grown in normal human fetal lung fibroblast cells MRC-5 (ATCC: CCL-171) or Huh7 (*12*) and HCoV-OC43 in African green monkey kidney cells

BS-C-1 (ATCC: CCL-26) or human rhabdomyosarcoma cell line RD-151 (ATCC: CCL-136).

2. This protocol can be adapted using other sources of antibodies; e.g.:

   a. HCoV-229E: Murine monoclonal antibodies specific for the nucleocapsid protein (N) of HCoV-229E (Chemicon International) *(13)*: Clone 401-4A.
   b. HCoV-OC43: Murine monoclonal antibody to HCoV-OC43 (Chemicon international) *(14)*: clone 541-8f or 542-7D.
   c. An ascites fluid from mouse MAb O.4.3 *(15)*, directed against the surface glycoprotein (S) of HCoV-OC43.
   d. An ascites fluid from mouse MAb 4E11.3 directed against the nucleocapsid (N) protein of the serologically related hemagglutinating encephalomyelitis virus of pigs *(10)*.

3. HRP conjugate is sensitive to CO_2 and will lose its activity.
4. There are Excel spreadsheets to calculate $TCID_{50}$/ml on the Internet at, e.g., http://ubik.microbiol.washington.edu/protocols/bl3/tcid50.htm

Acknowledgments

This work was supported mainly by grant MT-9203 from the Canadian Institutes of Health Research (Institute of Infection and Immunity).

References

1. Hamre, D., and Procknow J. J. (1966) A new virus isolated from the human respiratory tract. *Proc. Soc. Exp. Biol. Med.* **121**, 190–193.
2. McIntosh, K., Becker W. B., and Chanock R. M. (1967) Growth in suckling mouse brain of "IBV-like" viruses from patients with upper respiratory tract disease. *Proc. Natl. Acad. Sci. USA* **58**, 2268–2273.
3. Tyrrell, D. A. J., and Bynoe, M. L. (1965) Cultivation of a novel type of common-cold virus in organ cultures. *Brit. Med. J.* **1**, 1467–1470.
4. McIntosh K. (2004) Coronavirus infection in acute lower respiratory tract disease of infants. *J. Infect. Dis.* **190**, 1033–1041.
5. Collins A. R., and Sorensen, O. (1986) Regulation of viral persistence in human glioblastoma and rhabdomyosarcoma cells infected with coronavirus OC43. *Microb. Pathog.* **1**, 573–582.
6. Talbot P. J., Ekandé, S., Cashman, N. R., Mounir, S., and Stewart, J. N. (1993) Neurotropism of human coronavirus 229E. *Adv. Exp. Med. Biol.* **342**, 339–346.
7. Arbour, N., Ekandé S., Côté, G., Lachance, C., Chagnon, F., Tardieu, M., Cashman, N. R., and Talbot, P. J. (1999) Persistent infection of human oligodendrocytic and neuroglial cell lines by human coronavirus 229E. *J. Virol.* **74**, 3326–3337.

8. Arbour, N., Côté, G., Lachance, C., Tardieu, M., Cashman, N. R., and Talbot, P. J. (1999) Acute and persistent infection of human neural cell lines by human coronavirus OC43. *J. Virol.* **73**, 3338–3350.
9. Tompkins, W. A., Watrach, A. M., Schmale, J. D., Schultz, R. M., and Harris, J. A. (1974) Cultural and antigenic properties of newly established cell strains derived from adenocarcinomas of the human colon and rectum. *J. Natl. Cancer Inst.* **52**, 1101–1110.
10. Bonavia, A, Arbour, N., Wee Yong, V., and Talbot, P. J. (1997) Infection of primary cultures of human neural cells by human coronavirus 229E and OC43. *J. Virol.* **71**, 800–806.
11. Karber G. (1931) Beitrag zar Kollktiven Behandlung Pharmakologischer Reiherersuche. *Arch. Exp. Pathol. Pharmakol.* **162**, 480–483.
12. Tang, B. S., Chan, K. H., Cheng, V. C., Woo, P. C., Lau, S. K., Lam, C. C., Chan, T. L., Wu, A. K., Hung, I. F., Leung, S. Y., and Yuen, K. Y. (2005) Comparative host gene transcription by microarray analysis early after infection of the Huh7 cell line by severe acute respiratory syndrome coronavirus and human coronavirus 229E. *J. Virol.* **79**, 6180–6193.
13. Che, X. Y., Qiu, L.W., Liao, Z.Y., Wang, Y. D, Wen, K., Pan, Y. X., Hao, W., Mei, Y. B., Cheng, V. C., and Yuen, K.Y. (2005) Antigenic cross-reactivity between severe acute respiratory syndrome-associated coronavirus and human coronaviruses 229E and OC43. *J. Infect. Dis.* **191**, 2033–2037.
14. Gerna, G., Campanini, G., Rovida, F., Percivalle, E., Sarasini, A., Marchi, A., and Baldanti, F. (2006) Genetic variability of human coronavirus OC43- , 229E- , and NL63-like strains and their association with lower respiratory tract infections of hospitalized infants and immunocompromised patients. *J. Med. Virol.* **78**, 938–949.
15. Butler, N., Pewe, L., Trandem, K., and Perlman, S. (2006) Murine encephalitis caused by HCoV-OC43, a human coronavirus with broad species specificity, is partly immune-mediated. *Virology* **347**, 410–421.

9

The Preparation of Chicken Tracheal Organ Cultures for Virus Isolation, Propagation, and Titration

Brenda V. Jones and Ruth M. Hennion

Abstract

Chicken tracheal organ cultures (TOCs), comprising transverse sections of chick embryo trachea with beating cilia, have proved useful in the isolation of several respiratory viruses and as a viral assay system, using ciliostasis as the criterion for infection. A simple technique for the preparation of chicken tracheal organ cultures in glass test tubes, in which virus growth and ciliostasis can be readily observed, is described.

Key words: tracheal organ culture; ciliostasis; respiratory virus; viral assay

1. Introduction

Tracheal organ cultures (TOCs) have been used for the study of a number of respiratory tract pathogens (*1*). The first human coronavirus (HCoV) was isolated using human ciliated embryonal trachea (*2*), and studies on persistent infection with Newcastle disease virus (*3*), isolation of the Hong Kong variant of influenza A2 virus (*4*), and studies on the pathogenicity of mycoplasmas (*5*) using TOCs have all been reported. More recently, TOCs have been used in studies on the pathogenicity and induction of protective immunity by a recombinant strain of infectious bronchitis virus (IBV) (*6*).

Tracheal organ cultures derived from 20-day-old chicken embryos are reported to be as sensitive as 9-day-old embryonating eggs for the isolation and titration of IBV (*7*), and are more sensitive than TOCs from chickens up to 31 days of

From: *Methods in Molecular Biology, vol. 454: SARS- and Other Coronaviruses,*
Edited by: D. Cavanagh, DOI: 10.1007/978-1-59745-181-9_9, © Humana Press, New York, NY

age, with complete ciliostasis, the criterion for infection, being observed 3 days after incubation.

With the ease of production and the proven usefulness of TOCs in virus isolation and in studies on pathogenicity and immunization strategies, it may be worth considering their more widespread use for research into respiratory tract viruses. The method described below, based on that previously reported *(5)*, utilizes chicken embryo TOCs on a rolling culture tube assembly, where TOCs are capable of maintaining ciliary activity for longer periods than in static cultures. Debris accumulating within the TOC rings is reduced, making observation of ciliary activity easier.

2. Materials

2.1. Preparation of Tracheal Sections

1. 19- to 20-day-old embryonating eggs from SPF flock.
2. Tissue chopper: the following method assumes the use of a McIlwain mechanical tissue chopper (Mickle Laboratory Engineering Co., Gomshall, UK).
3. Sterile curved scissors (small).
4. Sterile scissors (large).
5. Sterile forceps.
6. Sterile Whatman filter paper discs 55 mm diameter (*see* **Note 1**).
7. 70% industrial methylated spirits (IMS).
8. Double-edged razor blades.
9. Eagles Minimum Essential Medium with Earles salts, L-glutamine and 2.2 g/liter sodium bicarbonate (MEM) (Sigma-Aldrich).
10. Penicillin + streptomycin (100,000 U of each per ml).
11. 1 M HEPES buffer prepared from HEPES (free acid) and tissue culture grade water, sterilized in an autoclave at 115°C for 20 min.
12. Sterile Bijou bottles.
13. Sterile 100- and 150-mm-diameter Petri dishes.

2.2. Culture of Tracheal Sections

1. Tissue culture roller drum capable of rolling at approximately 8 revolutions/h at 37°C.
2. Associated rack suitable for holding 16-mm tubes on roller drum.
3. Sterile, extra-strong rimless soda glass tubes, 150 mm long × 16 mm outside diameter, suitable for bacteriological work (VWR International) (*see* **Note 2**).
4. Sterile silicone rubber bungs 16 mm diameter at wide end, 13 mm diameter at narrow end, and 24 mm in length (VWR International) (*see* **Note 3**).
5. Inverted microscope (60–100× magnification).

3. Method

To calculate the number of embryonated eggs required for an assay assume each trachea will yield 17 to 20 rings. Expect to lose up to 20% of the cultures during the preliminary incubation step, owing to damage to the rings or spontaneous cessation of ciliary activity.

3.1. Preparation of Tracheal Sections

1. Prepare culture medium by the addition of HEPES buffer, penicillin, and streptomycin solution to MEM, to give final concentration of 40 mM HEPES and 250 U/ml penicillin and streptomycin.
2. On a clean workbench spray the top of the eggs with 70% IMS (*see* **Note 4**).
3. Using curved scissors remove the top of the shell, lift the embryo out by the wing, and place it in a 150-mm Petri dish.
4. Sever the spinal cord just below the back of the head and discard the egg and yolk sac (*see* **Note 5**).
5. Position the embryo on its back and, using small forceps and scissors, cut the skin the full length of the body, ending under the beak. Care must be taken not to damage the underlying structures.
6. Locate the trachea and, using small scissors and forceps, dissect it away from the surrounding tissues (*see* **Note 6**).
7. Cut the trachea at the levels of the carina and larynx and remove it from the embryo, placing the tissue in a Bijou bottle containing culture medium (*see* **Note 7**).
8. Repeat steps 3–7 for all available embryos.
9. Place one trachea at a time on a disc of filter paper and, using two pairs of fine forceps, gently remove as much fat as possible (*see* **Note 8**).
10. Place the cleaned tracheas in a 100-mm Petri dish containing culture medium.
11. Swab the McIlwain tissue chopper with 70% IMS.
12. Place two filter paper discs on top of the PVC cutting table disc and slide the assembled discs under the cutting table clips on the tissue chopper.
13. Raise the chopping arm of the tissue chopper and attach the razor blade.
14. Position the arm over the center of the cutting table (*see* **Note 9**).
15. Place tracheas on to the filter paper under, and perpendicular to, the raised blade (*see* **Note 10**).
16. Adjust the machine to cut sections 0.5–1.0 mm thick and activate the chopping arm.
17. Once the arm has stopped moving, discard the first few rings from each end of the cut tracheas; then, with a scalpel, scrape the remaining rings into a 150-mm Petri dish containing culture medium.
18. With a large-bore glass Pasteur pipette gently aspirate the medium to disperse the cut tissue into individual rings.
19. Repeat steps 12–18 until all the tracheas have been sectioned (*see* **Note 11**).

3.2. Culture of Tracheal Sections

1. With a large-bore glass Pasteur pipette dispense one TOC ring together with approximately 0.5 ml of culture medium into a glass tube (*see* **Note 12**).
2. Seal with silicone bung and check visually that each tube contains one complete ring (*see* **Note 13**).
3. Put the tubes in a roller tube rack and place on the roller apparatus, set to roll at a rate of about 8 revolutions/h, at approximately 37°C. Leave the tubes rolling for 1 to 2 days (*see* **Note 14**).
4. Check each tube culture for complete rings and the presence of ciliary activity, using a low-power inverted microscope.
5. Discard any tubes in which less than 60% of the luminal surface has clearly visible ciliary activity.
6. The remaining tubes may be used for viral assays (*see* **Note 15**).

4. Notes

1. Batches of sterile Whatman filter papers can be prepared by interleaving individual discs with slips of grease-proof paper and placing them in a glass Petri dish. Wrap the dish in aluminum foil and sterilize in a hot air oven (160°C for 1 h).
2. Batches of sterile tubes can be prepared by placing them, open end down, in suitable sized lidded tins lined with aluminum foil. Sterilize in a hot air oven as above.
3. Batches of sterile silicone rubber bungs can be prepared by placing them, narrow end down, in shallow, lidded tins. Sterilize by autoclaving.
4. Preparation of TOCs can be performed on the open laboratory bench after cleaning the surfaces with 70% IMS or any other suitable disinfectant.
5. Care must be taken at this stage not to damage the trachea.
6. The trachea can be identified by the presence of transverse ridges seen down its length owing to the underlying rings of cartilage.
7. The carina and larynx can be identified by the increased diameter at the ends of the trachea.
8. To avoid damage to the trachea hold it as close to one end as possible with the first pair of forceps and use the second pair to strip away the fatty tissue.
9. At this stage gently lower the arm onto the cutting area disc, loosen the screw holding the blade slightly, check that the blade is aligned correctly (the full length of the blade must be in contact with cutting area), tighten the screw again, and raise the arm.
10. A maximum of five tracheas can be laid side by side on the cutting bed at any one time. Gently stretch each trachea as it is placed on the cutting area, and when all five are in the correct position, wet them with a few drops of culture medium.
11. It is important to use a fresh blade edge and paper discs for each set of five tracheas to be sectioned.

12. Check for damaged glass tubes at this stage, particularly around the rims. Discard any with cracks as these can fail when bungs are inserted, leading to injured fingers.
13. Make sure the tracheal rings are fully submerged in culture medium and not stuck on the wall of the tube. Discard any that appear ragged or incomplete.
14. The speed of the roller apparatus is quite slow. Check that the tube roller is actually moving before leaving the cultures to incubate.
15. A simple quantal assay for infectivity of IBV has been described by Cook et al. *(7)* and used extensively in our Institute. Five tubes of TOCs per tenfold serial dilution of virus gives sufficiently accurate results for most purposes. A simplification of the method of Cook et al. (1976), used for many years by Cavanagh and colleagues, is to add 0.5 ml of diluted virus per TOC tube without prior removal of the medium already in the tube. TOCs are scored as positive for virus when ciliary activity is completely abrogated. If a virus is poorly ciliostatic, its presence can be demonstrated using indirect immunofluorescence, with the TOCs conveniently not fixed *(8)*.

References

1. McGee, Z.A., and Woods, M. L. (1987) Use of organ cultures in microbiological research. *Ann. Rev. Microbiol.* **41**, 291–300.
2. Tyrell, D. A. J., and Bynoe, M. L. (1965) Cultivation of novel type of common-cold virus in organ cultures. *Br. Med. J.* **5448**, 1467–1470.
3. Cummiskey, J. F., Hallum, J. V., Skinner, M. S., and Leslie, G. A. (1973) Persistent Newcastle disease virus infection in embryonic chicken tracheal organ cultures. *Infect. Immun.* **8** (4), 657–664.
4. Higgins, P. G., and Ellis, E. M. (1972) The isolation of influenza viruses. *J. Clin. Path.* **25**, 521–524.
5. Cherry, J. D., and Taylor-Robinson, D. (1970) Large-quantity production of chicken embryo tracheal organ cultures and use in virus and mycoplasma studies. *Appl. Microbiol.* **19**(4), 658–662.
6. Hodgson, T.,Casais, R., Dove, B., Britton, P. and Cavanagh, D (2004) Recombinant infectious bronchitis coronavirus Baudette with the spike protein gene of the pathogenic M41 strain remains attenuated but induces protective immunity. *J. Virol.* **78**(24), 13804–13811.
7. Cook, J. K. A., Darbyshire, J. H., and Peters, R. W. (1976) The use of chicken tracheal organ cultures for the isolation and assay of infectious bronchitis virus. *Arch. Virol.* **50**, 109–118.
8. Bhattacharjee, P. S., Naylor C. J., and Jones R. C. (1994) A simple method for immunofluorescence staining of tracheal organ cultures for the rapid identification of infectious bronchitis virus. *Avian Pathol.* **23**, 471–480.

10

Isolation and Propagation of Coronaviruses in Embryonated Eggs

James S. Guy

Abstract

The embryonated egg is a complex structure comprising an embryo and its supporting membranes (chorioallantoic, amniotic, yolk). The developing embryo and its membranes provide the diversity of cell types that are needed for successful replication of a wide variety of different viruses. Within the family *Coronaviridae*, the embryonated egg has been used as a host system primarily for two group 3 coronaviruses, infectious bronchitis virus (IBV) and turkey coronavirus (TCoV), but it also has been shown to be suitable for pheasant coronavirus. IBV replicates well in the embryonated chicken egg, regardless of the inoculation route; however, the allantoic route is favored as the virus replicates extensively in chorioallantoic membrane and high titers are found in allantoic fluid. TCoV replicates only in embryo tissues, within epithelium of the intestines and bursa of Fabricius; thus amniotic inoculation is required for isolation and propagation of this virus. Embryonated eggs also provide a potential host system for studies aimed at identifying other, novel coronavirus species.

Key words: embryonated egg; allantoic; amniotic; chicken; turkey; isolation; propagation; diagnosis; detection.

1. Introduction

Embryonated eggs are utilized as a laboratory host system for primary isolation and propagation of a variety of different viruses, including the group 3 coronaviruses, infectious bronchitis virus (IBV), turkey coronavirus (TCoV), and pheasant coronavirus *(1,2,3,4)*. They have been extensively utilized for propagation of these viruses for research purposes and, in the case of IBV, for

From: *Methods in Molecular Biology, vol. 454: SARS- and Other Coronaviruses*,
Edited by: D. Cavanagh, DOI: 10.1007/978-1-59745-181-9_10, © Humana Press, New York, NY

commercial production of vaccines. In addition, embryonated eggs provide a potential host system for studies aimed at identifying other, novel coronavirus species.

The embryonated egg comprises the developing embryo and several supporting membranes that enclose cavities or "sacs" within the egg *(5)*. The shell membrane lies immediately beneath the shell; this is a tough fibrinous membrane that forms the air sac in the blunt end of the egg. In contrast to the shell membrane, chorioallantoic, amniotic, and yolk membranes comprise largely epithelium and represent potential sites of coronaviral replication. The chorioallantoic membrane (CAM) lies directly beneath the shell membrane; this is a highly vascular membrane that serves as the respiratory organ of the embryo. The CAM is the largest of the embryo membranes, and it encloses the largest cavity within the egg, the allantoic cavity; in the embryonated chicken egg, this cavity contains approximately 5–10 ml of fluid, depending upon the stage of embryonation. The amniotic membrane encloses the embryo and forms the amniotic cavity; in the embryonated chicken egg, this cavity contains approximately 1 ml of fluid. The yolk sac is attached to the embryo and contains the nutrients the embryo utilizes during development.

The developing embryo and its membranes (CAM, amniotic, yolk) provide the diversity of cell types that are needed for successful replication of a wide variety of different viruses. Embryonated eggs may be inoculated by depositing virus directly onto the CAM, or by depositing virus within allantoic, amniotic, and yolk sacs *(6)*. For group 3 coronaviruses, inoculation of eggs by allantoic or amniotic routes has been shown to provide these viruses with access to specific cell types that support their replication *(2– 4)*. IBV replicates in a variety of epithelial surfaces of the chicken including the respiratory tract, gastrointestinal tract, kidney, and oviduct *(7)*. In the embryonated chicken egg, IBV replicates well regardless of inoculation route; however, the allantoic route is favored as the virus replicates extensively in epithelium of the CAM and high titers are shed into allantoic fluid *(8)*. A pheasant coronavirus has been isolated and propagated in embryonated chicken eggs using procedures similar to those utilized for IBV (allantoic route inoculation) *(3)*. In contrast, TCoV is an enterotropic virus that replicates only in intestinal epithelium and bursa of Fabricius of chickens and turkeys *(1,4,9)*. The enterotropic nature of TCoV is also observed in the embryonated egg; the virus replicates only in embryonic intestines and bursa of Fabricius, sites that are reached only via amniotic inoculation.

2. Materials

2.1. Collection of Samples for Egg Inoculation

1. Dulbecco's modified Eagle's medium (DMEM) supplemented with 1% fetal bovine serum (FBS) and antibiotics (penicillin [1000 U/ml], gentamicin

[0.05 mg/ml], amphotericin B [5 µg/ml]). Adjust pH to 7.0–7.4. Tryptose phosphate broth and other cell culture basal media (minimal essential medium, RPMI 1640, etc.) may be substituted for DMEM.

2. Sterile cotton-tipped swabs are used for collection of antemortem samples (e.g., respiratory secretions, feces, etc.) from older birds. Type 4 Calgiswabs (Puritan Medical Products) are preferred for young birds.
3. Sterile Whirlpak[R] bags (Fisher Scientific) are used for the collection of tissues.

2.2. Egg Inoculation and Incubation

1. Fertile eggs are obtained, preferably from specific-pathogen-free (SPF) flocks (Charles River/SPAFAS). Alternatively, fertile eggs may be used that are from healthy flocks that are free of antibody to the virus of interest (*see* **Note 1**).
2. Disinfectant: 70% ethanol containing 3.5% iodine and 1.5% sodium iodide.
3. A vibrating engraver (Fisher Scientific) or drill (Dremel) is used to prepare holes in the egg shells. Prior to use, disinfect the tip of the engraving tool/drill to prevent contamination of the egg.
4. Plastic cement, glue, tape, or nail varnish are used to seal holes in the egg shells after inoculation.
5. Egg candlers are available from a variety of commercial sources.
6. A suitable egg incubator is needed; these are available from a variety of commercial sources. Commercially available egg incubators generally are equipped with a heat source, a humidifier, and a timer-based, mechanical turning system.

2.3. Collection of Specimens from Inoculated Eggs

1. Sterile scissors and forceps.
2. Sterile pipettes or 5-ml syringes with 1-in., 18-gauge needles.
3. Sterile plastic tubes; 12 × 75-mm snap-cap tubes or microcentrifuge tubes.

3. Methods

Embryonated chicken and turkey eggs are extensively utilized for isolation and propagation of IBV and TCoV, respectively *(2,4)*. These same eggs and techniques may be useful for amplification of other coronaviruses, and this has been demonstrated with isolation and propagation of pheasant coronavirus in embryonated chicken eggs *(3)*. Embryonated eggs of other avian species may be utilized; these are inoculated essentially as described for chicken and turkey eggs, primarily by making adjustments in the length of time embryos are incubated before inoculation. Embryonated chicken eggs are inoculated by the allantoic route at approximately the middle of the 21-day embryonation period, at 8–10 days of embryonation; they are inoculated by the amniotic route late in the incubation period, at 14–16 days of embryonation. Turkey and duck eggs have a

28–day embryonation period and generally are inoculated by the allantoic route at 11–14 days of embryonation, and by the amniotic route at 18–22 days of embryonation.

Embryonated eggs are incubated at a temperature of 38°–39°C with a relative humidity of 83–87%. They should be turned several times a day to ensure proper embryo development and to prevent development of adhesions between the embryo and its membranes. Fertile eggs may be stored for brief periods with minimal loss of viability *(10)*. Ideally, fertile eggs are stored at a temperature of 19°C with a relative humidity of approximately 70%. Alternatively, eggs may be stored at room temperature, but these should be tilted at 45°, and daily alternated from side to side to minimize loss of embryo viability.

Methods for detection of coronaviruses in inoculated embryonated eggs include electron microscopy, immunohistochemistry, and reverse transcriptase-polymerase chain reaction (RT-PCR) procedures *(2,4,11,12)*. Electron microscopy is a particularly useful tool as this method depends solely on morphologic identification of the virus and does not require specific reagents *(13)*. The characteristic electron microscopic morphology of coronaviruses allows their presumptive identification in embryonic fluids (e.g., allantoic fluid) or embryo intestinal contents. A variety of immunohistochemical and RT-PCR procedures has been developed for detection of coronaviruses. These same procedures may be useful for detection of novel coronaviruses owing to antigenic and genomic similarities among coronaviruses, particularly those within the same antigenic group *(2,4,9,11,12,14–16)*.

3.1. Collection of Samples for Egg Inoculation

1. Place swabs used to collect clinical samples such as respiratory secretions and feces into 2–3 ml of DMEM supplemented with FBS and antibiotics.
2. Collect tissues using an aseptic technique and place in clean, tightly sealed bags (Whirlpak bags).
3. Chill clinical samples immediately after collection and transport them to the laboratory with minimal delay. Samples may be shipped on ice, dry ice, or with commercially available cold packs (*see* **Note 2**).

3.2. Preparation of Samples for Egg Inoculation

1. Use a vortex mixer to expel material from swabs; then remove and discard the swab. Clarify by centrifugation (1000–2000 × *g* for 10 min) in a refrigerated centrifuge. Filter, if needed, through a 0.45-μm filter, and store at –70°C (*see* **Note 3**).
2. Prepare tissues and feces as 10–20% suspensions in DMEM supplemented with FBS and antibiotics. Homogenize tissues using a mortar and pestle, Ten Broeck homogenizer, or Stomacher[R] (Fisher). Clarify tissue and fecal suspensions by

centrifugation (1000–2000 × *g* for 10 min) in a refrigerated centrifuge; this removes cellular debris and most bacteria. Filter, if needed, through a 0.45-μm filter, and store at –70°C (*see* **Note 3**).

3.3. Allantoic Sac Inoculation

1. Embryonated chicken eggs at 8–10 days of embryonation are commonly used; eggs from other avian species may be used by making adjustments in the ages at which embryos are inoculated. Turkey and duck eggs (28-day embryonation period) are generally inoculated by this route at 11–14 days of embryonation.
2. Place eggs in an egg flat with the air cell up. Candle eggs to ensure viability and mark the edge of the air cell using a pencil.
3. Disinfect the area marked on the shell and drill a small hole just above the mark so that the hole penetrates the air cell, but not the portion of the egg below it.
4. A 1-ml syringe with a 25-gauge, 0.5-in. (12-mm) needle is used to inoculate the eggs. Insert the needle up to the hub while holding the syringe vertically and inject 0.1–0.3 ml of inoculum into the allantoic cavity.
5. Seal the holes and return the eggs to the incubator.
6. Incubate the eggs for 3–7 days.
7. Evaluate the embryos and the allantoic fluid for presence of virus as described below.

3.4. Amniotic Sac Inoculation (Method A)

1. Fertile embryonated eggs are inoculated late in the incubation period. Chicken eggs are inoculated at 14–16 days of embryonation; turkey and duck eggs are inoculated at 18–22 days of embryonation.
2. Candle eggs to ensure embryo viability. Place eggs in an egg flat with the air cell up. Disinfect the shell at the top of the egg, over the center of the air cell. Drill a small hole through the shell at the center of air cell using a vibrating engraver.
3. Use a 1-ml syringe with a 22-gauge, 1.5-in. (38-mm) needle to inoculate chicken, duck, and turkey embryos. Insert the needle up to the hub while holding the syringe vertically and inject 0.1–0.2 ml of inoculum into the amniotic cavity (*see* **Note 4**).
4. Seal the holes and return the eggs to the incubator.
5. Inoculated embryos are generally examined for the presence of the virus after incubation for 2–5 days.
6. Evaluate the inoculated embryos for the presence of the virus as described below.

3.5. Amniotic Sac Inoculation (Method B)

1. Fertile embryonated chicken eggs are inoculated, as above, at 14–16 days of embryonation; turkey and duck eggs at 18–22 days of embryonation. Candle the eggs and mark the general location of the embryo (*see* **Note 5**).

2. Place eggs in an egg flat with the air cell up. Disinfect the shell at the top of the egg, over the center of the air cell. Drill a small hole through the shell at the center of the air cell.

3. Use a 1-ml syringe with a 22-gauge, 1.5-in. (38-mm) needle to inoculate chicken, duck, and turkey embryos. Inoculate the eggs in a darkened room, as the embryo must be visualized for this method of amniotic inoculation. Hold the egg against an egg candler and insert the needle into the egg and toward the shadow of the embryo. As the tip of the needle approaches the embryo, a quick stab is used to penetrate the amniotic sac. Penetration of the amniotic sac may be verified by moving the needle sideways; the embryo should move as the needle moves (*see* **Note 4**).

4. Seal the holes and return the eggs to the incubator.

5. Inoculated embryos are generally examined for the presence of the virus after incubation for 2–5 days.

3.6. Collection of Allantoic Fluid from Eggs Inoculated by the Allantoic Route

1. Candle eggs once daily after inoculation. Discard all eggs with embryos that die within the first 24 h after inoculation (*see* **Note 6**).

2. Collect allantoic fluid from all the eggs with embryos that die more than 24 h after inoculation and from eggs with embryos that survive through the specified incubation period. Eggs with live embryos following the specified incubation period are refrigerated for at least 4 h, or overnight, prior to collection of allantoic fluid (*see* **Note 7**).

3. Place eggs in an egg flat with the air cell up. Disinfect the portion of the egg shell that covers the air cell, and use sterile forceps or fine scissors to crack and remove the egg shell over the air cell.

4. Use forceps to gently dissect through the shell membrane and CAM to expose the allantoic fluid. Use forceps to depress the membranes within the allantoic cavity so that allantoic fluid pools around the tip of the forceps. Use a pipette or syringe with a needle to aspirate the fluid. Place fluid in sterile 12 × 75-mm snap-cap tubes or other vials. Store at –70°C.

5. Examine the allantoic fluid for the presence of coronavirus using electron microscopy, immunohistochemistry, or RT-PCR (*see* **Note 8**).

3.7. Collection of Embryo Tissues from Eggs Inoculated by the Amniotic Route

1. Candle eggs once daily after inoculation. Discard all eggs with embryos that die within the first 24 h after inoculation (*see* **Note 6**).

2. Examine all eggs with embryos that die more than 24 h after inoculation and eggs with embryos that survive through the specified incubation period (*see* **Note 9**).

3. Euthanize embryos by placing eggs in a plastic bag or plastic bucket filled with carbon dioxide gas, or refrigerate (4°C) overnight. Alternatively, embryos may be euthanized by cervical dislocation upon removal from the eggs using the handles of a pair of scissors (*see* **Note 10**).
4. Place eggs in an egg flat with the air cell up. Disinfect the portion of the egg shell that covers the air cell, and use sterile forceps or fine scissors to crack and remove the egg shell over the air cell.
5. Use forceps to dissect through the shell membrane and CAM.
6. Grasp the embryo with sterile forceps and gently remove it from the egg.
7. Remove selected tissues and/or intestinal contents from the embryo for coronavirus detection using electron microscopy, immunohistochemistry,, or RT-PCR.

4. Notes

1. Fertile eggs from non-SPF flocks may be used; however, the presence of antibodies may interfere with isolation and propagation, and the presence of egg-transmitted infectious agents may result in contamination of any viruses obtained with these eggs.
2. If dry ice is use, samples must be placed in tightly sealed containers to prevent inactivation of viruses from released carbon dioxide.
3. The supernatant fluid should be filtered if the specimen is feces or other sample that is probably contaminated with high concentrations of bacteria. Filtration of samples will reduce virus titer, and should be used only when necessary.
4. The accuracy of delivering an inoculum into the amniotic sac using this method may be checked by injecting a dye such as crystal violet (0.2% crystal violet in 95% ethanol), then opening the eggs and determining the site of dye deposition.
5. The principal disadvantage of method B is the difficulty of visualizing the embryo, particularly in embryonated eggs having a dark shell color (e.g., turkey eggs, brown chicken eggs).
6. Embryo deaths that occur less than 24 h after inoculation are generally due to bacterial contamination, toxicity of the inoculum, or injury.
7. Refrigeration kills the embryo and causes the blood to clot. This prevents contamination of allantoic fluid with blood.
8. Multiple passages in embryonated eggs may be necessary for initial isolation of coronaviruses; allantoic fluid is used as inoculum for passages. Embryos at each passage should be evaluated for gross lesions. For IBV, embryo-lethal strains generally result in embryos with cutaneous hemorrhage; non-embryo-lethal strains result in stunting, curling, clubbing of down, or urate deposits in the mesonephros of the kidney. Virus replication in embryonated eggs may not be associated with readily detectable embryo lesions.
9. TCoV rarely results in embryo mortality; for this virus only those eggs with live embryos following the specified incubation period are examined. The possibility of embryo-lethal viruses should not be overlooked.

10. The method of euthanasia employed will depend upon the method used to detect virus in inoculated embryos. Fresh tissues are required if immunohistochemistry is to be employed; for this, embryos should be euthanized by cervical dislocation or exposed briefly to carbon dioxide gas.

References

1. Adams, N. R., and Hofstad, M. S. (1970) Isolation of transmissible enteritis agent of turkeys in avian embryos *Avian Dis.* **15**, 426–433.
2. Cavanagh, D., and Naqi, S. A. (2003) Infectious bronchitis. In: Saif, Y. M., Barnes, H. J., Fadly, A., Glisson J. R., McDougald, L.R., and Swayne, D. E. (eds.) *Diseases of Poultry*, 11th Ed. Iowa State University Press, Ames, pp. 101–120.
3. Gough, R. E., Cox, W. J., Winkler, C. E., Sharp, M. W., and Spackman, D. (1996) Isolation and identification of infectious bronchitis virus from pheasants *Vet. Rec.* **138**, 208–209.
4. Guy, J. S. (2003) Turkey coronavirus enteritis. In: Saif, Y. M., Barnes, H. J., Fadly, A., Glisson, J. R., McDougald, L. R., and Swayne, D. E., (eds.) *Diseases of Poultry*, 11th Ed. Iowa State University Press, Ames, pp. 300–308.
5. Hawkes, R. A. (1979) General principles underlying laboratory diagnosis of viral infections. In: Lennette, E. H, and Schmidt, N. J. (eds.) *Diagnostic Procedures for Viral, Rickettsial and Chlamydial Infections*, 5th Ed. American Public Health Association, Washington, DC, pp. 1–48.
6. Senne, D. A. (1998) Virus propagation in embryonating eggs. In: Swayne, D. E., Glisson, J. R., Jackwood, M. W., Pearson, J. E., and Reed, W. M. (eds.) *A Laboratory Manual for Isolation and Identification of Avian Pathogens*, 4th Ed., American Association of Avian Pathologists, Kennett Square, PA, pp. 235–240.
7. Cavanagh, D. (2003) Severe acute respiratory syndrome vaccine development: experiences of vaccination against avian infectious bronchitis virus *Avian Pathol.* **32**, 567–582.
8. Jordan, F. T. W., and Nassar, T. J. (1973) The combined influence of age of embryo, temperature and duration of incubation on the replication and yield of avian infectious bronchitis virus in the developing chick embryo. *Avian Pathol.* **2**, 279–294.
9. Guy, J. S. (2000) Turkey coronavirus is more closely related to avian infectious bronchitis virus than to mammalian coronaviruses: a review *Avian Pathol.* **29**, 207–212.
10. Brake, J., Walsh T. J., Benton, C. E., Petitte, J. N, Meyerhof, R., and Penalva, G. (1997). Egg handling and storage *Poult. Sci.* **76**, 144–151.
11. Jonassen, C. M., Kofstad, T., Larsen, I. L., Lovland, A., Handeland, K., Follestad, and A., Lillehaug, A. (2005) Molecular identification and characterization of novel coronaviruses infecting graylag geese (*Anser anser*), feral pigeons (*Columba livia*) and mallards (*Anas platyrhynchos*). *J. Gen. Virol.* **86**, 1597–1607.
12. Stephensen, C. B., Casebolt, D. B., and Gangopadhyay, N. N.(1999) Phylogenetic analysis of a highly conserved region of the polymerase gene from 11 coronaviruses and development of a consensus polymerase chain reaction assay *Virus Res* **60**, 181–189.

13. McNulty, M. S., Curran, W. L., Todd, D., and McFerran, J. B. (1979) Detection of viruses in avian faeces by direct electron microscopy *Avian Pathol.* **8**, 239–247.
14. Cavanagh, D., Mawditt, K., Welchman, D. B, Britton, P., and Gough, R.E. (2002) Coronaviruses from pheasants (*Phasianus colchicus*) are genetically closely related to coronaviruses of domestic fowl (infectious bronchitis virus) and turkeys. *Avian Pathol.* **31**, 181–193.
15. Guy, J. S., Barnes, H. J., Smith, L. G., and Breslin, J. (1997) Antigenic characterization of a turkey coronavirus identified in poult enteritis and mortality syndrome-affected turkeys. *Avian Dis.* **41**, 583–590.
16. Cavanagh, D. (2005) Coronaviruses in poultry and other birds *Avian Pathol.* **23**, 439–448.

11

Large-Scale Preparation of UV-Inactivated SARS Coronavirus Virions for Vaccine Antigen

Yasuko Tsunetsugu-Yokota

Abstract

In general, a whole virion serves as a simple vaccine antigen and often essential material for the analysis of immune responses against virus infection. However, to work with highly contagious pathogens, it is necessary to take precautions against laboratory-acquired infection. We have learned many lessons from the recent outbreak of severe acute respiratory syndrome (SARS). In order to develop an effective vaccine and diagnostic tools, we prepared UV-inactivated SARS coronavirus on a large scale under the strict Biosafety Level 3 (BSL3) regulation. Our protocol for large-scale preparation of UV-inactivated SARS-CoV including virus expansion, titration, inactivation, and ultracentrifugation is applicable to any newly emerging virus we might encounter in the future.

Key words: SARS-CoV; coronavirus; UV-inactivation; vaccine antigen; VERO E6 cells; plaque purification; concentration of virions.

1. Introduction

The outbreak of severe acute respiratory syndrome (SARS) occurred in southern China in November 2002. The causative agent was a novel coronavirus originating from wild animals in a live market [see reviews in (1,2)]. From November 2002 to July 2003, a cumulative total of more than 8098 probable SARS cases with more than 774 deaths were reported in 26 countries (http://www.wpro.who.int/sars/). However, after the epidemic appeared to be over, a laboratory-acquired SARS outbreak occurred in China in late March and mid-April, 2004, with one fatality. Although the outbreak was finally

From: *Methods in Molecular Biology, vol. 454: SARS- and Other Coronaviruses,*
Edited by: D. Cavanagh, DOI: 10.1007/978-1-59745-181-9_11, © Humana Press, New York, NY

contained, the biosafety procedure in the laboratory conducting SARS research with live virus was cause for serious concern (http://www.cdc.gov/ncidod/sars/sarsprepplan.html). Thus, these cases from China highlighted the risk of SARS transmission from laboratory-acquired infection and were a good reminder of the need to handle infectious SARS coronavirus (SARS-CoV) with great care.

When the rapid development of a vaccine against an emerging disease such as SARS is required, a whole virion preparation is a simple vaccine antigen, if the safety issue is overcome [review in *(3)*]. We have demonstrated that subcutaneously administered UV-inactivated SARS-CoV elicits a high level of IgG-type neutralizing antibodies and weak T-cell responses in mice *(4)*. A whole virion is also useful, and sometimes necessary, for the analysis of immune response in vaccinated animals. Using UV-inactivated SARS-CoV, we established several monoclonal antibodies that recognize the spike (S) or nucleocapsid (N) proteins *(5)*, which is quite useful not only for basic research but also for the clinical diagnosis of SARS-CoV infection. Here we describe our protocol for the large-scale preparation of UV-inactivated SARS-CoV virion under the strict Biosafety Level 3 (BSL3) regulation.

2. Materials

2.1. BSL3 Laboratory Facilities and Equipment

The BSL3 laboratory in our institute comes close to the detailed description in the *WHO Laboratory Biosafety Manual* 3rd Edition, 2004. It is located in an isolated area on the third basement floor. The BSL3 laboratory area is further separated into several laboratories according on the pathogen being handled. As an example, the basic outline of BSL3 facilities and equipment in a laboratory working with SARS-CoV and other airborne viruses is described briefly below.

1. Double doors and a controlled ventilation system that maintains a directional airflow into the laboratory with a visual monitoring device. The air supply is high-efficiency particulate air, HEPA filtered, and exhaust from the laboratory is discharged through an HEPA filter to the outside of the building.
2. Anteroom, pass box, and hand-washing station with hands-free controls.
3. Biological safety cabinet (class II, type B), autoclave, and centrifuge with capped (safety) bucket are specially designed. Basically, any liquid waste is sterilized by the autoclave of the BSL laboratory liquid-waste-treating system and drained into a storage tank for secondary decontamination.
4. Personal protective equipment (PPE) comprises disposable gloves, particulate filter masks, arm covers, closed-toed footwear, foot covers, and head coverings in addition to solid-front gowns and ethanol spray. These are located in the anteroom. Other supporting materials are kept in a common corridor space or a storage room.

2.2. Titration of SARS-CoV

1. VERO E6 cells (ATCC #CRL-1586).
2. Eagle's Minimal Essential medium (MEM) with nonessential amino acids supplemented with 10% FCS, 2 mM L-glutamine, 1.0 mM sodium pyruvate, 1.5 g/liter sodium bicarbonate, and antibiotics for VERO E6 cell culture.
3. SeaPlaque low melting agarose (FMC, 4%): dissolve 4 g agarose in ddH$_2$O 100 ml in a microwave oven, store at room temperature.
4. 2X MEM.
5. Crystal violet.
6. Formalin (33% conc.) diluted to 3.3% with PBS.
7. Dulbecco's MEM for virus production.

2.3. UV-Irradiation and Ultracentrifuge

1. Ultraclear tube (Beckman #344058, 25 mmC).
2. 15-UV lamp (365-nm wavelength, 1350 μW/cm^2 at 15 cm distance).
3. Polyethylene glycol PEG6000 (#169-0915, Wako Pure Chemical Industries, Ltd., Osaka, Japan).
4. Sterile and UV-resistant container.

2.4. Concentration of UV-Inactivated SARS-CoV

1. Beta-cyclodextrin (Fluka # 28707, Sigma-Aldrich): dissolve 1 g in 20 ml ddH$_2$O to make a 5% solution.
2. Centrifugal concentrator 150 K Apollo 30 ml (Orbital Bioscience Inc., MA).
3. BCA protein assay kit (#23227, 23225, Pierce Co., IL).

3. Methods

3.1. Practice in BSL3 Laboratory

Based on the WHO guidelines, our institute has provided a biosafety control laboratory with several staff members and has developed a specific code of practice for use in biosafety laboratories. Anyone intending to work with microbes must take a training program for practice in BSL1, 2, 3 laboratories and only those who have registered and taken additional courses are allowed to walk into a BSL3 laboratory. We have a laboratory supervisor in each BSL3 laboratory unit and a subsupervisor for each working group for class 3 pathogens to control access and deal with any biosafety issues. During practice, it is important to avoid aerosol generation as much as possible.

3.2. Titration of SARS-CoV

For titration of SARS-CoV, either a plaque titration or tissue culture infectious dose (TCID) assay is utilized. TCID assay is simple and easy, but a plaque assay is usually more accurate than a TCID assay.

3.2.1. Plaque Titration

1. Seed VERO E6 cells to a six-well plate with 10%FCS-DMEM at ~0.5–1 × 10^6/well. (If cells are collect from a confluent T75 flask, two plates can be seeded and one used a day later.)
2. When wells become confluent, replace the medium with 0.5 ml of 1% FCS-DMEM.
3. Add serially diluted (~tenfold) SARS-CoV with 1% FCS-DMEM, 100 μl to each well.
4. Incubate for 1 h at 37°C in a humidified CO_2 incubator.
5. During incubation, melt 4% low melting agarose using a microwave oven (*see* **Note 1**).
6. Mix 4% low melting agarose (1/4 volume) with FCS-MEM[#] without agarose (3/4 volume) and keep it in a water bath at 42°C to make 10 ml 1% agarose solution, 5 ml of 2X EM, 1 ml of FCS, and 1.5 ml of sterile water.
7. Remove medium and overlay 1 ml of 1% low melting agarose in 10% FCS-MEM as quickly as possible.
8. Incubate for 48 h.
9. Add 10% formalin/PBS to kill the virus, and leave the plate for 10 min.
10. Remove formalin agarose and add 10% formalin/PBS containing 0.1% crystal violet.
11. Leave the plate overnight and soak it in warm water to remove agarose completely.
12. Dry the plates.

3.2.2. Tissue Culture Infectious Doses (TCID)

1. Seed VERO E6 cells to 96-well plate with 10%FCS-DMEM at ~0.5 × 10^5/well.
2. When wells become confluent, replace the medium with 50 μl of 1% FCS-DMEM.
3. Prepare serial dilutions of the virus (~fivefold) and add 50 μl to the plate, in duplicate.
4. Incubate for ~48–72 h at 37°C in a humidified CO_2 incubator.
5. When cytopathic effect (CPE) appears, add 2 or 3 drops of 10% formalin/PBS to each well and leave the plate for a few minutes.
6. Invert the plates onto a stack of towel papers to remove the formalin/PBS.
7. Tap out any remaining liquid and remove the bubbles.
8. Add 2 or 3 drops of 10% formalin/PBS containing 0.1% crystal violet.
9. Wash out the dye under tap water.
10. Dry the plates.

The CPE will appear as clear wells, whereas the cells that survived are colored blue as shown in **Fig. 1**.

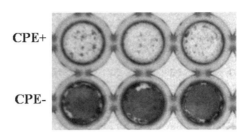

CPE+

CPE-

Fig. 1. CPE at 48 h in VERO E6 cells infected with SARS-CoV. Confluent VERO E6 cells cultured in a 96-well plate are infected with SARS-CoV (1000 TCID) and incubated for 2 days. When CPE was observed microscopically, 10% formalin/PBS, followed by 10% formalin/PBS containing 0.1% crystal violet was added to each well. Upper row: infected; lower row: uninfected culture, in triplicate.

3.3. Large-Scale Preparation of SARS-CoV

1. Prepare confluent VERO E6 cell culture of 8X T225 filter-capped flasks in 1% FCS-DMEM, 50 ml each or 10X T175 flasks, 40 ml each.
2. Inoculate SARS-CoV stock virus ($\sim 1 \times 10^7$ TCID/ml, 1 ml) in each flask.
3. Incubate for 24 h at 37°C in a humidified CO_2 incubator (*see* **Note 2**).
4. Collect supernatant and transfer to 10×50-ml conical tubes (40 ml each).
5. Centrifuge at 3000 rpm for 10 min.
6. Transfer supernatant into a sterile and UV-resistant container placed on a stand under the UV-lamp apparatus (*see* **Note 3**).
7. UV irradiation for 20 min at 15 cm distance (1350 μW $\times$ 20 min $\times$ 60 sec = 1620 mJ/cm^2 at 365 nm) (*see* **Note 4**).
8. Keep UV-irradiated supernatant at 4°C.

(If we do the preparation twice a week, total volume would be 800 ml)

3.4. PEG Precipitation Stock (Once a Week)

1. Centrifuge 8000 rpm for 30 min (<200 ml in 250-ml PP tube.)
2. Transfer cleared supernatants to disposable plastic bottles (500 ml) containing PEG 6000 (36 g) and NaCl (13.14 g) to final 8% and 0.5 M, respectively.
3. Adjust the total volume up to 450 ml with PBS.
4. Dissolve completely by stirring for more than 4 h in a refrigerator.
5. Centrifuge at 8000 rpm for 30 min (<200 ml in 250 ml PP tube) to collect PEG pellets.
6. Remove liquid completely with a kimwipe and resuspend pellets with PBS ($\sim$1 ml/ bottle, total $\sim$5 ml).
7. Transfer the virion solution to 15-ml conical tubes and keep frozen until ultracentrifugation.

(For 2-liter culture, the total volume should be less than 15 ml)

8. Keep frozen at –80°C.

3.5. Purification of Virions with 20–60% Discontinuous Sucrose Gradient Centrifugation

1. In an ultraclear tube (38.5 ml total volume), prepare the following sucrose gradient with 60% sucrose, 8 ml, and 20% sucrose, 10 ml.
2. Overlay ~15 ml UV-irradiated virion solution recovered from 2 liters of culture.
3. Centrifuge at 24,000 rpm for 2 h at 4°C using SW30.1 rotor (or SW28).
4. Collect the band of virions at the 60% and the 20% interface (2 to 4 ml).
5. Aliquot and keep frozen at –80°C until the loss of infectivity is confirmed. (*see* **Note 5**)

3.6. Check of Purified Virions

The concentration of purified virions is determined by BCA assay with BSA as a standard. However, we found that the concentration is overestimated by this method, probably because of the large amount of sucrose. When the medium of concentrated virion was changed to PBS, the BCA assay gave a reasonable titer of virion concentration (*see* **Note 5**). It is better to confirm the amount of virion by 10% SDS-PAGE. Based on a major band of N protein as shown in **Fig. 2**, we

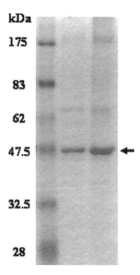

Fig. 2. SDS-PAGE analysis of purified SARS-CoV. Purifed virions were run on 10% SDS-PAGE, prefixed and Coomassie Brilliant Blue staining was carried out. The arrow indicates the N protein, which was confirmed by Western blot using anti-N monoclonal antibody (SKOT 9) *(5)*. In lanes 1 and 2, respectively, 3 and 5 μg of virion were loaded.

can estimate the approximate amount of virion. Generally, 1 to 2 mg of virions is obtained from 2 liters of culture.

4. Notes

1. It is better not to use an autoclave to melt the agarose, as repeated autoclaving damages its gelling activity.
2. CPE is not apparent at this point in time. However, the yield of cell-free virus is good enough [personal communication with Dr. Shigeru Morikawa (the First Department of Virology, National Institute of Infectious Diseases (NIID), Tokyo)].
3. Select a container for UV irradiation of a total of 400 ml supernatant; the liquid should be less than 2.5 cm high to ensure inactivation of the virus.
4. Alternatively, we know that UV irradiation of purified virions (0.5 ml, $>10^9$ PFU) plated in a Falcon dish (ϕ35 mm) for 10 min using UVP Model UVG-11 (4-W UV-lamp, 254-nm wavelength; 1520 μW/cm^2) completely eliminates viral infectivity (600 mJ/cm^2 at 254 nm).
5. When the loss of infectivity of purified virion is confirmed by TCID assay, one can handle UV-inactivated virion at BSL2. In order to increase the safety of this UV-inactivated SARS-CoV, we inactivated the virion vaccine using both UV and formalin. The purified UV-inactivated virion was treated with formalin at a final concentration of 0.002% and left in a safety cabinet overnight to allow the formalin to evaporate naturally. Then, in order to avoid any risk of residual formalin, virion was diluted 20-fold with PBS and concentrated in the presence of 0.5% β-cyclodextrin using a centrifugal concentrator (see below). This doubly inactivated SARS-CoV vaccine elicited comparable immune responses with UV-inactivated SARS-CoV vaccine in mice (*6*).
6. To concentrate purified virions or recombinant N protein, a centrifugal concentrator was used. However, we realized that more than 90% of virion or recombinant N protein was lost during this procedure. This is probably due to the sticky nature of viral proteins on the filter membrane. We tested blocking reagents of membrane filter and found that β-cyclodextrin worked quite well.

Acknowledgment

I thank Dr. Shigeru Morikawa for providing his original protocol for UV irradiation and purification of SARS-CoV. I also acknowledge my collaborators, Dr. Masamichi Ohshima (Department of Immunology, NIID) and Mr. Hirotaka Takagi (Biosafety Laboratory, NIID) for their practical help in developing this protocol, and Dr. Katuaki Shinohara (Biosafety Laboratory, NIID) for reviewing the part of this manuscript relating to biosafety.

References

1. Holmes, K.V. (2003) SARS coronavirus: a new challenge for prevention and therapy. *J. Clin. Invest.* **111**, 1605–1609.
2. Peiris, J. S., Guan, Y., and Yuen, K. Y. (2004) Severe acute respiratory syndrome. *Nature Med.* **10**, S88–97.
3. Tsunetsugu-Yokota, Y., Ohnishi, K., and Takemori, T. (2006) Severe acute respiratory syndrome (SARS) coronavirus: application of monoclonal antibodies and development of an effective vaccine. *Rev. Med. Virol.* **16**, 117–131.
4. Takasuka, N., Fujii, H., Takahashi, Y., Kasai, M., Morikawa, S., Itamura, S., Ishii, K., Sakaguchi, M., Ohnishi, K., Ohshima, M., Hashimoto, S., Odagiri, T., Tashiro, M., Yoshikura, H., Takemori, T., and Tsunetsugu-Yokota, Y. (2004) A subcutaneously injected UV-inactivated SARS coronavirus vaccine elicits systemic humoral immunity in mice. *Int. Immunol.* **16**, 1423–1430.
5. Ohnishi, K., Sakaguchi, M., Kaji, T., Akagawa, K., Taniyama, T., Kasai, M., Tsunetsugu-Yokota, Y., Oshima, M., Yamamoto, K., Takasuka, N., Hashimoto, S., Ato, M., Fujii, H., Takahashi, Y., Morikawa, S., Ishii, K., Sata, T., Takagi, H., Itamura, S., Odagiri, T., Miyamura, T., Kurane, I., Tashiro, M., Kurata, T., Yoshikura, H., and Takemori, T. (2005) Immunological detection of severe acute respiratory syndrome coronavirus by monoclonal antibodies. *Jpn. J. Infect. Dis.* **58**, 88–94.
6. Tsunetsugu-Yokota, Y., Ato, M., Takahashi, Y., Hashimoto, S.-I., Kaji, T., Kuraoka, M., Yamamoto, K., Yamomoto, T., Ohshima, M., Ohnishi, K., and Takemori, T. (2007) Formalin-treated UV-inactivated SARS coronavirus vaccine retains its immunogenicity and promotes Th2-type immune responses. *Jpn. J. Infect. Dis.* **60**, 106–112.

III

Structure of Coronaviruses Analyzed by Electron Microscopy

12

Purification and Electron Cryomicroscopy of Coronavirus Particles

Benjamin W. Neuman, Brian D. Adair, Mark Yeager, and Michael J. Buchmeier

Abstract

Intact, enveloped coronavirus particles vary widely in size and contour, and are thus refractory to study by traditional structural means such as X-ray crystallography. Electron microscopy (EM) overcomes some problems associated with particle variability and has been an important tool for investigating coronavirus ultrastructure. However, EM sample preparation requires that the specimen be dried onto a carbon support film before imaging, collapsing internal particle structure in the case of coronaviruses. Moreover, conventional EM achieves image contrast by immersing the specimen briefly in heavy-metal-containing stain, which reveals some features while obscuring others. Electron cryomicroscopy (cryo-EM) instead employs a porous support film, to which the specimen is adsorbed and flash-frozen. Specimens preserved in vitreous ice over holes in the support film can then be imaged without additional staining. Cryo-EM, coupled with single-particle image analysis techniques, makes it possible to examine the size, structure and arrangement of coronavirus structural components in fully hydrated, native virions. Two virus purification procedures are described.

Key words: coronavirus; cryo-EM; virus purification; density gradient centrifugation

1. Introduction

The success of electron cryomicroscopy (cryo-EM) is highly dependent on the quality of the prepared sample. Experience has shown that in order to be useful cryo-EM virion preparations should contain approximately 10^{10} infectious particles per milliliter in a relatively pure suspension. This requires at minimum

From: *Methods in Molecular Biology, vol. 454: SARS- and Other Coronaviruses,*
Edited by: D. Cavanagh, DOI: 10.1007/978-1-59745-181-9_12, © Humana Press, New York, NY

a 100-fold concentration step, since coronavirus growth seldom surpasses 10^8 PFU/ml. Moreover, the coronavirus envelope and the protein-mediated connections between the membrane and the ribonucleoprotein are sensitive to mechanical and osmotic disruption. The viral spike proteins (S) are sensitive to ectodomain shearing and, if cleaved, to shedding of the amino-terminal S1 subunit. The method of concentration and purification described here is designed to minimize these disruptive forces and has been used successfully with several coronaviruses and other enveloped viruses.

Sample adsorption is sensitive to viscosity, and both the alternative serum-free purification protocol and the sample loading protocol provide potential solutions for high-viscosity preparations. These protocols are designed to assist the investigator in preparing virus samples for cryo-EM. From this point, microscopy and image analysis techniques will vary with the type of instrument and the nature of the investigation. It is hoped that these techniques will facilitate further examination of coronavirus, torovirus, arterivirus, and ronivirus supramolecular architecture.

2. Materials

2.1. Virus Purification and Concentration

1. Dulbecco's Modified Eagle's Medium supplemented with 10% fetal bovine serum, antibiotics, 10 mM HEPES buffer, pH 7.
2. Precipitation reagents: polyethylene glycol-8000, white flake type (PEG-8000 Ultra for Molecular Biology, available from Fluka) and NaCl, crystalline, high quality.
3. HEPES-saline: 9 g NaCl (0.9% final), 10 ml of 1 M HEPES, 990 ml purified water, pH adjusted to 7.0, vacuum-sterilized through a 0.2-μm pore size membrane.
4. 3X HEPES-saline: 27 g NaCl, 30 ml of 1 M HEPES, 930 ml purified water, pH adjusted to 7.0.
5. 50% (w/w) sucrose: 50 g sucrose, 50 g HEPES-saline, vacuum-sterilized through a 0.2-μm pore size membrane. Dilute with HEPES-saline to prepare 10, 20, and 30% sucrose solutions.
6. 25% neutral buffered formalin: 10 ml of formalin (37–40% formaldehyde), 5 ml of 3X concentrated HEPES-saline.
7. Centrifuges and rotors: A low-speed centrifuge and rotor with a capacity of ≥1 liter (Sorvall GSA, e.g.), and a high-speed centrifuge and rotor with a total capacity ≥100 ml (Beckman SW28, e.g.).

2.2. Serum-Free Virus Purification and Concentration

1. Virus production serum-free medium (VP-SFM; Gibco/Invitrogen) supplemented with antibiotics and glutamine.
2. All other reagents and equipment as in **Section 2.1.**

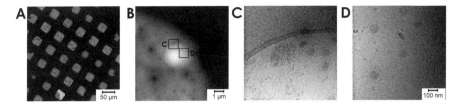

Fig. 1. Cryo-EM of feline coronavirus (FCoV). Arrays of holes containing vitrified FCoV particles can be seen at left (A), along with a close-up image of a single hole (B). The viral envelope, spikes, and internal components are visible at higher magnification (C) and (D). Automated image collection was performed using Leginon.

2.3. Quality Control EM

1. Formvar/carbon coated 200-mesh or 300-mesh EM grids (available from Ted Pella).
2. Plasma glow discharge unit (available from Emitech).
3. Negative stain: 2 g uranyl acetate (available from Electron Microscopy Sciences), 100 ml purified water, filtered through a 0.2-μm membrane, stored in a brown glass bottle away from light.
4. HEPES-buffered saline (**Section 2.1**, step 5).
5. Parafilm M (American Can Company).
6. Fine forceps for EM grid manipulation (available from Ted Pella).
7. Filter paper (Whatman 3 M).

2.4. Sample Loading and Grid Freezing

1. Holey carbon grids (Quantifoil holey film with circular holes; Quantifoil Micro Tools) (**Fig. 1**).
2. Cryogenic reagents: Liquid nitrogen and an ethane gas cylinder fitted with a flow regulator and a fine nozzle.
3. Ethane plunge device: large liquid nitrogen reservoir, smaller ethane liquefaction chamber made from a thermally conductive material, and a drop-release specimen holder (available as a set from Electron Microscopy Sciences).

3. Methods

As the quality of the virus preparation is the most important component of the cryo-EM technique, two purification protocols are listed below. Either can yield cryo-EM-quality coronavirus, but the method in **Section 3.1** is generally preferred because the serum proteins function as a "carrier" during the PEG precipitation step. Serum-free purification (**Section 3.2**) can be used when virus preparations tend to produce overly viscous purified virion solutions. For best results, the purification process should be completed in 1 day, and the virus loaded onto holey grids no later than 24 h after purification.

3.1. Virus Purification and Concentration

1. This method is suitable for most coronaviruses that grow well in cultured cells and has been used successfully with SARS-CoV, FcoV, and MHV *(1)* in addition to several types of arenavirus (B. N., unpublished data). For the purpose of this protocol, it is assumed that SARS-CoV is being prepared on Vero-E6 cells.
2. Culture Vero-E6 cells in DMEM to approximately 70–90% confluency (*see* **Note 1**).
3. Inoculate with SARS-CoV at a multiplicity of approximately 3 PFU/cell.
4. Remove the inoculum after 1 h and replace it with fresh medium.
5. Remove and discard the culture medium 24 h after inoculation. Replace it with fresh DMEM (*see* **Note 2**).
6. Collect the cell culture supernatant 48 h after inoculation. Store a small sample for plaque assay titration (*see* **Note 3**).
7. Transfer the supernatant to centrifuge bottles with screw caps, noting the total volume. Pellet cellular debris at $10,000 \times g$, 4°C for 20 min. It is best to use a high-capacity rotor at this stage to minimize preparation time.
8. During the centrifugation, prepare fresh screw-cap centrifuge bottles containing 10 g of dry PEG-8000 and 2.2 g of NaCl per 100 ml of culture medium to be added. Alternatively, prepare a large conical flask with sufficient PEG-8000 and NaCl to bring the entire volume of virus-containing medium to a final concentration of 10% PEG-8000, 2.2% NaCl. Chill HEPES-saline and neutral-buffered formalin on ice for later use (*see* **Note 4**).
9. After centrifugation (step 6), an off-white or yellow pellet of cell debris will be visible. Decant the supernatant into the centrifuge bottles or conical flask prepared earlier with PEG and NaCl.
10. Swirl the mixture gently until the additives are fully dissolved. Add a clean stir bar and incubate at 4°C for a further 30 min with gentle stirring.
11. Transfer the solution to centrifuge bottles, if you have not done so already Collect the PEG-precipitated protein fraction by centrifugation for 30 min at approximately $10,000 \times g$.
12. During the centrifugation, prepare three-step 10% - 20% - 30% sucrose gradients in tubes appropriate to the high-speed centrifuge rotor. If using the Beckman SW-28 rotor, use ~8 ml for each step, leaving ~10 ml for sample loading and balancing (*see* **Note 5**).
13. Decant and discard supernatants immediately to minimize sample loss. A large opaque white pellet should be present in each of the flasks following centrifugation. Swirl the pellet in cold HEPES-saline until dissolved. Avoid using a pipette at this step, if possible. It is critical that the PEG pellets be completely resuspended before proceeding to the next step (*see* **Note 6**).

14. Overlay the PEG-protein solution carefully onto the sucrose gradients. Balance with remaining sample or additional HEPES-saline. Pellet the virions through the sucrose cushions by centrifugation at 100,000 × *g* for 90 min at 4°C.

15. Decant and discard supernatants immediately. Invert the empty tubes on an absorbent surface for 5 min and tap gently to wick away any remaining sucrose solution. Resuspend the virion pellet in as small a volume of HEPES-saline as possible (typically 100–200 µl). Do not use a pipette to resuspend the virus, as this may shear spikes and damage fragile viral envelopes (*see* **Note 7**).

16. When the pellet has been resuspended, the HEPES-saline will turn somewhat milky in color. Use a P-1000 pipette tip with the pointed end removed to gently transfer the virus suspension to a cryovial. Discard any insoluble material. Set aside a small sample for diagnostic purposes.

17. At this stage, the virus should be monodisperse, and can be inactivated if desired. Treatment with a 1% final concentration of ice-cold neutral buffered formalin, 2-propiolactone treatment, or gamma-irradiation should all yield intact, inactivated EM-quality particles. Samples for cryo-EM may be stored at 4°C for up to 24 h (*see* **Note 8**).

3.2. Serum-Free Virus Purification and Concentration

1. This alternative method is suitable for purification of viruses that grow to lower titers. Serum-free culture and preparation can also be used to remedy solutions that fail for cryo-EM owing to high viscosity, nonviral protein contamination, or large amounts of insoluble material. Percentage recovery will often be lower than with the protocol described in **Section 3.1** (*see* **Note 9**).

2. Perform steps 2 to 11 in **Section 3.1**, substituting VP-SFM for DMEM starting at the time of inoculation.

3. The PEG-protein pellets should be white and may be quite small. Decant and discard the supernatants immediately. Resuspend by swirling gently in 10 ml HEPES-saline (*see* **Note 10**).

4. Perform steps 12 to 15 in **Section 3.1.** The final translucent pellet may be small and quite difficult to see.

3.3. Quality Control

1. The relative infectivity of the final preparation should be directly assessed by plaque assay or similar means as a retrospective measure of quality. However, the decision to freeze grids for cryo-EM must be taken immediately. The quickest way to assess sample quality is by standard transmission EM of a negatively stained sample. Ideal samples for cryo-EM should contain a high density of virions with intact spikes and little nonviral material. This protocol describes how to prepare a negatively stained coronavirus specimen for transmission EM.

2. Plasma glow discharge several formvar/carbon-coated EM grids prior to use to make the support surface more hydrophilic (*see* **Note 11**).

3. Trace several lines on a piece of parafilm M with the dull end of a pair of forceps to create channels for droplets of negative stain and wash buffers. On the parafilm, place one ~20 µl droplet of 2% uranyl acetate and up to three droplets of HEPES-saline along the grid. The number of HEPES-saline washes will depend on the viscosity of the virus sample being imaged (*see* **Note 12**).

4. Place 3–5 µl of the viral sample on each grid, and adsorb for 5 min.

5. Float each grid on each droplet of HEPES-saline for 30 sec. Do not reuse buffer or stain droplets to avoid sample contamination.

6. Float each grid on a uranyl acetate droplet for 1 min.

7. Touch the edge of the grid to the filter paper several times to remove excess stain (*see* **Note 13**).

8. Samples should be visualized immediately, but can be stored for longer periods under a vacuum.

3.4. Sample Loading and Grid Freezing

1. This procedure describes a protocol for vitrifying coronavirus specimens.

2. Plasma glow discharge holey carbon grids before use.

3. Fill the outer coolant reservoir of the vitrification apparatus with liquid nitrogen. This will be used to cool ethane to liquid form, and ultimately a semisolid "slush" that will be used to cryopreserve the virus sample.

4. Wait until the temperature of the inner cooling chamber has stabilized, and the liquid nitrogen is no longer boiling. Dispense ethane into the inner cooling chamber. Wear protective clothing and eye protection at this stage, since ethane splatter can be dangerous. Fill the ethane chamber without allowing any liquid nitrogen to splash into the inner chamber. Stir occasionally until the consistency of the liquid ethane changes as it begins to freeze. If the ethane freezes at any point, liquefy it by adding fresh ethane.

5. Center a pair of EM forceps in the plunge device over the inner ethane chamber. Check that the grid will be plunged well below the level of the ethane slush. Adjust the level of ethane if necessary.

6. Pick up a holey carbon grid by the edge using the forceps and slot the forceps into the plunge device.

7. Adsorb 3–5 µl of sample onto the carbon-coated (dull) side of the grid.

8. Blot the edge of the grid against the filter paper to remove excess volume (*see* **Note 14**).

9. Immediately plunge the grid into the ethane slush (*see* **Note 15**).

10. Detach the forceps, still holding onto the grid below the surface of the ethane. Avoid touching the walls of the ethane chamber.

11. Quickly transfer the sample from the ethane to the outer liquid nitrogen reservoir.

12. Place the grid into a grid box while still under the liquid nitrogen. Avoid bumping the grid on the walls of the liquid nitrogen reservoir. The grid must be stored and handled under liquid nitrogen from this point forward to remain vitrified.

4. Notes

1. Plan to prepare at least $\sim 10^{10}$ PFU of SARS-CoV. Assuming an average SARS-CoV titer of $\sim 10^7$ PFU/ml, 1 liter or more of infectious cell culture medium will be needed. One-well plates (available from Nalgene) can be a cost- and space-effective alternative to standard tissue culture flasks for large-scale virus preparation.

2. Early time points containing little virus can be discarded. Purification efficiency is strongly dependent on concentration.

3. Virus can be collected at 2-h intervals beginning 48 h after inoculation to increase the final yield. However, for best results, each virus sample should be processed immediately, rather than collected and stored.

4. For viruses with cleaved spike proteins, such as MHV, the amount of NaCl added should be reduced to 1 g/100 ml to minimize damage to the spikes.

5. Alternatively, virus can be banded at the interface of a 30–50% two-step sucrose gradient.

6. In general, rapid resuspension, cold temperature, and the minimization of mechanical stress will all improve the quality of the preparation. Soluble proteins contained in the pellet may alter the color to a yellowish hue.

7. The translucent virion pellet may be difficult to see. Whether or not the pellet is visible, attempt to resuspend immediately. If a pellet is present, the viscosity of the added HEPES-saline will increase noticeably upon resuspension.

8. Inactivation techniques should be validated beforehand. Both formaldehyde and 2-propiolactone can lose effectiveness over time. Tris buffers will react with formaldehyde, and should be avoided. HEPES buffer is therefore recommended for use throughout the purification process.

9. The growth of SARS-CoV, FcoV, and MHV is not affected by short-term treatment with VP-SFM as outlined here. However, cells do grow more slowly in VP-SFM as compared to DMEM, and thus VP-SFM is not recommended for the initial cell culture step.

10. If left in contact with the supernatant, serum-free PEG pellets will resuspended much more quickly than serum-containing pellets, and the sample may be lost.

11. Grids that continually sink in the saline or stain droplets likely have suffered extensive damage to the support surface and should not be used.

12. The purpose of glow discharge treatment is to impart a charge to the carbon-coated side of the grid, which can otherwise be quite hydrophobic. Therefore, glow discharge grids with the carbon side, which will appear slightly dull, facing upward.

13. The effects of the electron beam on large amounts of residual stain can cause a "blowout" of the carbon/formvar support surface. It is therefore important to remove all excess stain before visualization.

14. The vitrified ice supporting the specimen must be very thin (~ 1 μm or less) to permit clear imaging.

15. Sample blotting is unfortunately an inexact process. The thickness of the sample ice can be adjusted by altering blotting time, which typically ranges from 0.5 to

about 15 sec. Ice thickness and quality will have to be assessed by EM (**Fig. 1**), so it is recommended that several grids be prepared at once using different blot times.

Acknowledgments

Detailed notes left by Dr. Thomas Gallagher were invaluable in refining the purification procedure. Funding for this work was provided by the NIH/NIAID contract "Functional and Structural Proteomics of SARS Coronavirus" HHSN266200400058C and by the Pacific-Southwest Regional Center of Excellence AI-065359. Some of the work presented here was conducted at the National Resource for Automated Molecular Microscopy, which is supported by the National Institutes of Health though the National Center for Research Resources' P41 program (RR17573).

References

1. Neuman, B. W., Adair, B. D., Yoshioka C., et al. (2006) Supramolecular architecture of severe acute respiratory syndrome coronavirus revealed by electron cryomicroscopy. *J. Virol.* **80**, 7918–7928.
2. Stadler, K., Roberts, A., Becker, S., et al. (2005) SARS vaccine protective in mice. *Emerg. Infect. Dis.* **11**, 1312–1314.
3. Beniac, D. R., Andonov, A., Grudeski, E., and Booth, T. F. (2006) Architecture of the SARS coronavirus prefusion spike. *Nat. Struct. Mol. Biol.* **13**, 751–752.
4. Suloway, C., Pulokas, J, Fellmann, D., et al. (2005) Automated molecular microscopy: the new Leginon system. *J. Struct. Biol.* **151**, 41–60.

IV

EXPRESSION OF CORONAVIRUS PROTEINS AND CRYSTALLIZATION

13

Production of Coronavirus Nonstructural Proteins in Soluble Form for Crystallization

Yvonne Piotrowski, Rajesh Ponnusamy, Stephanie Glaser, Anniken Daabach, Ralf Moll, and Rolf Hilgenfeld

Abstract

For biophysical investigations on viral proteins, in particular for structure determination by X-ray crystallography, relatively large quantities of purified protein are necessary. However, expression of cDNAs coding for viral proteins in prokaryotic or eukaryotic systems is often not straightforward, and frequently the amount and/or the solubility of the protein obtained are not sufficient. Here, we describe a number of protocols for production of nonstructural proteins of coronaviruses that have proven to be efficient in increasing expression yields or solubilities.

Key words: nonstructural proteins; human coronavirus; expression; solubility; crystallization.

1. Introduction

The number of viral outbreaks has increased dramatically in recent years. Since 1996, the world has seen at least one major outbreak of a new virus or of a new variant of a known virus each year. Vaccines are hardly capable of stopping such outbreaks, because even with advances in their development, they will not be available for 6 to 12 months following the first appearance of a new pathogen. Immediate containment of viral outbreaks will therefore depend on quarantine and antiviral drugs. Unfortunately, no drug treatment is available for most known

From: *Methods in Molecular Biology, vol. 454: SARS- and Other Coronaviruses,*
Edited by: D. Cavanagh, DOI: 10.1007/978-1-59745-181-9_13, © Humana Press, New York, NY

viral diseases of humans, let alone for newly emerging ones. We believe that in view of this situation, it is necessary to develop lead compounds with activity against all major families of viruses, both those that infect humans and those that so far have been restricted to animals but may cross the species barrier by zoonotic transmission. Ultimately, we should aim at discovering antiviral compounds that exhibit a relatively broad activity against a range of new viruses should they emerge.

In order to discover such leads for antiviral compounds, a sophisticated approach is necessary. Random screening for antivirals may have its merits, but the case of HIV has demonstrated that this approach has not led to a single marketed drug in the past 20 years, whereas no fewer than 26 antiretroviral drugs that have been rationally designed are on the market or in late development phases. Therefore, we follow a structure-based approach to discover new compounds with activity against RNA viruses. Methods applied include *de-novo* design, virtual screening, and structure-guided medicinal chemistry, all of which require a detailed knowledge of the structure and function of the viral target enzymes.

Coronaviruses such as human coronavirus 229E (HCoV 229E) and HCoV OC43 are responsible for variants of the common cold, while SARS-CoV caused the outbreak of severe acute respiratory syndrome in 2003, which killed more than 800 people *(1)*. The human coronavirus NL63, first described in March 2004, has been identified as the causative agent of respiratory disease in very young children and in immunocompromised adults *(2)*. Furthermore, HCoV-NL63 infection has been associated with laryngotracheitis (croup) *(3)* and Kawasaki disease, although the latter has been contested by a recent study *(4)*.

The emergence of coronaviruses during the last decade and the wide variety of the diseases they cause—from the relatively harmless common cold to potentially lethal SARS—demonstrate the necessity of understanding the replication and transcription of these viruses, in order to find drugs that will interfere with these processes.

The 16 nonstructural proteins (Nsp) of the coronaviruses are required for genomic RNA synthesis (replication) and subgenomic RNA synthesis (transcription). They are encoded by the replicase gene, which comprises two open reading frames, ORF1a and ORF1b. ORF1a encodes the replicative polyprotein 1a (pp1a). The larger pp1ab results from a (−1) ribosomal frameshift during translation, which occurs just upstream of the ORF1a stop codon. Thus, pp1ab is the translation product of ORF1a with a large extension resulting from the ORF1b-encoded part. The polyproteins are processed into individual polypeptides by the virus-encoded proteases, one or two papain-like cysteine proteases (domains of Nsp3) and the main protease (Nsp5). Even though the function of a few nonstructural proteins is known, e.g., Nsp5 being the main protease, Nsp13 being a helicase and Nsp12 being an RNA-dependent RNA polymerase *(1,5)*, the process of replication and transcription has not yet been elucidated. It is expected that this problem will be tackled by structural and functional studies of individual nonstructural proteins and of the complexes they form with one another and with nucleic acids.

X-ray crystallography and nuclear magnetic resonance spectroscopy are the most important methods for determination of the three-dimensional structures of proteins at high resolution. Today, more than 85% of protein structures are elucidated by X-ray crystallography. For crystallization, rather elevated concentrations of the protein under study are needed (5–20 mg/ml) and the sample should be as pure as possible (>95%). To achieve this, an efficient expression system and a reliable purification protocol have to be established. Here we present a few protocols for expression in *Escherichia coli* of genes coding for nonstructural proteins of human coronaviruses NL63 and 229E and for increasing the solubility of the recombinant protein. We also describe the purification of proteins carrying hexahistidine or GST (glutathione-S transferase) affinity tags. Furthermore, we present a preliminary crystallization experiment that has been shown to give initial hints at crystallization conditions.

2. Materials

2.1. Cultivation

1. The culture medium is prepared by dissolving 15.5 g of Invitrogen's 2× tryptone-yeast (TY) powder (1.6% tryptone, 1% yeast extract, 0.5% NaCl) in 1 liter distilled water followed by autoclaving.
2. To prepare a stock solution of ampicillin, 1 g of ampicillin (Na-salt) is dissolved in 5 ml glycerol and 5 ml distilled water and filtered through a 0.45-μm filter. The solution is stored at –20°C.
3. The 1-M isopropyl-β,D-thiogalactopyranoside (IPTG) stock solution is prepared in 10 ml distilled water, filtered through a 0.45-μm filter, and stored at –20°C.

2.2. Sodium Dodecylsulfate Polyacrylamide Gel Electrophoresis (SDS-PAGE)

1. a. Laemmli solution A: 30 g acrylamide, 0.3 g N′,N′-bismethylenacrylamide, filled up to 100 ml with distilled water, filtered. Store at 4°C.
   b. Laemmli solution B: 18.15 g Tris-HCl, filled up with distilled water to 100 ml, pH 8.8. Store at 4°C.
   c. Laemmli solution C: 10 g SDS dissolved in distilled water in a total volume of 100 ml.

   d. Laemmli solution D: 3 g Tris-HCl, filled up with distilled water to a final volume of 50 ml, pH 6.8. Store at 4°C.

   e. 10% (w/v) ammonium persulfate solution (APS): Dissolve 1 g APS in 10 ml distilled water and filter through a 0.45-μm filter. Store the solution at 4°C.

   f. N,N,N′,N′-tetramethylethylenediamine (TEMED). Store at 4°C.

   g. 0.1% (w/v) SDS solution: 100 mg SDS dissolved in distilled water in a total volume of 100 ml.

2. Laemmli sample buffer: 1.5 ml Laemmli solution D, 2 ml Laemmli solution C, 2 ml glycerol 87% (w/v), 3.1 ml distilled water, 0.2 ml of a 1% (w/v) bromophenol blue solution (in ethanol), freshly add one-tenth of the β-mercaptoethanol to the mixture.

3. Mini gel running buffer: 1 g SDS, 3 g Tris, 14.4 g glycine, filled up to 1 liter with distilled water.

4. Coomassie staining solution: 0.01% (w/v) Brilliant Blue R 250 (Sigma), 50% (v/v), methanol, 10% (v/v) acetic acid.

5. Destaining solution: 40% (v/v) methanol, 10% (v/v) acetic acid.

2.3. Solubility Screen [after (6)]

1. Washing buffer: 10 mM Tris-HCl, pH 8.5, 100 mM NaCl, 1 mM EDTA (*see* **Note 1**).

2. Lysozyme solution: Dissolve 15 mg lysozyme in 1 ml of distilled water.

3. Solubility screen buffers for His-tagged protein (*see* **Note 2**):

   a. 100 mM Tris, 10% (v/v) glycerol, pH 7.6

   b. 100 mM Tris, 50 mM LiCl, pH 7.6

   c. 100 mM HEPES, 50 mM $(NH_4)_2SO_4$, 10% (v/v) glycerol, pH 7.0

   d. 100 mM HEPES, 100 mM KCl, pH 7.0

   e. 100 mM Tris, 50 mM NaCl, 10% (v/v) isopropanol, pH 8.2

   f. 100 mM K_2HPO_4/KH_2PO_4, 50 mM $(NH_4)_2SO_4$, 1% (v/v) Triton X-100, pH 6.0

   g. 100 mM triethanolamine, 100 mM KCl, 10 mM DTT, pH 8.5

   h. 100 mM Tris, 100 mM sodium glutamate, 10 mM DTT, pH 8.2

   i. 250 mM K_2HPO_4/KH_2PO_4, 0.1% (w/v) CHAPS, pH 6.0

   j. 100 mM triethanolamine, 50 mM LiCl, 5 mM EDTA, pH 8.5

   k. 100 mM sodium acetate, 100 mM glutamine, 10 mM DTT, pH 5.5

   l. 100 mM sodium acetate, 100 mM KCl, 0.1% (v/v) n-octyl-β-D-glucoside, pH 5.5

   m. 100 mM HEPES, 1 M $MgSO_4$, pH 7.0

   n. 100 mM HEPES, 50 mM LiCl, 0.1% (w/v) CHAPS, pH 7.0

   o. 100 mM K_2HPO_4/KH_2PO_4, 2,5 mM $ZnCl_2$, pH 4.3

   p. 100 mM Tris, 50 mM NaCl, 5 mM calcium acetate, pH 7.6

   q. 100 mM triethanolamine, 50 mM $(NH_4)_2SO_4$, 10 mM $MgSO_4$, pH 8.5

   r. 100 mM Tris, 100 mM KCl, 2 mM EDTA, 1% (v/v) Triton X-100, pH 8.2

   s. 100 mM sodium acetate, 1 M $MgSO_4$, pH 5.5

    t. 100 mM Tris, 2 M NaCl, 0.1% (v/v) n-octyl-β-D-glucoside, pH 7.6
    u. 100 mM Tris, 1 M (NH$_4$)$_2$SO$_4$, 10 mM DTT, pH 8.2
    v. 100 mM sodium acetate, 50 mM LiCl, 5 mM calcium acetate, pH 5.5
    w. 100 mM HEPES, 100 mM sodium glutamate, 5 mM DTT, pH 7.0
    x. 100 mM triethanolamine, 100 mM sodium glutamate, 0.02% (v/v) n-octyl-β-D-glucoside, 10% (v/v) glycerol, pH 8.5
    y. 100 mM Tris, 50 mM NaCl, 100 mM urea, pH 8.2
    z. 100 mM triethanolamine, 100 mM KCl, 0.05% (w/v) dextran sulfate, pH 8.5
  aa. 100 mM K$_2$HPO$_4$/KH$_2$PO$_4$, 50 mM (NH$_4$)$_2$SO$_4$, 0.05% (w/v) dextran sulfate, pH 6.0
  bb. 100 mM HEPES, 50 mM LiCl, 0.1% (w/v) deoxycholate, pH 7.0
  cc. 100 mM Tris, 100 mM KCl, 0.1% (w/v) deoxycholate, 25% (v/v) glycerol, pH 7.6
  dd. 100 mM potassium acetate, 50 mM NaCl, 0.05% (w/v) dextran sulfate, 0.1% (w/v) CHAPS, pH 5.5

4. Solubility screen buffers for glutathione-S-transferase (GST) fusion protein (*see* **Note 3**): Basic buffer is PBS buffer (140 mM NaCl, 2.7 mM KCl, 10 mM Na$_2$HPO$_4$, 1.8 mM KH$_2$PO$_4$, adjusted to pH 7.3 with HCl) containing:

    a. 1% Tween 20.
    b. 1% Triton X-100.
    c. 2 mM EDTA.
    d. 10% glycerol.
    e. negative control.
    f. 5 mM DTT.
    g. 0.1% n-octyl-β-D-glucopyranoside 0.1%.
    h. 0.17 mM dodecylmaltoside.

2.4. Large-Scale Purification

2.4.1. Purification of the His-Tagged Protein

1. Binding buffer: Solubility screen buffer + 20 mM imidazole (*see* **Note 4**).
2. Elution buffer: Solubility screen buffer + 500 mM imidazole (*see* **Note 4**).
3. Hen egg-white lysozyme, lyophilized.
4. Complete protease inhibitor cocktail tablet (EDTA-free) (Roche).
5. HisTrap FF 1-ml column (Amersham Biosciences).
6. Amicon ultrafiltration cell.

2.4.2. Purification of the GST-Fusion Protein

1. 1× PBS buffer (140 mM NaCl, 2.7 mM KCl, 10 mM Na$_2$HPO$_4$, 1.8 mM KH$_2$PO$_4$, pH 7.3) + additive (*see* **Note 5**).
2. Cleavage buffer: 50 mM Tris-HCl, 150 mM NaCl, 1 mM EDTA, 1 mM DTT, pH 7.5.

3. Elution buffer: 50 mM Tris-HCl, 20 mM reduced glutathione, pH 8.0.
4. Hen egg-white lysozyme, lyophilized.
5. Complete protease inhibitor cocktail tablet (EDTA-free) (Roche).
6. PreScission protease: 10 mg/ml.
7. GSTrap FF 5-ml column (Amersham Biosciences).
8. Amicon ultrafiltration cell.

2.5. Preliminary Crystallization Experiment

1. A pregreased 48-well plate from Hampton Research (VDX48 plate with sealant, cat.-no. HR3-275) and 12-mm-diameter circle cover slides.
2. Buffers: 1 M citric acid pH 4.0, 1 M citric acid pH 5.0, 1 M MES pH 6.0, 1 M HEPES pH 7.0, 1 M Tris pH 8.0, 1 M bicine pH 9.0, adjusted to the desired pH with HCl or NaOH (*see* **Note 6**).
3. Precipitant solutions: 4 M $(NH_4)_2SO_4$, 50% polyethylene glycol (PEG) 6000.

3. Methods

The more that is known about the target protein prior to the onset of experiments, the better. Homologous sequence alignments and secondary-structure prediction programs are helpful tools from the outset. They can provide a hint as to whether to clone the whole gene into the expression vector or to truncate it because of potentially unfolded regions, or to even go for larger parts of polyproteins rather than individual proteins. An example is shown in **Fig. 1** (*see* **Note 7**).

In this chapter, we present a guide to the production of sufficient amounts of soluble protein and describe it for some nonstructural proteins of HCoV NL63 and HCoV 229E. We do not describe cloning procedures, as they are common knowledge in almost every laboratory working in the field. However, we do provide a recipe for a preliminary crystallization experiment.

We start with various expression constructs of the target and the search for suitable expression conditions. All expression experiments are performed in *E. coli* strains. The cultivation experiments are done on a small scale (Section 3.3, step 2) in order to test as many parameters as possible. We show that the expression yield can differ using different temperatures, times of induction, and amounts of inductor.

Having found the conditions for optimal expression, we cultivate on a larger scale (Section 3.3, step 3), because more cells are needed for the subsequent solubility screen. The solubility screen (Section 3.4.) demonstrates that the choice of the purification tag that was made during the cloning procedure at the very beginning, as well as the use of additives, can have a huge effect on the solubility of the target protein. Furthermore, co-purification of two or more individual

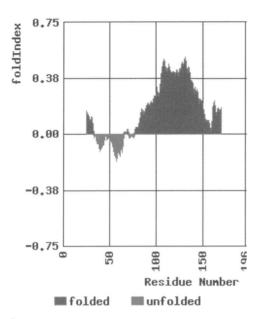

Fig. 1. FoldIndex© *(7)* prediction for HCoV-229E Nsp8. Peaks below 0.00 on the Y-axis show regions predicted to be unfolded; peaks above 0.00 indicate folded regions.

proteins can improve the solubility dramatically and might indicate that these proteins function as a complex, which we demonstrate with some figures.

Finally, we demonstrate that investing to obtain well-soluble protein is worth the effort. We perform a preliminary crystallization screen (Section 3.6.) using ammonium sulfate and PEG. Even this initial screen can sometimes yield crystals if the protein is well purified and reasonably soluble. Such crystals can usually be optimized and used in structure determination by X-ray crystallography.

3.1. SDS-PAGE

1. The SDS-PAGE instruction presented here uses the Hoefer SE250 gel system, but it can be easily modified to suit any other system.
2. Prepare a 0.75-mm-thick 15% gel by pipetting 2.78 ml Laemmli solution A, 1.38 ml Laemmli solution B, 55 μl Laemmli solution C, 1.28 ml water, 55 μl 10% APS, and 3.3 μl TEMED. Pour the gel and leave some space for the stacking gel. To avoid evaporation, carefully pour 0.1% (w/v) SDS on top of the nonpolymerized gel solution. Wait 20–30 min, until the gel is polymerized.
3. Decant the 0.1% (w/v) SDS solution.
4. For the stacking gel, pipette 1 ml Laemmli solution A, 50 μl Laemmli solution C, 1.25 ml Laemmli solution D, 2.65 ml water, 50 μl APS, and 5 μl TEMED. Pour the stacking gel on top of the separating gel. Insert a comb with the necessary number

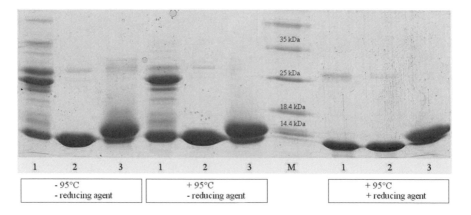

Fig. 2. SDS-PAGE of Nsp9 (14 kDa) of HCoV 229E (wildtype [containing one disulfide bridge in the dimer (1)]; mutant [Cys mutated to Ala] (2)], and SARS-CoV (3) under different conditions; reducing agent = β-mercaptoethanol; M = protein molecular-weight marker (Fermentas).

of pockets into the stacking gel. Wait 20–30 min, until the gel is polymerized. Carefully remove the comb after polymerization (*see* **Note 8**).

5. Attach the gel sandwich to the core that is fixed to the lower chamber in the case of the Hoefer SE250 gel system. Fill the sample wells and every upper and lower buffer chamber with running buffer. Make sure that the lower electrode is completely submerged.

6. For sample preparation, mix an equal volume of protein solution with the Laemmli sample buffer according to the size of the wells in the stacking gel. Heat the samples for 5 min at 95°C and centrifuge for 1 min at ~16,000× *g*. Load the samples and the molecular weight marker into the wells (*see* **Note 9; Fig. 2**).

7. The gel electrophoresis is carried out for ~1 h (until the blue dye front is almost running out of the gel) at 20 mA/gel.

8. Transfer the gel to a basin containing the Coomassie staining solution and incubate at room temperature for 30 min.

9. Discard the staining solution in a separate waste container and destain the gel by incubating it with destaining solution at room temperature until the background is clear.

3.2. Cell Lysis (Under Denaturing Conditions) (see Note 10)

1. Take a 1-ml sample from the culture and measure the optical density at 660 nm (OD_{660}).

2. Centrifuge the sample in a 1.5-ml reaction tube at ~16,000× *g*, 4°C for 5 min and discard the supernatant.

3. Resuspend the pellet in 67 µl Laemmli solution D (SDS-PAGE solution).

4. Add 8 μl of 10% SDS-solution (Laemmli solution C), resulting in a concentration of 1% SDS in the final volume.
5. Shake the tube for 10 min at 950 rpm and 95°C.
6. Centrifuge the tube at ~16,000× g, 4°C for 10 min.
7. Transfer the supernatant into a new 1.5-ml reaction tube.

3.3. Cultivation

3.3.1. Overnight Culture (5-ml Culture Volume) (see Note 11)

1. Add 5 μl of the ampicillin stock solution (100 mg/ml) to 5 ml 1 × TY medium (resulting in a final concentration of 100 μg/ml ampicillin).
2. Inoculate the 5 ml of 1 × TY/ampicillin medium with a single colony or a mixed colony (*see* **Note 12**).
3. Let the culture grow overnight at 37°C, shaking at 250 rpm.

3.3.2. Cultivation for Expression

1. Prepare a 2-ml overnight culture. Let the culture grow overnight at 37°C, shaking at 250 rpm.
2. In four parallel experiments (I, II, III, and IV), inoculate 40 ml each of fresh 1 × TY/ampicillin medium, using 400 μl of the overnight culture (*see* **Note 13**).
3. Grow the cells at 37°C, shaking at 180 rpm.
4. Remove 1 ml of culture from each cultivation ("preinduction sample") for cell lysis under denaturing conditions—when the OD_{660} reaches 0.5 for experiments I and II and 1.0 for III and IV.
5. Induce the expression by adding 40 μl to I and II and 20 μl to III and IV of the 1-M IPTG stock solution to the remaining cultures directly after the removal of the preinduction sample, resulting in a final concentration of 1 mM and 0.5 mM IPTG, respectively. Then shift cultures I and III to 20°C and leave II and IV at 37°C, shaking at 180 rpm.
6. Expression is carried out for 4 h. After this, a 1-ml sample of each culture is taken ("postinduction sample"), the OD_{660} is determined and cell lysis is performed under denaturing conditions (*see* **Note 14**).
7. Analyze the samples by SDS-PAGE. Take 10 μl of each sample and adjust to an OD_{660} value of 0.5, e.g., 10 μl of the preinduction sample OD_{660} = 0.5 (experiments I and II), or 5 μl (filled up to 10 μl with distilled water) of the preinduction sample OD_{660} = 1.0 (experiments III and IV). One example of the influence of the parameters tested is shown in **Fig. 3** (*see* **Note 15**).

3.3.3. Cultivation for Solubility Screen

1. Prepare a 5-ml overnight culture. Let the culture grow overnight at 37°C, shaking at 250 rpm.

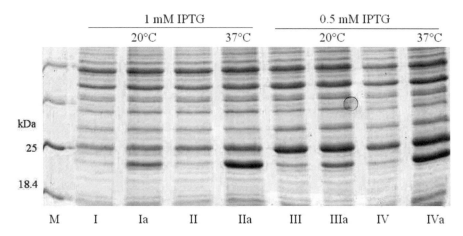

Fig. 3. Expression of the HCoV-NL63 Nsp3 X-domain with the vector pEXP1-DEST in BL21 (DE3) pLysS, containing a sequence encoding a 6xHis-tag (total molecular mass of 23 kDa). M = protein molecular-weight marker (Fermentas); I: 1 mM IPTG, 20°C, II: 1 mM IPTG, 37°C; III: 0.5 mM IPTG, 20°C, IV: 0.5 mM IPTG, 37°C; before (I–IV) and after induction (Ia–IVa).

2. Centrifuge the overnight culture at ∼5200 × g, at room temperature for 10 min. Discard the supernatant (*see* **Note 16**).
3. Resuspend the pellet in a few ml of fresh 1 × TY/ampicillin medium.
4. Transfer the resuspended cells to 500 ml 1 × TY medium, containing 500 μl ampicillin stock solution.
5. Grow the cells at 37°C, shaking at 170 rpm, until they reach an OD_{660} = 0.5 (*see* **Note 17**).
6. Induce the expression of the desired gene by adding 500 μl of 1-M IPTG to get a final concentration of 1 mM IPTG.
7. Let the cultures grow for 4 h at 20°C and 37°C, respectively.
8. Harvest the cells by centrifugation for 15 min in a suitable beaker at ∼7200 × g and 4°C. Discard the supernatant, but leave about 25 ml to resuspend the pellet. Transfer the suspension to a 50-ml Sarstedt tube.
9. Centrifuge the suspension again at ∼5200 × g, 4°C for 15 min. The pellets can be stored at −20°C for later use.

3.4. Solubility Screen [Modified after (6)]

1. Resuspend and wash the pellet of the 500-ml cultivation with 30 ml of washing buffer (*see* **Notes 18 and 19**).
2. Divide the 30-ml suspension into 30 × 1.5-ml reaction tubes, each containing 1 ml of the resuspended cells.
3. Centrifuge the tubes at ∼16,000 × g, 4°C for 10 min and discard the supernatant.

4. If the protein is produced with a His-tag, resuspend the cells in 1 ml of one of the lysis buffers listed in Section 2.3 (point 3). If the gene of interest is expressed as a GST fusion protein, resuspend the pellet in 1 ml of one of the buffers given in Section 2.3 (point 4).
5. To each tube add 10 µl of lysozyme stock solution (resulting in a final concentration of 0.15 mg/ml lysozyme) and vortex the solution for a few seconds.
6. Lyse the cells on ice by 1-min sonication [Branson Sonifier Cell Disruptor B15 (output control 4, % duty cycle 50)].
7. After sonication, centrifuge the tubes at ~16,000 × g, 4°C for 30 min.
8. Transfer the supernatant into a new 1.5-ml reaction tube.
9. Analyze 10 µl of each of the 30 crude cell extracts by SDS-PAGE.
10. As controls, the complete extracts made by cell lysis under denaturing conditions before and after induction of gene expression (resulting in a final OD_{660} = 0.5) are analyzed in parallel by SDS-PAGE. The influence of the purification tag on the solubility of the protein is demonstrated in **Fig. 4**. While the protein produced with the 6 × His-tag is more or less insoluble, the protein fused to GST shows high solubility. In contrast, **Fig. 5** illustrates that instead of changing the construct, additives can also increase the amount of soluble protein (*see* **Note 20**).

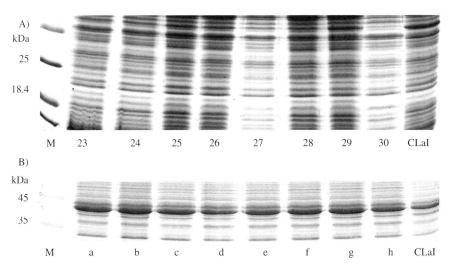

Fig. 4. Solubility screen of HCoV-NL63 Nsp3 X-domain produced with a 6xHis-tag (A, 23 kDa) and as a GST-fusion protein (B, 46 kDa). M = protein molecular-weight marker (Fermentas): (A) 23–30: buffers as listed in Section 2.3 (point 3); (B) a–h: buffers as listed in Section 2.3 (point 4); CLaI: cell lysis under denaturing conditions after induction.

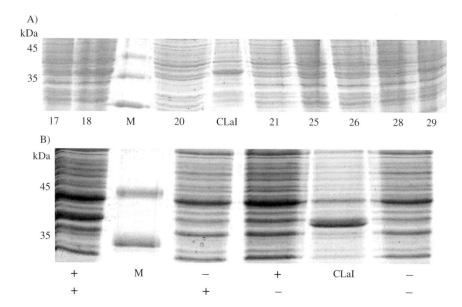

Fig. 5. Solubility screen of HCoV-NL63 Nsp1 as a GST-fusion protein (40 kDa) based on the described method for His-tagged proteins: (A) solubility screen with selected buffers (Section 2.3, point 3); (B): test for the necessity of additives based on buffer 18, upper labels: Triton X-100, lower labels: EDTA (+ in presence, – in absence); M: protein molecular-weight marker (Fermentas); CLaI: cell lysis under denaturing conditions after induction.

3.5. Large-Scale Purification (see Note 21)

3.5.1. Purification of the His-Tagged Protein

1. Resuspend the cells of a 1-liter cultivation in 25 ml binding buffer that contains 10 mg lysozyme and one-half of a tablet of complete protease inhibitor cocktail.
2. Incubate the solution for 30 min on ice.
3. Break the cells using a French press at 25,000 psi (*see* **Note 22**).
4. Perform an ultracentrifugation for 1 h at ~150,000 × *g* and 4°C.
5. Filter the supernatant through a 0.45-μm filter.
6. Apply the supernatant using a peristaltic pump with a flow of 1 ml/min on a HisTrap FF 1-ml column that has been equilibrated with the binding buffer (*see* **Note 23**).
7. Wash the column with binding buffer (1 ml/min) until the OD_{280} reaches a steady baseline.
8. Elute the protein with a flow of 1 ml/min using a step gradient of 20 mM (binding buffer) to 500 mM imidazole (elution buffer) over 50 column volumes (*see* **Note 24**). Collect the flow-through in 2-ml fractions.
9. Analyze the fractions that show an absorbance at 280 nm by SDS-PAGE.

10. Pool the fractions of highest purity and dialyze the solution overnight against an appropriate buffer at 4°C (*see* **Note 25**).
11. Concentrate the protein solution to 5–20 mg/ml using an Amicon ultrafiltration cell (*see* **Note 26**).

3.5.2. Purification of the GST-Fusion Protein

1. Resuspend the cells of a 1-liter cultivation in 25 ml 1 × PBS buffer + additive (Section 2.4.2., point 1) containing 10 mg lysozyme and one-half of a tablet of complete protease inhibitor cocktail.
2. Incubate the solution for 30 min on ice.
3. Break the cells using a French press at 25,000 psi (*see* **Note 22**).
4. Centrifuge for 1 h at ~150,000 × g and 4°C.
5. Filter the supernatant through a 0.45-μm filter.
6. Apply the supernatant using a peristaltic pump with a flow of 0.5 ml/min on a GSTrap FF 5-ml column that has been equilibrated with 1 × PBS buffer (*see* **Note 23**).
7. Wash with 1X PBS buffer (2 ml/min) until the OD$_{280}$ reaches a steady baseline.
8. Wash the GSTrap 5-ml column with ten column volumes (2 ml/min) cleavage buffer.
9. Prepare the PreScission protease mix by mixing 50 μl PreScission protease with 4.95 ml cleavage buffer (*see* **Note 27**).
10. Load the PreScission protease mix onto the column.
11. Seal the column and incubate overnight at 4°C.
12. Elute the protein with 20 ml cleavage buffer (0.5 ml/min). Collect the flow-through in 0.5-ml fractions.
13. Remove the GST protein and the PreScission protease, which is itself fused to GST, by washing the column with 2 ml/min elution buffer. Collect the flow-through in 2-ml fractions.
14. Analyze the fractions containing the protein under study, the GST, and the PreScission protease by SDS-PAGE.
15. Pool the fractions containing the protein of highest purity and concentrate to 5–20 mg/ml using an Amicon ultrafiltration cell (*see* **Note 26**).

3.6. Preliminary Crystallization Experiment

1. Fill the wells of the pregreased Hampton research 48-well plate (VDX48 plate with sealant) with 250 μl of the reservoir solution according to the pipetting scheme shown in **Table 1**, using a 1-M stock solution of the buffer, a 4-M stock solution of $(NH_4)_2SO_4$, and 50% PEG 6000 (*see* **Note 28**).
2. Centrifuge the solution of the purified protein for 10 min at 16,000× g and 4°C.
3. Pipette 1 μl of the protein solution on a circular cover slide (12 mM diameter) and add 1 μl of the reservoir solution of the corresponding well.

Table 1
Pipet Scheme for the Preliminary Crystallization Experiment with Amonium Sulfate and PEG 6000

	$H_2O \searrow$	(NH$_4$)$_2$SO$_4$				PEG 6000			
		0.8 M 50 µl	1.6 M 100 µl	2.4 M 150 µl	3.2 M 200 µl	5% 25 µl	10% 50 µl	20% 100 µl	30% 150 µl
		1	2	3	4	5	6	7	8
pH 4.0 25 µl	A	175 µl	125 µl	75 µl	25 µl	200 µl	175 µl	125 µl	75 µl
pH 5.0 25 µl	B	175 µl	125 µl	75 µl	25 µl	200 µl	175 µl	125 µl	75 µl
pH 6.0 25 µl	C	175 µl	125 µl	75 µl	25 µl	200 µl	175 µl	125 µl	75 µl
pH 7.0 25 µl	D	175 µl	125 µl	75 µl	25 µl	200 µl	175 µl	125 µl	75 µl
pH 8.0 25 µl	E	175 µl	125 µl	75 µl	25 µl	200 µl	175 µl	125 µl	75 µl
pH 9.0 25 µl	F	175 µl	125 µl	75 µl	25 µl	200 µl	175 µl	125 µl	75 µl

4. Turn the cover slide upside down and place it on top of the well.
5. Carefully press on the cover slide to fix it onto the well.
6. Store the plate at 12°C (*see* **Note 29**).
7. Initially, check the plate daily, then weekly, and later monthly (*see* **Note 30**).

4. Notes

1. All buffers should be filtered through a 0.45-µm filter and stored at 4°C to decrease the risk of contamination.
2. The buffers for the His-tagged protein are chosen according to the literature *(6)*. These buffers vary in composition (salts, additives) and pH stabilizing the protein and/or improving solubility.
3. According to the *GST Gene Fusion System Handbook* (Amersham Biosciences), PBS is used as a buffer for purification of the fusion protein. Additives are selected with properties of the protein in mind, i.e., isoelectric point, hydropathy, etc.
4. The buffer that gave the best result in the solubility screen (Section 3.4) is the basic buffer. Thus, the binding buffer contains the basic buffer plus 20 mM imidazole and the elution buffer consists of the basic buffer plus 500 mM imidazole. The basic buffer resulting from the solubility screen should be compatible with the Ni^{2+} sepharose, e.g., it should not contain EDTA, as this would strip the Ni^{2+} from the column.
5. The buffer used here for cell lysis is the $1 \times$ PBS buffer, containing the additive that gave the best result in the solubility screen (Section 3.4).
6. The calculated amounts of the substances for the stock solutions are dissolved in distilled water and filtered through a 0.45-µm filter. For solutions that cannot be filtered because of high viscosity, e.g., the 50% PEG 6000 solution, the powder can be dissolved in prefiltered distilled water. The pH of the buffers is adjusted to the desired pH with NaOH or HCl.
7. In the case of Nsp8 of HCoV 229E, the FoldIndex© program *(7)* predicts an unfolded region at the N terminus of the protein (**Fig. 1**). Actually, this part is involved in the interaction with Nsp7, as revealed by the crystal structure of the Nsp7:Nsp8 complex of SARS-CoV *(8)*.
8. Be sure that the gel gets polymerized; otherwise artifact bands will be observed when the protein is subjected to gel electrophoresis.
9. Depending on the protein, heating and mercaptoethanol reduction may influence the electrophoretic mobility. If the observed molecular mass differs from the one expected from the polypeptide sequence, other temperature conditions should be tried. SDS-PAGE under nonreducing, compared to reducing, conditions gives a hint concerning disulfide bonds present in the protein, as shown in **Fig. 2**.
10. During the cell lysis under denaturing conditions, the cells are totally destroyed and the proteins are unfolded. Thus, this sample represents the positive control on an SDS-PAGE gel, displaying the total amount of proteins in the cell.

11. The overnight culture is done to have an optimal growth of the cells. After the overnight incubation, the cells are adapted to the medium and strong enough to be transferred into a larger volume of medium. If the colony from the agar plate is directly transferred to a larger volume, it takes a longer time until the cells reach the desired cell density and by then the risk of contamination increases.

12. The decision as to whether to pick a single colony or a mixture of colonies depends on the stability of the clones. Using a mixture of colonies can increase the chance of finding an expression clone. In the same way, it can increase the risk of a failed expression, as the cells can contain damaged plasmids. Thus, several single colonies have to be analyzed individually. Nevertheless, checking out ten single colonies can result in no expression, whereas one or two mixtures of colonies may include good expression clones.

13. It is not absolutely necessary to centrifuge the overnight culture in order to remove degraded ampicillin and expressed β-lactamase, as the volume of the overnight culture is really small.

14. To go for a higher expression rate, additional parameters can be tested, e.g., different *E. coli* strains [BL21(DE3), Tuner pLacI (DE3), C41 (DE3)] and media (Luria Bertani, rich medium), as well as different incubation times.

15. From comparison with the intensity of the host proteins, the amount of the target protein produced can be estimated. Weak bands of the heterologously produced protein in a high background of *E. coli* proteins are difficult to detect. In any case, correct expression has to be confirmed by Western blot methods (works well for crude lysates, if the protein of interest shows up with a high yield), N-terminal sequencing, and mass spectrometry analysis (for purified recombinant protein).

16. Unlike cultivation for the expression analysis (Section 3.3.2), it is recommended to centrifuge the overnight culture to remove the degraded ampicillin and the expressed β-lactamase to avoid contamination during cultivation.

17. The expression parameters, such as induction time (OD_{660}), amount of inducer, and duration of the expression, are adjusted to the results from the initial cultivation (see Section 3.3.2). Our example is from cultures I and II, shown in **Fig. 3**.

18. Even if a better expression condition is found at 20°C, a cultivation should also be performed at 37°C, in order to test whether the protein produced in lower yields is more soluble than the one obtained in higher amounts.

19. Starting from the first step, everything should be done on ice and at 4°C to avoid protein degradation. In addition, the use of protease inhibitors, for instance, complete protease inhibitor cocktail (Roche), may be advisable.

20. Sometimes much effort can be invested in finding a buffer for cell lysis to obtain soluble protein. Depending on the function of the viral protein in replication or transcription, it may need another nonstructural protein to form a complex for proper folding. **Figure 6** demonstrates the independent purifications of Nsp7 and Nsp8 from HCoV 229E. Whereas Nsp7 is soluble, Nsp8 is not. However, by combining the pellets and purifying the two together, the solubility of Nsp8 increases dramatically as shown in **Fig. 7**. Recently, the crystal structure of a

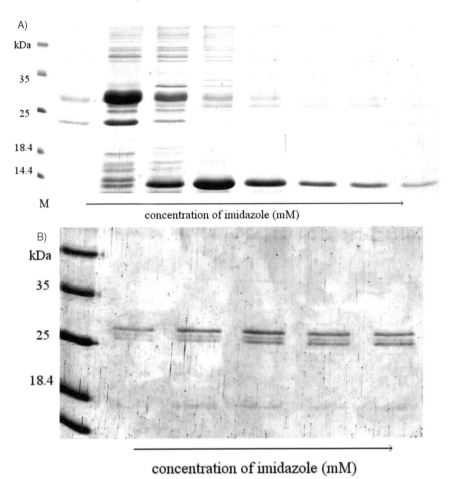

Fig. 6. Purification of His-tagged HCoV-229E Nsp7 (A) and Nsp8 (B). Elution of the proteins bound to a Ni²⁺-NTA affinity column with increasing concentrations of imidazole. Nsp7 (11 kDa) can be eluted by small amounts of imidazole, whereas Nsp8 (22 kDa) cannot.

hexadecameric (8:8) Nsp7:Nsp8 complex from the SARS coronavirus has been determined *(8)*.

21. Once the expression and cell-lysis conditions have been optimized to yield large amounts of soluble protein, biochemical and structural studies can be started. The protein is purified from a 1-liter cultivation (according to Section 3.3.3., with the necessary adjustments for the larger cultivation volume) using affinity chromatography: (i) *Affinity Chromatography: Principle and Methods Handbook*; (ii) *GST Gene Fusion System Handbook*: Amersham Biosciences). This

Piotrowski et al.

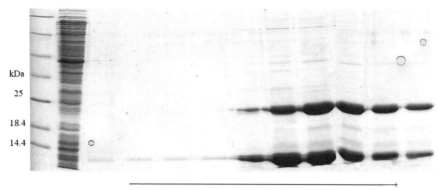

concentration of imidazole (mM)

Fig. 7. Purification of HCoV-229E Nsp7 and Nsp8 in combination. The His-tagged proteins are eluted from the Ni^{2+}-NTA affinity column with increasing concentrations of imidazole. Both Nsp7 (11 kDa) and Nsp8 (22 kDa) can be eluted in larger amounts than in the single experiments (compare **Fig. 6**).

can be followed by ion-exchange or size-exclusion chromatography to remove more impurities.

22. French press is just one possibility for breaking the cells. Sonication or several thaw/freeze cycles can also be done for cell disruption.
23. Instead of a peristaltic pump, an automatic device can be used as well. For instance, the Äkta Prime (Amersham Biosciences) is easy to use for simple purification of proteins on a laboratory scale.
24. If the purification can be done with automated equipment instead of a step gradient, the protein can be eluted using a linear gradient. This gradient may separate proteins with similar binding strengths.
25. It depends on the purity of the eluted protein whether the dialysis can be done directly afterward or if a second purification step (ion-exchange or size-exclusion chromatography) is advisable to remove more impurities. Experience shows that sometimes only the purest fraction will give high-quality crystals; therefore, inclusion of too much of the flanks of the protein peak should be avoided.
26. For concentrating the protein solution, methods other than ultrafiltration can be used, depending on what is established in the laboratory.
27. In the protocol given here, the GST fusion protein contains a cleavage site for PreScission protease. Other fusion constructs may require different proteases (e.g., factor Xa) for processing. In the example described here, the protein of interest is cleaved off from the GST fusion by the PreScission protease directly on the column. It is also possible to do the cleavage in a reaction tube after elution of the GST fusion protein with the elution buffer. The fractions containing the GST fusion protein are pooled and dialyzed against the cleavage buffer before starting the cleavage. After overnight incubation, the solution is applied onto the GSTrap 5-ml column and washed with cleavage buffer. The flow-through

contains the protein. Furthermore, cleavage of the fusion protein is not always required. A number of target-fusion crystal structures have been reported. In some cases, fusion proteins can be useful tools for crystallization *(9,10)*.

28. In addition to the mentioned preliminary crystallization experiment with ammonium sulfate and PEG 6000 other precipitants can be tested as well, for instance, sodium chloride and 2-methyl-2,4-pentanediol (MPD) *(11)*. There are commercial crystallization kits available. Sigma-Aldrich offers crystallization kits (basic and extension) for proteins that are based on the original screening protocol of Jancarik and Kim *(12)*, which utilizes crystallization conditions that have worked before for various proteins and allows testing of a large range of buffers, pH, additives, and precipitants. The protocol can be used for setting up the crystallization experiment by hand, or alternatively, by pipetting robots that allow screening of a larger number of possible crystallization conditions (high-throughput experiment) with less volume of the protein solution and less effort.

29. Temperature also has an influence on the crystallization of the protein. If there is enough protein solution available, other temperatures such as 4°C, 25°C, or 37°C should be tested.

30. **Figure 8** shows crystals of the HCoV-NL63 Nsp3 X-domain obtained from the preliminary crystallization experiment (Section 3.6). A solution containing 1 μl protein (7 mg/ml) and 1 μl reservoir solution was equilibrated against 250 μl reservoir solution of well B7 (20% PEG 6000, 0.1 M citric acid pH 5.0). The crystal diffracted to 3.5 Å resolution. Within this screen, crystals could be obtained if PEG 6000 (10–20%) was used as precipitant in a range of pH 7.0 to 9.0, whereas $(NH_4)_2SO_4$ had no effect on the crystallization of the protein. The reservoir solution has been optimized and crystals have been grown that

Fig. 8. Crystals of HCoV-NL63 Nsp3 X-domain. The reservoir solution contains 20% PEG 6000, 0.1 M citric acid pH 5.0.

A)

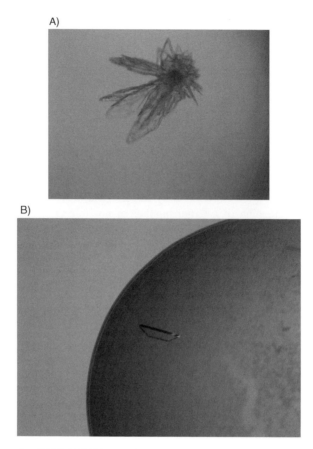

B)

Fig. 9. Crystals of HCoV-229E Nsp9: (A) First crystallization trials yield intergrown crystals (B) Addition of 5% MPD to the reservoir buffer results in single crystals.

diffract to 1.8 Å resolution. Optimization of the crystallization conditions can also include additives such as 2-methyl-2,4-pentanediol (MPD). **Figure 9** shows intergrown crystals of Nsp9 of HCoV 229E and a single crystal grown in the presence of 5% MPD. Keep in mind, that each component of the reservoir solution can either promote or hinder the crystallization experiment. So, the choice of the buffer, salts, or organic additives should be very carefully considered and investigated *(13)*.

Acknowledgments

The work described here is being supported, in part, by the VIZIER ("Comparative Structural Genomics of Viral Enzymes Involved in Replication") project of the European Commission (contract no. LSHG-CT-2004-511960; http://www.vizier-europe.org), and by SEPSDA (contract no. SP22-CT-2004-003831, www.sepsda.eu). Support by the Sino-German Center on the Promotion of

Research, Beijing, is gratefully acknowledged. R. H. thanks the Fonds der Chemischen Industrie for continuous support. We thank L. van der Hoek and J. Ziebuhr for the cDNA.

References

1. Groneberg, D. A., Hilgenfeld, R., and Zabel, P. (2005) Molecular mechanisms of severe acute respiratory syndrome (SARS). *Respir. Res.* **6**, 8–23.
2. van der Hoek, L., Pyrc, K., and Berkhout, B. (2006) Human coronavirus NL63, a new respiratory virus. *FEMS Microbiol. Rev.* **30**, 760–773.
3. van der Hoek, L., Sure, K., Ihorst, G., Stang, A., Pyrc, K., Jebbink, M. F., Petersen, G., Forster, J., Berkhout, B., and Überla, K. (2005) Croup is associated with the novel coronavirus NL63. *PLoS Med.* **2**, 764–770.
4. Dominguez, S. R., Anderson, M. S., Glodé, M. P., Robinson, C. C., and Holmes, K. V. (2006) Blinded case-control study of the relationship between human coronavirus NL63 and Kawasaki syndrome. *J. Infect. Dis.* **194**, 1697–1701.
5. Stadler, K., Masignani, V., Eickmann, M., Becker, S., Abrignani, S., Klenk, H. D., and Rappuoli, R. (2003) SARS—Beginning to understand a new virus. *Nature Rev. Microbiol.* **1**, 209–218.
6. Lindwall, G., Chau, M.-F., Gardner, S. R., and Kohlstaedt, L. A. (2000) A sparse matrix approach to the solubilization of overexpressed proteins. *Protein Eng.* **13**, 67–71.
7. Prilusky, J., Felder, C. E., Zeev-Ben-Mordehai, T., Rydberg, E. H., Man, O., Beckmann, J. S., Silmann, I., and Sussman, J. L. (2005) FoldIndex©: a simple tool to predict whether a given protein sequence is intrinsically unfolded. *Bioinformatics* **21**, 3435–3438.
8. Zhai, Y., Sun, F., Li, X., Pang, H., Xu, X., Bartlam, M., and Rao, Z. (2005) Insights into SARS-CoV transcription and replication from the structure of the nsp7-nsp8 hexadecamer, *Nature Struct. Mol. Biol.* **12**, 980–986.
9. Smyth, D. R., Mrozkiewicz, M. K., McGrath, W. J., Listwan, P., and Kobe, B. (2003) Crystal structures of fusion proteins with large-affinity tags. *Protein Sci.* **12**, 1313–1322.
10. Hogg, T., and Hilgenfeld, R. (2007) Protein crystallography in drug design. In: Tayler, J. B., and Triggle, D. J. (eds.) *Comprehensive Medicinal Chemistry II.* Elsevier, Amsterdam, pp. 875–900.
11. McPherson, A. (1999) Strategies and special approaches in growing crystals. In *Crystallization of Biological Macromolecules.* Cold Spring Harbor Laboratory Press, New York, pp. 271–329.
12. Jancarik, J., and Kim, S.-H. (1991) Sparse matrix sampling: a screening method for crystallization of proteins. *J. Appl. Cryst.* **24**, 409–411.
13. Mesters, J. R. (2007) Practical protein crystallization. In: Drenth, J. (ed.) *Principles of Protein X-Ray Crystallography* 3rd Ed.. Springer, New York, pp. 297–304.

V

RAISING ANTIBODIES TO CORONAVIRUS PROTEINS

14

Generating Antibodies to the Gene 3 Proteins of Infectious Bronchitis Virus

Amanda R. Pendleton and Carolyn E. Machamer

Abstract

Infectious bronchitis virus (IBV), a group 3 coronavirus, produces three proteins (IBV E, IBV 3a, and IBV 3b) from subgenomic mRNA 3 during infection. IBV E, a viral envelope protein, plays a role in virus budding, possibly by altering membrane morphology at the virus assembly site. In addition to this role, IBV E may also function as a viroporin, although no data from infected cells have confirmed this possibility definitively. Conversely, the IBV 3a and IBV 3b proteins are nonstructural proteins. These proteins are dispensable for replication in cell culture, but are thought to be important for infection of the natural host. This chapter details methods for generating and screening antibodies to these gene 3 proteins. Antibodies were raised in rabbits following inoculation with IBV-specific peptides and GST fusion proteins, and were screened by immunofluorescence, radioimmunoprecipitation, and immunoblotting.

Key words: infectious bronchitis virus, 3a, 3b, E; antibodies; coronavirus; accessory proteins.

1. Introduction

Coronaviruses have large, positive-sense, single-stranded RNA genomes (*1*). During infection, four structural proteins are expressed: spike (S), membrane (M), nucleocapsid (N), and envelope (E). The viral envelope E protein is thought to alter membrane morphology at the virus assembly site, thus promoting virus budding (*2–10*). Additionally, more infectious bronchitis virus (IBV) E protein is produced in infected cells than is incorporated into virions, suggesting that IBV E may have a function apart from its role in virus budding (*3*). Consistent

From: *Methods in Molecular Biology, vol. 454: SARS- and Other Coronaviruses,*
Edited by: D. Cavanagh, DOI: 10.1007/978-1-59745-181-9_14, © Humana Press, New York, NY

with this idea, studies have indicated that the E protein from several coronaviruses may act as a viroporin *(11–14)*. However, no conclusive experiments in infected cells have confirmed this possibility. In addition to the coronavirus structural proteins, several other open reading frames (ORFs) located between the structural protein ORFs produce proteins of undetermined function during infection *(15)*. These proteins are thought to be nonstructural, accessory proteins that are essential for productive virus infection in the natural host, but dispensable for virus growth in cell culture *(16–23)*. Therefore, study of these accessory proteins, in addition to study of the coronavirus structural proteins, may more fully elucidate pathogenesis mechanisms and lead to the development of more effective antiviral drugs.

IBV, a group 3 coronavirus that infects chickens, produces two nonstructural proteins, 3a and 3b, from ORFs located between the S and E genes *(24)*. Both of these proteins are translated from subgenomic mRNA 3, with IBV 3a translation initiated via the first AUG, and the downstream IBV 3b translation initiated via leaky ribosomal scanning *(25)*. Subgenomic mRNA 3 also produces the IBV E protein through an internal ribosomal entry site. We have developed antibodies specific to each of these IBV gene 3 proteins for use in characterization and functional studies. IBV E was found to localize to Golgi membranes via its C-terminal tail *(3,26)*. IBV E also promoted the release of virus-like particles through the interaction of its C-terminal tail with the IBV M protein *(27)*. Studies on the IBV 3a protein demonstrated that one pool of IBV 3a localized cytoplasmically, while another pool associated with membranes at the smooth endoplasmic reticulum (ER) *(28)*. The short length of IBV 3a appeared to preclude efficient association with signal recognition particle (SRP), causing inefficient co-translational insertion of IBV 3a into membranes. Thus, IBV may limit 3a levels at smooth ER membranes by a novel mechanism.

Conversely, IBV 3b localized to the nucleus in mammalian cells when using a vaccinia virus expression system *(18,29)*. However, IBV 3b was undetectable via microscopy of IBV-infected or transiently transfected mammalian cells *(29)*. Surprisingly, IBV 3b, which was readily detected in chicken cells, localized to the cytoplasm with apparent nuclear exclusion. The half-life of IBV 3b was greatly reduced in mammalian cells as compared to chicken cells. This rapid turnover in mammalian cells was proteasome-dependent, whereas turnover in avian cells was proteasome-independent. Thus, the importance of using cells derived from the natural host when studying coronavirus nonstructural proteins is highlighted. This chapter details methods for generating polyclonal antibodies to the IBV gene 3 proteins, including protocols for generating peptides or fusion proteins for immunizing rabbits, prescreening of nonimmunized rabbit sera, general rabbit immunization procedures, and screening sera from immunized rabbits.

2. Materials

2.1. Preparing Peptides for Injection into Rabbits

1. Protein sequence

2.2. Preparing Fusion Proteins for Injection into Rabbits

1. Plasmid DNA: pGEX-2T (Amersham) (encodes the GST gene) and pGEX-2T with the gene of interest cloned in frame with the GST gene (*see* **Note 1**).
2. Competent BL21 *Escherichia coli* bacteria from the GST gene fusion system (Amersham). Store in 250-μl aliquots at –80°C. Thaw only once after aliquoting.
3. YT-amp broth: 0.8% (w/v) BactoTryptone (BD Biosciences, San Diego, CA), 1% (w/v) Bacto yeast extract (BD Biosciences), and 0.5% (w/v) NaCl in dH$_2$O. Autoclave and add 80 μg/ml ampicillin when the solution has cooled to below 65°C. Store at 4°C (*see* **Note 2**).
4. LB-amp plates: 1% (w/v) BactorTryptone, 1% (w/v) Bacto yeast extract, 0.5% (w/v) NaCl, 0.1 M NaOH, and 1.3% BactoAgar (BD Biosciences) in dH$_2$O. Autoclave and add 80 μg/ml ampicillin when the solution has cooled to below 65°C. Pour into Petri dishes and store at 4°C when the agar has solidified.
5. Inducing solution (1000X): 100 mM isopropyl-β-D-thiogalactopyranoside (IPTG) (Amersham) in dH$_2$O. Make fresh before each use.
6. Phosphate buffered saline (PBS). Store at room temperature.
7. PBSP: Protease inhibitors (Sigma, catalog number P8340) are added to PBS right before use.
8. Glutathione sepharose beads: Glutathione sepharose 4B (Amersham). Store at 4°C.
9. Elution solution: 10 mM reduced glutathione (Sigma), 50 mM Tris-HCl (pH 8) in dH$_2$O. Make fresh when needed.
10. Loading sample buffer (LSB) (4X): 200 mM Tris-HCl (pH 6.8), 8% (w/v) SDS, 60% (v/v) glycerol, and 0.02% (w/v) bromophenol blue in dH$_2$O. Store at 4°C. Since the buffer solidifies at 4°C, it has to be heated briefly at 65°C before use.
11. 2-mercaptoethanol. Store at room temperature.
12. Destain solution: 40% (v/v) methanol and 10% (v/v) acetic acid in dH$_2$O. Store at room temperature.
13. Coomassie solution: 0.25% (w/v) Coomassie brilliant blue R (Sigma) in destain solution. Store at room temperature.

2.3. Prescreening Rabbit Sera before Injection of Peptides or Proteins

1. Sera from four nonimmunized rabbits.
2. Phosphate buffered saline (PBS). Store at room temperature.
3. All materials listed in Sections 2.13 and 2.14 if prescreening sera using SDS-PAGE and immunoblotting.
4. All materials listed in Section 2.11 if prescreening sera using indirect immunofluorescence microscopy.

2.4. Rabbit Immunization

1. A minimum of 1.75 mg of peptide or a minimum of 875 μg of GST fusion protein with a concentration of no less than 0.25 mg/ml.

2.5. Screening Sera from Immunized Rabbits

2.5.1. Screening Sera from Immunized Rabbits Using in vitro Transcribed and Translated Protein

1. All of the materials listed in Section 2.10.
2. All of the materials listed in Section 2.12, except for the [^{35}S]-methionine-cysteine and the methionine and cysteine-free DMEM.

2.5.2. Screening Sera from Immunized Rabbits Using Other Methods

1. All of the materials listed in Sections 2.7 (IBV infection), Section 2.8 (vTF7-3 infection/transfection), or Section 2.9 (transient transfection), depending on the preferred methods of screening.
2. All of the materials listed in Sections 2.13 and 2.14 if screening sera using immunoblotting.
3. All of the materials listed in Section 2.12 if screening by immunoprecipitation.
4. All of the materials listed in Section 2.11 if screening by indirect immunofluorescence microscopy.

2.6. Affinity Purification of Sera

1. Kit equilibration buffer #1 from the Reduce-Imm Immobilized Reductant Kit (Pierce).
2. Kit equilibration buffer #2 from the Reduce-Imm Immobilized Reductant Kit (Pierce).
3. DTT solution: 10 mM dithiothreitol in kit equilibration buffer #1. Make this solution fresh before using.
4. Diluted peptide: Dissolve 5–10 mg of peptide in kit equilibration buffer #1. Make this solution fresh before using.
5. Diluted protein: Dissolve 5–6 mg of GST fusion protein in kit equilibration buffer #2. Make this solution fresh before using.
6. Ellman's Reagent (Pierce)
7. Reduce-Imm kit reductant column (Pierce)
8. SulfoLink coupling column (Pierce)
9. SulfoLink equilibration buffer (Pierce): 0.1 M sodium phosphate and 5 mM EDTA [pH 6.0] in dH$_2$O.
10. SulfoLink coupling buffer (Pierce): 5 mM EDTA and 50 mM Tris-HCl [pH 8.5] in dH$_2$O.

11. Cysteine buffer: 15.8 mg of L-cysteine • HCl in 2 ml of SulfoLink coupling buffer.
12. Wash buffer (Pierce): 1.0 M NaCl and 0.5% (w/v) NaN$_3$ in dH$_2$O.
13. 0.05% NaN$_3$:0.05% (w/v) NaN$_3$ in dH$_2$O. Store at 4°C.
14. 0.05% degassed NaN$_3$: aspirate 0.05% NaN$_3$ for 5 min.
15. Storage buffer: 10 mM EDTA, 0.05% NaN$_3$, 50% (v/v) glycerol in dH$_2$O.
16. Rabbit sera.
17. WB 1: 0.2% deoxycholic acid and 10 mM Tris-HCl [pH 7.4] in dH$_2$O. Store at 4°C.
18. WB 2: 0.5 M NaCl and 10 mM Tris-HCl [pH 7.4] in dH$_2$O. Store at 4°C.
19. Elution solution: 4 M MgCl$_2$ in dH$_2$O. Store at 4°C.
20. BSA: 30% bovine serum albumin (Sigma). Store at 4°C.
21. Dialysis bags (Pierce–10 K MWCO): Make sure that bags are knotted and clipped before beginning the elution in step 5 of Section 3.6.
22. Phosphate buffered saline (PBS) (pH 7.4). Store at room temperature.

2.7. IBV Infection

1. The Vero-adapted Beaudette strain of IBV is used to infect Vero cells *(30)*. The egg-adapted strain of IBV (American Type Tissue Culture VR-22) is used to infect DF1 cells. Work with IBV should be done using Biosafety Level 2 (BL-2) precautions, including the use of biological safety cabinets, autoclaving of all disposable items that have been in contact with virus, and inactivation of all virus-containing solutions with bleach. Store virus in 1-ml aliquots at –80°C. Do not subject to freeze/thaw more than twice.
2. Serum-free DMEM: High-glucose Dulbecco's modified Eagle's medium (Invitrogen Life Technologies, Carlsbad, CA). Store at 4°C.
3. Cells: Vero (African green monkey kidney epithelial cells) and DF1 (UMN-SAH/DF1, immortalized chicken embryo fibroblast cells).
4. Normal growth medium: 10% (v/v) fetal calf serum (FBS) (Atlanta Biologicals, Norcross, GA) and 0.1 mg/ml Normocin (Invivogen, San Diego, CA) in serum-free DMEM. Store at 4°C.

2.8. vTF7-3 Infection/Transfection

1. vTF7-3: vaccinia virus encoding T7 RNA polymerase *(31)*. Store in 200-μl aliquots at –80°C. Do not subject to freeze/thaw more than five times. Work with vaccinia virus should be done using BL-2 precautions. These precautions include the use of biological safety cabinets, autoclaving of all disposable items that have been in contact with virus, and inactivation of all virus-containing solutions with bleach. Additionally, a virus-only laminar flow hood can be used to prevent the accidental contamination of cell lines with virus for routine tissue culture work. Finally, institutional or governmental regulations on researcher vaccination and

virus containment should be consulted before beginning any projects involving vaccinia virus.

2. Plasmid DNA: Gene of interest must be cloned behind the T7 RNA polymerase promoter.
3. Serum-free DMEM: High glucose Dulbecco's modified Eagle's medium (Invitrogen). Store at 4°C.
4. Cells: Vero (African green monkey kidney epithelial cells) and DF1 (UMNSAH/DF1, immortalized chicken embryo fibroblast cells).
5. Opti-MEM® I (Invitrogen). Store at 4°C.
6. Fugene 6 (Roche Molecular Biochemicals, Indianapolis, IN). Store at 4°C.
7. Normal growth medium: 10% (v/v) fetal calf serum (FBS) (Atlanta Biologicals) and 0.1 mg/ml Normocin (Invivogen) in serum-free DMEM. Store at 4°C.

2.9. Transient Transfection of Cells

1. Cells: Vero (African green monkey kidney epithelial cells) or DF1 (UMNSAH/DF1, immortalized chicken embryo fibroblast cells) (*see* **Note 3**).
2. Plasmid DNA
3. Fugene 6 (Roche): Store at 4°C.
4. Opti-MEM® I (Invitrogen): Store at 4°C.

2.10. In vitro *Transcription and Translation*

1. TNT master mix from the quick-coupled transcription/translation T7 system (Promega Corporation, Madison, WI): Store in 40-μl aliquots at –80°C. Only use the aliquots once after thawing them.
2. Plasmid DNA: Gene of interest must be cloned behind the T7 RNA polymerase promoter.
3. [^{35}S]-Redivue-methionine (Amersham): Store at 4°C.

2.11. Indirect Immunofluorescence Microscopy

1. Phosphate buffered saline (PBS) [pH 7.4] (10X): 0.1 M NaH_2PO_4, 0.1 M Na_2HPO_4, and 1.5 M NaCl in dH_2O: Store at room temperature.
2. 3% paraformaldehyde solution: Add 3 g of paraformaldehyde to 80 ml of dH_2O and 50 μl of 5 M NaOH. Heat the solution to 50°C until the paraformaldehyde is dissolved. Add 10 ml of 10X PBS and bring the total volume up to 100 ml with dH_2O. Check with pH paper (*see* **Note 4**) that the pH is 7.4. Store in 15-ml aliquots at –20°C.
3. PBS/Gly: 10 mM glycine and 0.02% (w/v) NaN_3 in 1X PBS. Store at room temperature.
4. Permeabilization solution: 0.5% (v/v) TritonX-100 in PBS/Gly. Make fresh before each use.

5. Primary and secondary antibody dilution buffers: 1% (v/v) bovine serum albumin (BSA) (Sigma-Aldrich Co., St. Louis, MO) in PBS/Gly. Make fresh before each use.
6. Mounting solution: 0.1 M N-propyl gallate in glycerol. Store at room temperature protected from light (*see* **Note 5**).

2.12. *Immunoprecipitation*

1. [^{35}S]-methionine-cysteine: pro-mix (Amersham). Store in 35-μl aliquots at –80°C.
2. Methionine and cysteine-free DMEM (Invitrogen)
3. Phosphate buffered saline (PBS). Store at room temperature.
4. Detergent solution: 62.5 mM EDTA, 50 mM Tris-HCl [pH 8], 0.4 % deoxycholic acid (CalBiochem, La Jolla, CA), and 1% nonidet P40 substitute (Fluka Chemie AG, Buchs, Switzerland) in dH$_2$O. Store at 4°C. Add protease inhibitors (Sigma, cat. no. P8340) immediately before each use.
5. 10% SDS: 10% (w/v) sodium dodecylsulfate (SDS) in dH$_2$O. Store at room temperature.
6. 2-mercaptoethanol (BioRad). Store at room temperature.
7. WP buffer 1: 2% (w/v) SDS and 1% (v/v) 2-mercaptoethanol in dH$_2$O. Make fresh before use.
8. WP buffer 2: 20 mM Tris-HCl [pH 8], 1 M NaCl, 1% (v/v) TritonX-100, and 0.02% (w/v) NaN$_3$. Make fresh before use.
9. Washed pansorbin (to reduce background binding): 25 ml of standardized pansorbin cells (Calbiochem) are mixed with 5 ml of 10% SDS and 270 μl of 2-mercaptoethanol. Cells are then taken through a heating procedure [heating to 100°C for 15 min, centrifugation at 7840 × *g* for 7 min at 4°C, and resuspension in 30 ml of WP buffer 1]. This heating procedure is then repeated once. Cells are washed five times in 30 ml of a WP buffer 2 before being resuspended in 25 ml of WP buffer 2 and stored in 1-ml aliquots at –20°C (*see* **Note 6**).
10. Primary antibodies: Undiluted rabbit serum
11. RIPA buffer: 10 mM Tris-HCl [pH 7.4], 0.1% (w/v) SDS, 1% (w/v) deoxycholine, 1% (v/v) nonidet P 40 substitute, and 150 mM NaCl in dH$_2$O. Store at 4°C.
12. Loading sample buffer (LSB) (4X): 200 mM Tris-HCl [pH 6.8], 8% (w/v) SDS, 60% (v/v) glycerol, and 0.02% (w/v) bromophenol blue in dH$_2$O. Store at 4°C. Since the buffer solidifies at 4°C, it has to be heated briefly at 65°C before use.

2.13. *SDS Polyacrylamide Gel Electrophoresis (SDS-PAGE)*

1. Ethanol (the Warner-Graham Company, Cockeysville, MD). Store at room temperature in a flammables cabinet.

2. Acrylamide: 40% acrylamide (BioRad Laboratories, Hercules, CA), 2% bis-acrylamide (BioRad), 30% polyacrylamide:bis (37.5:1 with 2.6% C, BioRad). Store at 4°C. Note that acrylamide is a neurotoxin before it polymerizes, so gloves should be worn when handling these chemicals.
3. TEMED (BioRad): Store at room temperature.
4. Separating gel buffer (4X): 1.5 M Tris-HCl [pH 8.8] in dH$_2$O. Store at room temperature.
5. 10% SDS: 10% (w/v) sodium dodecylsulfate (SDS) in dH$_2$O. Store at room temperature.
6. 30% APS: 30% (w/v) ammonium persulfate (APS) in dH$_2$O. Store at 4°C. Make no more than 2 ml of this solution at a time, as it will generally last for only 3 months when stored at this temperature.
7. Stacking gel buffer (4X): 0.5 M Tris-HCl [pH 6.8] in dH$_2$O. Store at room temperature.
8. Layering buffer: 0.2% (w/v) SDS in dH$_2$O. Store at room temperature.
9. Running buffer (5X): 125 mM Tris, 1 M glycine, and 0.5% (w/v) SDS in dH$_2$O. Store at room temperature.
10. Loading sample buffer (LSB) (4X): 200 mM Tris-HCl [pH 6.8], 8% (w/v) SDS, 60% (v/v) glycerol, and 0.02% (w/v) bromophenol blue in dH$_2$O. Store at 4°C. Since the buffer solidifies at 4°C, it has to be heated briefly at 65°C before use.
11. 2-mercaptoethanol (BioRad). Store at room temperature.
12. Molecular weight markers: Precision Plus Protein™ All Blue Standards for immunoblotting (BioRad) or [^{14}C] molecular weight markers (Amersham Pharmacia Biotech, Inc., Piscataway, NJ) for radioactive gels. Store at –20°C.

2.14. Immunoblotting

1. Cooled transfer buffer: 25 mM Tris, 0.2 M glycine, 15% (v/v) MeOH in dH$_2$O. Store at 4°C. Buffer can be used six times before being discarded.
2. 3 MM Chr chromatography paper (Whatman Inc., Florham Park, NJ).
3. Polyvinylidene fluoride (PVDF) membrane (Millipore Corporation, Billerica, MA) (*see* **Note 7**).
4. Methanol: Store at room temperature in a flammables cabinet.
5. TTBS: 150 mM NaCl, 0.05% (v/v) Tween-20, and 10 mM Tris-HCl [pH 7.4] in dH$_2$O. Store at 4°C.
6. Blocking buffer: 5% (w/v) nonfat dry milk in TTBS. Make fresh before each use.
7. Primary antibody dilution buffer: 1% (w/v) nonfat dry milk in TTBS. Make fresh before each use.
8. Secondary antibodies: 1:10,000 dilution of HRP-conjugated donkey anti-rabbit IgG antibodies (Amersham) in blocking buffer. Make fresh before each use.
9. HRP substrate: ECL™ Western Blotting Detection Reagents (Amersham). Store at 4°C.

3. Methods

3.1. Preparing Peptides for Injection into Rabbits

1. Protein sequence must be analyzed to determine likely antigenic epitopes. First, hydropathy plots should be created to identify putative transmembrane segments in the protein. Peptides corresponding to transmembrane regions should not be used for immunizing rabbits since they may not allow the antibody access to the protein in some assays. One helpful website for creating hydropathy plots is found at http://ca.expasy.org/tools/protscale.html. Proteins should also be analyzed for the presence of domains that are highly conserved in many different proteins, such as coiled-coil, protein targeting, or DNA-binding motifs. Peptides corresponding to these types of motifs should not be used for immunizing rabbits as they may generate antibodies that recognize several different proteins. A list of helpful websites for predicting these types of motifs is found at http://ca.expasy.org/tools/. We have had success using peptides (around 14 amino acids in length) that correspond to the unique C- or N-terminal regions of the IBV gene 3 proteins. However, the antigenicity of these different peptides has varied, as shown in **Fig. 1A**.

2. Peptides are synthesized using an automated synthesizer and purified to ≥90% purity (*see* **Note 8**). If the peptide region does not contain a cysteine, then a cysteine must be added during the synthesis process so that it can be coupled to keyhole limpet hemocyanin (*see* **Note 9**). A minimum of 1.75 mg of peptide is required for injection into rabbits in Section 3.4. We order commercially synthesized peptides from Boston Biomolecules, Inc. (Boston, MA), precoupled to keyhole limpet hemocyanin through a cysteine added to one end of the peptide. However, coupling can also be done using commercially available kits from companies such as Pierce Biotechnology, Inc. (*see* **Note 10**).

3.2. Preparing Fusion Proteins for Injection into Rabbits

1. GST fusion proteins are expressed and purified using the GST gene fusion system (Amersham). A minimum of 875 μg of protein with a concentration of no less than 0.25 mg/ml is required for injection into rabbits in Section 3.4 (*see* **Note 10**).

2. 50 μl of competent BL21 *E. coli* bacteria is thawed on ice before being incubated for 20 min on ice with 10 ng of plasmids encoding either GST (pGEX-2T) or GST fused to the protein of interest. Bacteria are then heated at 42°C for 1 min, placed on ice for 2 min, and plated on YT-amp plates overnight at 37°C.

3. One individual bacterial colony carrying the pGEX-2T plasmid is picked and grown overnight in 2 ml of YT-amp broth. Twelve individual bacterial colonies carrying the plasmid encoding the GST fusion protein are picked and grown overnight in 2 ml of YT-amp broth (*see* **Note 11**).

4. The next day each individual culture is diluted 1:10 in 2 ml of YT-amp broth and grown to an OD_{600} of approximately 0.6.

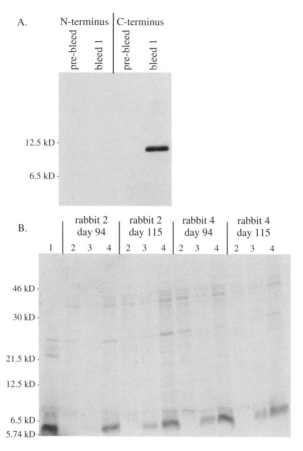

Fig. 1. Screening sera from immunized rabbits: (A) IBV-infected Vero cells (Section 3.7) were lysed at 15 h postinfection according to the protocol in Section 3.3, step 2. Sera from rabbits immunized with peptides corresponding to either the N-terminus or the C-terminus of IBV E were then screened using the immunoblotting protocol in Section 3.5.2. Results from a 1:1000 dilution of sera harvested on day 52 (bleed 1) are shown. The C-terminal peptide of IBV E was much more immunogenic than the N-terminal peptide. Although rabbits immunized with the N-terminal IBV E peptide eventually produced antibodies specific to IBV E, the titers were always lower than those of rabbits immunized with the C-terminal IBV E peptide. (B) Sera from rabbits immunized with GST-IBV 3b were screened using the protocol in Section 3.5.1. The responses of two different rabbits at two different times are shown. Lane 1: 10% input of *in vitro* transcribed and translated IBV 3b. Lane 2: immunoprecipitation of *in vitro* transcribed and translated IBV 3b using preimmune rabbit sera. Lane 3: immunoprecipitation of *in vitro* transcribed and translated IBV 3b using 1 µl of immunized rabbit sera. Lane 4: immunoprecipitation of *in vitro* transcribed and translated IBV 3b using 3 µl of immunized rabbit sera.

5. GST protein expression is induced by adding 2 µl of inducing solution to cultures and growing them for 48 h at 13.5°C in a shaking water bath (*see* **Notes 12** and **13**).

6. Bacteria are pelleted by centrifugation at 20,800 × *g* for 1 min and resuspended in microcentrifuge tubes in 200 µl of PBSP.

7. Bacteria are lysed by sonication (three 30-sec pulses on setting 4) using a Sonic Dismembrator Model 100 (Fisher) (*see* **Note 14**).

8. Cellular debris is pelleted by centrifugation at 20,800 × *g* for 5 min at 4°C. Samples should be kept at 4°C throughout the rest of the protocol to minimize protein degradation.

9. 25 µl of glutathioine sepharose beads per 200 µl of sonicated supernatant are prepared by washing beads twice in 500 µl of PBS to 25 µl of beads in microcentrifuge tubes. Beads are centrifuged at 1020 × *g* for 5 min at 4°C.

10. The supernatant protein is bound to glutathione sepharose beads by rotation overnight at 4°C.

11. Beads are washed three times with PBSP at 4°C, followed by centrifugation at 1020 × *g* for 5 min at 4°C after each wash.

12. GST proteins are eluted by rotating beads at 4°C for 24 h in the elution solution. This elution step is repeated and GST or GSTfusion protein eluates are pooled (*see* **Note 15**).

13. 20 µl of eluate is mixed with 2X LSB supplemented with 5% (v/v) 2-mercaptoethanol, heated at 95°C for 5 min, and subjected to SDS-PAGE as described in Section 3.13. Then 10, 5, and 1 µg of purified bovine serum albumin (BSA) is also run onto gels as standards for visually assessing eluate yield.

14. Gels are stained with Coomassie solution for 1 h at room temperature. Gels are then rinsed once with destain solution and incubated overnight at room temperature without rocking in destain solution.

15. Gels are soaked for 10 min in dH₂O and dried for 1.25 h at 80°C on a gel dryer. Representative data using this procedure with the GST-IBV3b fusion protein are shown in **Fig. 2**.

3.3. Prescreening Rabbit Sera before Injection of Peptides or Proteins

1. We have found that sera from some nonimmunized rabbits may recognize cell or viral proteins. Therefore, we prescreen the sera from several nonimmunized rabbits (generally twice as many rabbits as we use for immunization procedures) before immunizing them with peptides or fusion proteins. Any rabbits whose preimmune sera recognize proteins from infected or transiently transfected cells should not be used for generating antibodies.

2. For prescreening by immunoblotting, IBV-infected, vTF7-3 infected, or transiently transfected cells are washed once with PBS and lysed directly in their dishes with 2X LSB plus 5% (v/v) 2-mercaptoethanol. Noninfected or untransfected cells are also lysed. Approximately 100 µl of the LSB mixture is used per

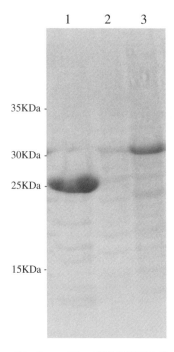

Fig. 2. Expression and purification of the GST-IBV 3b fusion protein. GST and GST-IBV 3b were expressed in bacteria and purified according to the protocol in Section 3.2, with 20 µl of the final eluate being subjected to SDS-PAGE and Coomassie staining. Lane 1 shows purified GST eluate. Lane 2 shows eluate from uninduced cultures containing plasmids that encoded GST-IBV 3b. Lane 3 shows purified GST-IBV 3b eluate.

35-mm dish. Once this mixture is added, a pipette tip is used to gently swirl the mixture across the cells. As the cells are lysed, the mixture becomes viscous owing to DNA and is then transferred to a microcentrifuge tube (samples can be stored at this point at –20°C for several weeks). Samples are then heated for 5 min at 95°C, until they are no longer viscous when the side of the tube is tapped. Samples are then subjected to SDS-PAGE (Section 3.13) and immunoblotting (Section 3.14). The primary antibodies used are a 1:1000 dilution of nonimmunized rabbit sera in the primary antibody dilution buffer. An example of the results produced from this protocol is shown in **Fig. 3**.

3. For prescreening by indirect immunofluorescence microscopy, IBV-infected, vTF7-3 infected, or transiently transfected cells can be processed according to Section 3.11. The primary antibodies used are a 1:500 dilution of nonimmunized rabbit sera.

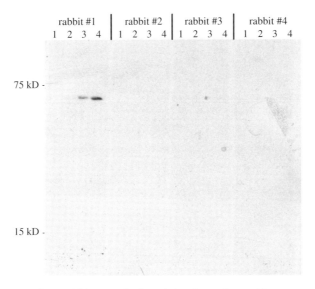

Fig. 3. Prescreening rabbit sera before injection of peptides or proteins. Before immunizing rabbits with the GST-IBV 3b fusion protein, four rabbit sera were tested for nonspecific recognition of proteins by SDS-PAGE and immunoblotting according to the protocol in Section 3.3. Lane 1: noninfected Vero cells. Lane 2: IBV-infected Vero cells (Section 3.7). Lane 3: vTF7-3 infected cells transfected with an empty plasmid. Lane 4: vTF7-3-infected cells transfected with a plasmid encoding IBV 3b behind the T7 RNA polymerase promoter (Section 3.8). Rabbits 2 and 4 were subsequently immunized with GST-IBV 3b. Rabbit 1 contains antibodies that recognize a vaccinia virus protein, a problem we have encountered occasionally.

3.4. Rabbit Immunization

1. Following prescreening in Section 3.10, two rabbits were chosen for immunization with peptides or fusion proteins. We use two rabbits per immunogen, since individual rabbit immunization responses may differ. The following procedure can be done commercially by a company such as Covance Research Products, Inc. (Denver, PA) (*see* **Notes 16** and **17**).
2. 21-day-old NZW rabbits are injected intradermally in the back with 250 µg of protein per rabbit combined with Freund's complete adjuvant (FCA).
3. Subsequent injections were done with 125 µg of protein combined with Freund's incomplete adjuvant (FIA) on the following days: 42 (subcutaneous nodal area injection), 63 (subcutaneous neck injection), 84 (subcutaneous and intramuscular injection), 105 (subcutaneous dorsal injection), 135 (subcutaneous nodal area injection), and 156 (subcutaneous dorsal injection) (*see* **Note 18**).
4. Approximately 5 ml of serum was harvested from rabbits on days 52 and 73 and approximately 20 ml of serum on days 94, 115, and 145. The terminal rabbit bleed on day 167 yielded approximately 50 ml. All sera are stored at –80°C.

3.5. Screening Sera from Immunized Rabbits

3.5.1. Screening Sera from Immunized Rabbits Using in vitro Transcribed and Translated Protein

1. Transcribe and translate the protein of interest *in vitro* as described in Section 3.10. An empty vector should be used as a control for the *in vitro* transcription and translation reaction (*see* **Note 19**).
2. Samples are diluted in 500 µl of detergent solution.
3. Samples are immunoprecipitated as in Section 3.12 starting with step 3. The primary antibodies used are 3 µl of undiluted sera from each harvest day from each rabbit. Always use preimmune sera (sera from the rabbit before immunization began) as a negative control. A representative result obtained from the method is shown in **Fig. 1B**.

3.5.2. Screening Sera from Immunized Rabbits Using Other Methods

1. Vero or DF1 cells are infected with IBV (Section 3.7). Noninfected cells should be used as a negative control. Alternatively, cells that are transiently transfected with plasmids encoding the protein of interest can be used (Section 3.8 or 3.9). In this case, cells transiently transfected with an empty vector should be used as a negative control (*see* **Note 19**).
2. Cells can then be subjected to immunoblotting (Section 3.14), immunoprecipitation (Section 3.12), or indirect immunofluorescence microscopy (Section 3.11) (*see* **Note 20**).
3. If screening by immunoblotting, primary antibodies are 1:1000, 1:2000, and 1:5000 dilutions of sera from each harvest day from each rabbit. Always use similar dilutions of preimmune sera as a negative control during screening.
4. If screening by immunoprecipitation, the primary antibodies used are 3 µl of undiluted sera from each harvest day from each rabbit. Always use preimmune sera as a negative control (*see* **Note 21**).
5. If screening by indirect immunofluorescence microscopy, primary antibodies are 1:50, 1:100, 1:200, 1:500, and 1:1000 dilutions of sera from each harvest day from each rabbit. Always use preimmune sera as a negative control during screening (*see* **Note 22**).

3.6. Affinity Purification of Sera

1. Occasionally, sera may show high background staining in negative control samples, and when this occurs, affinity purification of the sera may help reduce it. Affinity purification can be done with either the peptides or the GST fusion proteins that were used to immunize rabbits in Section 3.4. The protocol for affinity purification of sera using peptides is given below. Changes that must be made to this protocol when using fusion proteins rather than peptides are given in the Notes.

2. First, the cysteines in the purified peptides from Section 3.1 are reduced. To do this, we used the commercially available Reduce-Imm Immobilized Reductant Kit according to the manufacturer's instructions.

   a. The kit reductant column is washed with 10 ml of kit equilibration buffer #1 and activated by adding 10 ml of DTT solution.
   b. The column is then washed again in 20 ml of kit equilibration buffer #1 (*see* **Note 23**).
   c. 1 ml of diluted peptide is then added to the column (*see* **Note 24**).
   d. The column is eluted with 9 ml of kit equilibration buffer #1, and 1-ml fractions are collected (*see* **Note 25**).

3. Fractions are tested for the presence of free sulfhydryls using Ellman's Reagent.

   a. 50 μl of each fraction is transferred to a microcentrifuge tube containing 950 μl of kit equilibration buffer #1.
   b. 100 μl of Ellman's Reagent is then added to each tube.
   c. Tubes are incubated for 15 min at room temperature, and free sulfhydryl concentration is determined by absorbance measured at 412 nm.
   d. Fractions with similarly high absorbance measurements are pooled for use in subsequent steps.
   e. Additionally, the absorbance at 280 nm of the pooled fractions is measured, in order to calculate coupling efficiency during step 3.

4. To couple peptides to agarose beads, we used the commercially available SulfoLink kit according to the manufacturer's instructions.

   a. First, the coupling column is brought to room temperature, and the storage buffer is allowed to drain from the column after removing both top and bottom caps.
   b. The column is then equilibrated with 12 ml of SulfoLink equilibration buffer.
   c. The pooled peptide fractions from step 3d are then added to the column and the top cap of the column is replaced.
   d. The column is then rotated for 15 min at room temperature.
   e. Column-peptide mixtures are then incubated for an additional 30 min at room temperature without rotation.
   f. The buffer is allowed to drain and the columns are washed with 6 ml of coupling buffer.
   g. The absorbance at 280 nm of the column eluate is measured (*see* **Note 26**).
   h. Unbound column sites are then blocked by the addition of 2 ml of cysteine buffer.
   i. The column is then rotated for 15 min at room temperature, before being allowed to incubate without rotation for an additional 30 min at room temperature.
   j. All liquid is then drained from the column and the column is washed four times in 4 ml of wash buffer and three times in 4 ml of 0.05% NaN$_3$.
   k. If needed, the column can now be stored at 4°C. Usually, an additional 4–6 ml of 0.05% degassed NaN$_3$ is added to the column before storage to ensure that the column will not dry out during storage.

5. Protein-specific antibodies in the rabbit sera are then bound to the column, washed, and eluted.

    a. The storage liquid is removed from the column, and 2.5 ml of serum is added to the column and rotated overnight at 4°C.

    b. The column is allowed to drain and this flow-through is saved as it may contain antibodies that did not bind to the column (*see* **Note 27**).

    c. The column is then washed with 25 ml of ice-cold WB 1 and 25 ml of ice-cold WB 2. The column should be kept at 4°C during these washes.

    d. The column is eluted by adding 5 ml of elution solution. Three 1.5-ml fractions are collected and 5 µl of 30% BSA is immediately added.

    e. Fractions are transferred into dialysis bags and dialyzed three times for 24 h at 4°C in 1 liter of PBS (*see* **Note 28**).

    f. Fractions are separated into 100-µl aliquots and screened using procedures from Section 3.5. Aliquots should be stored at –80°C.

3.7. IBV Infection

1. IBV is brought to a total volume of 200 µl in serum-free DMEM per 35-mm dish such that the multiplicity of infection (MOI) is 1 (*see* **Note 29**).
2. Cells are washed once in serum-free DMEM. This virus dilution is then added to cells for 1 h at 37°C with rocking every 10 min. Virus is then removed and cells are incubated in normal growth medium for the appropriate amount of time postinfection (p.i.).

3.8. vTF7-3 Infection/Transfection

1. The vTF7-3 transient transfection system is used to express large amounts of proteins in a short period of time. In this system, cells are infected with a vaccinia virus that encodes the T7 RNA polymerase. Cells are then transfected with plasmids in which the gene of interest is cloned behind the T7 RNA polymerase promoter (*see* **Note 29**).
2. vTF7-3 is brought to a total volume of 200 µl in serum-free DMEM per 35-mm dish such that the MOI is 10.
3. This virus dilution is added to cells for 1 h at 37°C with rocking every 10 min.
4. While the virus is adsorbing to the cells, the transfection mixture is prepared. For each 35-mm dish, 6 µl of Fugene 6 is added to 94 µl of Opti-MEM. 2 µg of plasmid DNA is then added to this mixture and incubated at room temperature for 15 min (*see* **Note 3**).
5. At 1 h p.i., the virus dilution is removed from cells and 1 ml of regular growth medium is added to the cells.
6. The transfection mixture is added dropwise with swirling to the infected cells.

7. Cells are lysed for SDS-PAGE or processed for indirect immunofluorescence at approximately 4 h p.i.

3.9. Transient Transfection of Cells

1. For each 35-mm dish, 6 μl of Fugene 6 is added to 94 μl of Opti-MEM. 2 μg of plasmid DNA is then added to this mixture and incubated at room temperature for 15 min (*see* **Notes 3** and **29**)
2. The transfection mixture is then added dropwise to the cells with swirling.
3. Generally, cells are lysed for SDS-PAGE or processed for indirect immunofluorescence microscopy at approximately 24 h posttransfection.

3.10. In vitro *Transcription and Translation*

1. The TNT quick-coupled transcription/translation T7 system (Promega) is used for *in vitro* transcription and translation. This system utilizes the T7 RNA polymerase to drive gene expression; therefore, genes must be cloned into plasmids behind a T7 RNA polymerase promoter.
2. 0.5 μg of a plasmid DNA is incubated for 90 min at 30°C with 10 μl of the TNT master mix and 1 μl of [^{35}S]-Redivue-methionine. This reaction mix is usually used immediately. However, it can be stored at –20°C for 2 or 3 days if necessary.

3.11. Indirect Immunofluorescence Microscopy

1. Adherent cells that will be analyzed by indirect immunofluorescence microscopy are plated on sterilized, No.1, 22-mm square, precleaned, corrosive resistant, borosilicate coverslips (Fisher) in 35-mm dishes.
2. Cells are washed once in 1X PBS before being fixed in 3% paraformaldehyde solution for 10 min at room temperature (*see* **Note 22**).
3. Cells are washed quickly once with PBS/Gly before being washed a second time in PBS/Gly for 5 min at room temperature.
4. Cells are permeabilized in permeabilization solution for 3 min at room temperature (*see* **Note 22**).
5. Step 2 is repeated.
6. 40 μl of diluted primary antibodies is pipetted onto parafilm. Coverslips are then placed onto the primary antibodies, cell side down, for 20 min at room temperature (*see* **Note 30**).
7. Coverslips are placed back into the 35-mm dish cell side up and washed as in step 2.
8. Coverslips are incubated with diluted secondary antibodies as in step 6, except that they should be protected from light during this incubation.
9. Coverslips are put back into the 35-mm dish and washed as in step 2, except that they should be protected from light during washing.

10. A small drop of mounting solution is placed onto glass slides, and coverslips are placed, cell side down, onto this mounting solution. Excess mounting solution is removed by aspiration. Slides can be stored at 4°C, protected from light for several weeks. Slides are visualized on an Axioskop microscope (Zeiss, Thornwood, NY) with an attached Sensys charge-coupled device camera (Photometrics, Tucson, AZ) using IP Lab imaging software (Signal Analytics, Vienna, VA) (*see* **Note 31**).

3.12. Immunoprecipitation

1. Cells grown in 35-mm dishes are incubated with 0.085 mCi of $[^{35}S]$-methionine-cysteine in 500 μl of methionine and cysteine-free DMEM for 30 min at 37°C (*see* **Notes 32 and 33**)
2. Cells are rinsed once in 0°C PBS and then incubated for 10 min on ice in 500 μl of detergent solution.
3. Cell lysates are transferred to microcentrifuge tubes and centrifuged at full speed (20,800 × g) for 20 min at 4°C.
4. Cell supernatants are transferred to fresh microcentrifuge tubes, supplemented with 10 μl of 10% SDS, and precleared by adding 20 μl of washed pansorbin and rotating at 4°C for 15 min.
5. Samples are centrifuged at 20,800 × g for 1 min at room temperature, and supernatants are transferred to fresh microcentrifuge tubes containing primary antibodies (*see* **Note 21**). Samples are then rotated overnight at 4°C.
6. 15 μl of washed pansorbin is then added to samples and rotated for 20 min at 4°C.
7. Samples are centrifuged at 20,800 × g for 30 sec at room temperature to pellet the pansorbin with antigen-antibody complexes. Pellets are then washed three times in RIPA buffer. Following the final wash, the last drop of any remaining RIPA buffer is removed. Samples can be stored at this point for several weeks at –20°C.
8. Washed pansorbin pellets are resuspended in 20 μl of 2X LSB plus 5% (v/v) 2-mercaptoethanol, heated for 3 min at 95°C, and centrifuged at 20,800 × g for 1 min at room temperature.
9. The supernatants are subjected to SDS-PAGE as in Section 3.13.
10. Following SDS-PAGE, gels are washed for 10 min in dH$_2$O (*see* **Note 34**) and dried for 1.25 h on a gel dryer.
11. Gels are exposed either to film or to a K-HD imaging screen (Bio-Rad). Exposure to a K-HD imaging screen is followed by scanning with a personal molecular imager FX (Bio-Rad). Pixel intensity can then be determined with Quantity One software (Bio-Rad).

3.13. SDS Polyacrylamide Gel Electrophoresis (SDS-PAGE)

1. This protocol is for use with 16-cm gel plate systems similar to the Hoefer SE 400 or Hoefer SE 600 system; however, it can be easily adapted to other gel

systems. Gel plates, spacers, and combs should be wiped thoroughly clean with ethanol before they are assembled.

2. Prepare a 1.5-mm-thick, 17.5% low-bis polyacrylamide separating gel by mixing together 13.1 ml of 40% acrylamide, 1.5 ml of 2% bis-acrylamide, 7.5 ml of 4X separating gel buffer, 7.9 ml of dH$_2$O, and 300 µl of 10% SDS in a side-armed Erlenmeyer flask. Swirl gently to mix (*see* **Note 35**).

3. To pour a plug for the gel, remove a small volume (~2 ml) of the polyacrylamide mixture and place it in a disposable borosilicate glass tube (Fisher Scientific Research, Pittsburg, PA). Add 5 µl of TEMED and 5 µl of 30% APS. Vortex briefly. Using a glass Pasteur pipette, immediately pour this mixture between the gel plates.

4. Deaerate the remainder of the polyacrylamide mixture that is left in the side-armed Erlenmeyer flask for approximately 5 min.

5. Once the small volume from step 3 is polymerized, remove the side-armed flask from the vacuum and add 20 µl of TEMED and 50 µl of 30% APS to the poly-acrylamide mixture. Swirl gently to mix. Immediately pour this mixture between the gel plates, leaving room for the stacking gel.

6. Gently layer approximately 2 ml of layering buffer on top of the separating gel. The separating gel should take approximately 15 min to polymerize.

7. Pour off the layering buffer and remove any excess with a paper towel.

8. Prepare a stacking gel by mixing together 1.5 ml of 30% polyacrylamide:bis, 2.5 ml of 4X stacking gel buffer, 6 ml of dH$_2$O, and 100 µl of 10% SDS. Swirl gently to mix. Add 10 µl TEMED and 15 µl of 30% APS and immediately pour the polyacrylamide mixture between the gel plates. Immediately insert a comb. The stacking gel should polymerize within approximately 15 min.

9. Remove the comb, add 1X running buffer to the top and bottom chambers of the gel apparatus and, using a 20-gauge needle, rinse out the stacking gel wells, removing any bubbles caught between the plates in the bottom chamber.

10. Add 2X LSB supplemented with 5% (v/v) 2-mercaptoethanol to the samples. Heat the samples for 5 min at 95°C and load them into the stacking gel wells. One well should also contain molecular weight markers.

11. Run the gel on the continuous voltage setting at 50 mA for approximately 1.5 to 2 h until the dye front is about 1 in. from the bottom of the gel (*see* **Note 36**).

3.14. Immunoblotting

1. Experimental samples are subjected to SDS-PAGE according to Section 3.13.

2. A transfer cassette is assembled by first disassembling the gel apparatus. Two pieces of chromatography paper are layered in dH$_2$O onto the transfer cassette. A gel-sized piece of PVDF membrane (*see* **Note 7**) is briefly wetted in methanol and then layered on top of the chromatography paper. The gel is then layered on top of the PVDF membrane, followed by two additional pieces of chromatography paper. A 10-ml plastic pipette is then rolled across these layers to remove any bubbles trapped between them before closing the transfer cassette.

3. The transfer cassette is put into a transfer apparatus kept in a 4°C cold room, filled with enough cooled transfer buffer to cover the top of the PVDF membrane. The transfer cassette should be oriented so that the gel is closest to the cathode and the PVDF membrane is closest to the anode. Transfer then proceeds for 1 h at 75 V with constant stirring from a stir bar at the bottom of the transfer apparatus.

4. The membrane is then removed from the transfer apparatus and cassette and blocked for 30 min at room temperature in blocking buffer with rocking.

5. The membrane is then rinsed briefly twice in TTBS at room temperature and incubated with primary antibodies overnight at 4°C.

6. The membrane is washed twice for 15 min each in TTBS at room temperature, followed by two additional washes for 5 min each in TTBS at room temperature.

7. Membranes are incubated in secondary antibodies for 1 h at room temperature.

8. Step 6 is repeated.

9. The membrane is incubated for 1 min with HRP substrate. Excess substrate is removed by briefly patting the membrane with chromatography paper. The membrane is then wrapped in a small amount of saran wrap.

10. The membrane is either exposed to film for an appropriate amount of time or imaged using a VersaDoc Model 5000 imaging system (BioRad) with an attached cooled CCD AF Nikkor camera (Nikon, Inc., Melville, NY). Quantitation of images taken with the VersaDoc imaging system can then be performed using Quantity One software (BioRad).

4. Notes

1. We have noted the commercial source of our chemicals throughout this chapter. However, unless specifically stated otherwise, chemicals from other commercial sources are usually acceptable substitutes.

2. dH$_2$O: water having a resistivity of 18.2 MΩ-cm

3. Other cell types can be used with this transient transfection protocol. However, the protocol listed here is optimized for Vero and DF1 cells. To use this protocol with other cell types, the total amount of DNA and the DNA:Fugene ratio will have to be optimized. Additionally, other transfection reagents may give higher transfection efficiencies with different cell types.

4. Paraformaldehyde solutions will corrode pH electrodes; therefore, pH paper should always be used when testing the pH of paraformaldehyde solutions.

5. Other types of mounting solutions that contain fluorescence stabilizing chemicals can generally be used with this protocol. However, if secondary antibodies are conjugated to cyanine dyes (such as Cy2, Cy3, or Cy5), then mounting solutions that contain aromatic amines (such as phenylenediamine) should not be used.

6. As an alternative to using washed pansorbin, protein A-sepharose can be used to precipitate antigen-antibody complexes.

7. The IBV 3a and IBV 3b proteins do not bind well to nitrocellulose membranes. The IBV E protein binds to both nitrocellulose and PVDF membranes equally well.

8. Peptide purities greater than 90% can be obtained from commercial companies for an increased price. However, in our experience, 90%-pure peptides are sufficient for obtaining rabbit sera specific to the protein of interest.

9. If a peptide corresponding to the C-terminus of a protein is synthesized, then the extra cysteine should be added to the N-terminus of the peptide. On the other hand, if a peptide corresponding to the N-terminus of a protein is synthesized, then the extra cysteine should be added to the C-terminus of the peptide.

10. In our experience, ordering IBV 3a, IBV 3b, and IBV E peptides for injecting rabbits has been much more cost- and time-effective than expressing and purifying fusion proteins from bacteria. Expressing and purifying the IBV 3a and IBV 3b fusion proteins in bacteria have been very difficult, requiring optimization of many steps in the protocol listed in Section 3.2. Moreover, high yields of soluble IBV 3a fusion protein have not yet been obtained, likely owing to toxicity to bacteria.

11. In generating the GST-IBV 3b fusion protein, problems were consistently encountered during attempts to scale up protein expression and purification, so multiple small-scale productions were performed instead.

12. Multiple 1000X inducing solution concentrations (1 M, 0.1 M, 0.01 M, and 0.001 M IPTG) were tested for their ability to produce soluble GST-IBV 3b.

13. Multiple inducing temperatures and times were tested for their ability to produce soluble GST-IBV 3b, including 37°C for 2 h, 30°C for 2 h, 25°C for 2 h, 16°C for 24 h, 16°C for 48 h, 13.5°C for 24 h, 13.5°C for 48 h, 10°C for 24 h, and 10°C for 48 h. The temperature and time that best induced soluble protein was 13.5°C for 48 h. Additionally, one-half of the volume of water in the shaking water bath was added immediately before tubes were added, so that the water temperature gradually decreased over several minutes to 13.5°C.

14. Any frothing of samples during the sonication process increases the likelihood of denaturing GST fusion proteins so that they are unable to bind to glutathione sepharose beads in subsequent steps.

15. Multiple elution times were tested for their ability to yield high GST-IBV 3b concentrations, including 2 h, 4 h, 10 h, 24 h, and 48 h.

16. The same protocol is used for injecting rabbits with peptides, except that the micrograms of peptide used for each injection is doubled.

17. Rats can also be immunized to generate sera for co-localization experiments with rabbit sera. This protocol is also used for immunizing rats, except that 200 μg of protein is required for the initial injection and 100 μg of protein is required for all subsequent injections.

18. This injection schedule can be terminated early if rabbit sera show high enough titers following early injections.

19. In our experience, screening sera by *in vitro* transcription and translation of protein is essential. We first obtained anti-IBV 3b sera from injecting rabbits with

IBV 3b peptides. However, we screened these sera only by immunoblotting and indirect immunofluorescence microscopy of transiently transfected and infected Vero cells. No specific IBV 3b staining was seen during this screening; therefore, we assumed that anti-IBV 3b antibodies were not generated. This result led us to inject rabbits with purified GST-IBV 3b fusion proteins. Screening of these new sera using *in vitro* transcribed and translated IBV 3b demonstrated that high titers of anti-IBV 3b antibodies had been generated *(29)*. However, IBV 3b could still not be detected in Vero cells. Further experiments demonstrated that the extremely short half-life of IBV 3b in transiently transfected or infected Vero cells hampers its visualization in these cells. However, IBV 3b is readily detected in chicken cells since it is turned over differently in these cells. Subsequent experiments demonstrated that the antibody production using IBV 3b peptides showed similar staining patterns in chicken cells as the antibodies generated from GST-IBV 3b fusion proteins. IBV 3b was eventually detected by immunoprecipitation from Vero cells by increasing the dish size and the amount of radioactive label. Representative results are shown in **Fig. 4**.

20. Enzyme-linked immunosorbent assay (ELISA) methods can be used for screening sera. However, antibodies may show different binding affinities depending on the methods used. Therefore, screening sera using methods that will be used in future experiments is generally more informative.

21. The amount of primary antibody needed for immunoprecipitation reactions must be titered to ensure that antibodies are saturating. Therefore, immunoprecip-

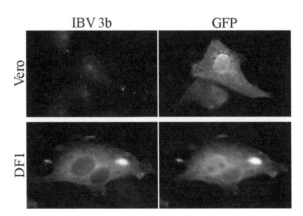

Fig. 4. Cell type may affect the expression of coronavirus accessory proteins. Plasmids encoding IBV 3b and green fluorescent protein (GFP) were transiently transfected into Vero and DF1 cells according to the protocol in Section 3.9. At 24 h posttransfection, cells were processed for indirect immunofluorescence microscopy according to the protocol in Section 3.11 with rabbit anti-IBV 3b and mouse anti-GFP primary antibodies. IBV 3b could not be detected in Vero cells, but was readily detected in DF1 cells. Subsequent experiments demonstrated that a greatly decreased IBV 3b half-life in Vero cells was responsible for the inability to visualize the protein in Vero cells *(29)*.

itation procedures are repeated with the supernatant to confirm that antibodies are in excess to antigen. We generally use 2 or 3 μl of serum for each immunoprecipitation.

22. Antibodies are often able to access their epitopes differently under different immunofluorescence microscopy fixation and permeabilization conditions. Therefore, alternate fixation and permeabilization conditions may have to be employed when screening antibodies: (i) concurrent fixation and permeabilization in PBS supplemented with 4% paraformaldehyde and 0.18% TritonX-100 for 10 min at room temperature; (ii) concurrent fixation and permeabilization in methanol for 5 min at –20°C; (iii) fixation in 3% paraformaldehyde for 10 min at room temperature followed by permeabilization in methanol of 5 min at –20°C; (iv) concurrent fixation and permeabilization in a 75% methanol, 25% acetic acid solution for 10 min at room temperature; (v) fixation in 3% paraformaldehyde for 10 min at room temperature followed by permeabilization in PBS supplemented with 0.05% saponin and 1% bovine serum albumin. For alternate protocol (v) all subsequent washes and incubations with antibodies should be done in PBS/Gly supplemented with 0.05% saponin and 1% bovine serum albumin.

23. Affinity purification using GST fusion proteins: wash the column with 10 ml of kit equilibration buffer #1, followed by an additional wash with 10 ml of kit equilibration buffer #2.

24. Affinity purification using GST fusion proteins: add 1 ml of the diluted protein to the column and incubate them together for 60 min at room temperature.

25. The column is eluted with 9 ml of kit equilibration buffer #2. 2-ml fractions are collected.

26. If the absorbance of the eluate is similar (given dilution factors) to the absorbance measured at the end of step 2, then the eluate should be reincubated with the coupling column and reeluted.

27. Affinity purification using GST fusion proteins: In order to remove antibodies that recognize GST, the entire procedure found in Section 3.6 can first be done using a column containing GST-coupled beads. The flow-through from this step is saved (it contains antibodies specific to the viral protein) and incubated with a different column that has the GST fusion protein coupled to it.

28. After fractions are collected, the column can be stored for reuse. To store the column it must first be washed with 10 ml of 0.02% (w/v) NaN$_3$ in PBS. 2 ml of storage buffer is then added and the column is stored at 4°C. Before reuse, the column should be washed once in elution buffer.

29. To obtain quality images by indirect immunofluorescence microscopy, cells should be infected or transfected when they are approximately 50% confluent. However, to obtain higher levels of protein expression when immunoblotting or immunoprecipitating, cells can be infected or transfected when they are as high as 70–90% confluent.

30. As a control for antibody specificity, sera from rabbits immunized with peptides can be preincubated with purified peptide before use in indirect immunofluores-

cence microscopy. 20 µl of serum is mixed with 20 µl of a solution containing 10 mg of peptide per 10 µl of dH_2O and rotated for 1 h at 4°C. 160 µl of a solution containing 1% (v/v) bovine serum albumin in PBS/Gly is then added to the antibody mixture and rotated overnight at 4°C. This antibody mixture is then diluted approximately 1:10 and used as the primary antibody in Section 3.11. Representative results of peptide-blocked versus unblocked antibody are shown in **Fig. 5**.

31. For visualization on an inverted microscope, coverslips can be sealed onto the slides by brushing boat sealer or nail polish around the edges of the coverslip.

32. For visualizing immunoprecipitated, unlabeled proteins, start the Section 3.12 protocol at step 2 and continue through step 9. Then perform the entire immunoblotting protocol. Rabbit primary antibodies can be used if the protein of interest is less than 20 kDa in size (approximately the size of the rabbit IgG light chains). Alternatively, primary antibodies from a different species can be used for immunoblotting.

33. Proteins with half-lives shorter than 30 min, such as IBV 3b when expressed in Vero cells *(29)*, may not be easily visualized in certain cell types. Therefore, cells may have to be plated in larger dishes and incubated with increased amounts of radioactivity.

34. For reproducible results, gels must be washed no longer than 10 min in dH_2O. Longer wash times allow IBV 3a and IBV 3b to diffuse out of the gel. Fixation of gels with 10% (w/v) trichloroacetic acid did not prevent diffusion of these proteins out of the gel in longer wash times.

35. Clear visualization of the IBV 3a and IBV 3b proteins is best achieved on 17.5% low-bis polyacrylamide gels. However, the slightly larger IBV E protein is easily visualized on 12 or 15% low-bis polyacrylamide gels. The same ratio of polyacrylamide:bis-polyacrylamide used when making 17.5% low-bis gels should also be used when making 12 or 15% low-bis gels.

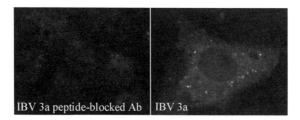

IBV 3a peptide-blocked Ab | IBV 3a

Fig. 5. Preincubation of sera with the peptides used to immunize rabbits can test antibody specificity. IBV-infected Vero cells (Section 3.7) were subjected to indirect immunofluorescence microscopy (Section 3.11) at 15 h postinfection. Sera from rabbits immunized with IBV 3a peptides were preincubated with either the IBV 3a peptide or dH_2O, as detailed in **Note 30**.

36. If the gel becomes too hot while electrophoresing, then the lanes at the edge of the gel run slower than those in the middle, causing a "smiling" effect. To minimize this possibility, fans and/or cold packs can be used to help cool gels.

Acknowledgments

We thank Emily Corse for the following: (i) designing the IBV 3a, IBV 3b, and IBV E peptides, (ii) prescreening rabbits before immunization with the IBV 3a, IBV 3b, and IBV E peptides, and (iii) screening anti-IBV E sera. This work was supported by National Institutes of Health grants GM42522 and GM64647.

References

1. Navas-Martin, S.R., and Weiss, S. (2004) Coronavirus replication and pathogenesis: Implications for the recent outbreak of severe acute respiratory syndrome (SARS), and the challenge for vaccine development. *J. Neurovirol.* **10**, 75–85.
2. Baudoux, P., Carrat, C., Besnardeau, L., Charley, B., and Laude, H. (1998) Coronavirus pseudoparticles formed with recombinant M and E proteins induce alpha interferon synthesis by leukocytes. *J. Virol.* **72**, 8636–8643.
3. Corse, E., and Machamer, C. E. (2000) Infectious bronchitis virus E protein is targeted to the Golgi complex and directs release of virus-like particles. *J. Virol.* **74**, 4319–4326.
4. Maeda, J., Maeda, A., and Makino S. (1999) Release of coronavirus E protein in membrane vesicles from virus-infected cells and E protein-expressing cells. *Virology* **263**, 265–272.
5. Vennema, H., Godeke, G. J., Rossen, J.W., et al. (1996) Nucleocapsid-independent assembly of coronavirus-like particles by co-expression of viral envelope protein genes. *EMBO J.* **15**, 2020–2028.
6. Fischer, F., Stegen, C.F., Masters, P.S., and Samsonoff ,W.A. (1998) Analysis of constructed E gene mutants of mouse hepatitis virus confirms a pivotal role for E protein in coronavirus assembly. *J. Virol.* **72**, 7885–7894.
7. Arbely, E., Khattari, Z., Brotons, G., Akkawi, M., Salditt, and T., Arkin, I. T. (2004) A highly unusual palindromic transmembrane helical hairpin formed by SARS coronavirus E protein. *J. Mol. Biol.* **34**, 769–779.
8. Raamsman, M.J., Locker, J.K., de Hooge, A., et al. (2000) Characterization of the coronavirus mouse hepatitis virus strain A59 small membrane protein E. *J. Virol.* **74**, 2333–2342.
9. Ortego, J., Escors, D., Laude, H., and Enjuanes, L. (2002) Generation of a replication-competent, propagation-deficient virus vector based on the transmissible gastroenteritis coronavirus genome. *J. Virol.* **76**, 11518–11529.
10. Kuo, L., and Masters, P. S. (2003) The small envelope protein E is not essential for murine coronavirus replication. *J. Virol.* **77**, 4597–4608.

11. Liao, Y., Lescar, J., Tam, J.P., and Liu, D.X. (2004) Expression of SARS-coronavirus envelope protein in *Escherichia coli* cells alters membrane permeability. *Biochem. Biophys. Res. Commun.* **325**, 374–380.

12. Liao, Y., Yuan, Q., Torres, J., Tam, J. P., and Liu, D. X. (2006) Biochemical and functional characterization of the membrane association and membrane permeabilizing activity of the severe acute respiratory syndrome coronavirus envelope protein. *Virology* **349**, 264–275.

13. Wilson, L., McKinlay, C., Gage, P., and Ewart, G. (2004) SARS coronavirus E protein forms cation-selective ion channels. *Virology* **330**, 322–331.

14. Madan, V., Garcia, Mde J., Sanz, M.A., and Carrasco, L. (2005) Viroporin activity of murine hepatitis virus E protein. *FEBS Lett.* **579**, 3607–3612.

15. Brown, T. D. K., and Brierley, I. (1995) The coronavirus nonstructural proteins. In: Siddell, S.G. (ed.) *The Coronaviridae*. New York: Plenum Press: pp. 191–217.

16. Youn, S., Leibowitz, J.L., and Collisson, E. W. (2005) *In vitro* assembled recombinant infectious bronchitis viruses demonstrate that the 5a open reading frame is not essential for replication. *Virology* **332**, 206–215.

17. Casais, R., Davies, M., Cavanagh, D., Britton, P. (2005) Gene 5 of the avian coronavirus infectious bronchitis virus is not essential for replication. *J. Virol.* **79**, 8065–8078.

18. Shen, S., Wen, Z. L., and Liu, D. X. (2003) Emergence of a coronavirus infectious bronchitis virus mutant with a truncated 3b gene: functional characterization of the 3b protein in pathogenesis and replication. *Virology* **311**, 16–27.

19. de Haan, C. A. , Masters, P.S., Shen, X., Weiss, S., and Rottier, P. J. (2002) The group-specific murine coronavirus genes are not essential, but their deletion, by reverse genetics, is attenuating in the natural host. *Virology* **296**, 177–189.

20. Haijema, B.J., Volders, H., and Rottier, P. J. (2004) Live, attenuated coronavirus vaccines through the directed deletion of group-specific genes provide protection against feline infectious peritonitis. *J. Virol.* **78**, 3863–3871.

21. Curtis, K.M., Yount, B., and Baric, R. S. (2002) Heterologous gene expression from transmissible gastroenteritis virus replicon particles. *J. Virol.* **76**, 422–434.

22. Ortego, J., Sola, I., Almazan, F., et al. (2003) Transmissible gastroenteritis coronavirus gene 7 is not essential but influences in vivo virus replication and virulence. *Virology* **308**, 13–22.

23. Sola, I., Alonso, S., Zuniga, S., Balasch, M., Plana-Duran, J., and Enjuanes, L.. (2003) Engineering the transmissible gastroenteritis virus genome as an expression vector inducing lactogenic immunity. *J. Virol.* **77**, 4357–4369.

24. Liu, D. X., Cavanagh, D, Green, P., and Inglis, S. C. (1991) A polycistronic mRNA specified by the coronavirus infectious bronchitis virus. *Virology* **184**, 531–544.

25. Liu, D. X., and Inglis, S.C. (1992) Internal entry of ribosomes on a tricistronic mRNA encoded by infectious bronchitis virus. *J. Virol.* **66**, 6143–6154.

26. Corse, E., and Machamer, C. E. (2002) The cytoplasmic tail of infectious bronchitis virus E protein directs Golgi targeting. *J. Virol.* **76**, 1273–1284.

27. Corse, E., and Machamer, C. E. (2003) The cytoplasmic tails of infectious bronchitis virus E and M proteins mediate their interaction. *Virology* **312**, 25–34.

28. Pendleton, A. R, and Machamer, C. E. (2005) Infectious bronchitis virus 3a protein localizes to a novel domain of the smooth endoplasmic reticulum. *J. Virol.* **79**, 6142–6151.

29. Pendleton, A. R., and Machamer, C. E. (2006) Differential localization and turnover of infectious bronchitis virus 3b protein in mammalian versus avian cells. *Virology* **345**, 337–345.
30. Machamer, C. E., and Rose, J. K. (1987) A specific transmembrane domain of a coronavirus E1 glycoprotein is required for its retention in the Golgi region. *J. Cell. Biol.* **105**, 1205–1214.
31. Fuerst, T.R., Niles, E. G., Studier, F. W., and Moss B. (1986) Eukaryotic transient-expression system based on recombinant vaccinia virus that synthesizes bacteriophage T7 RNA polymerase. *Proc. Natl. Acad. Sci. USA* **83**, 8122–8126.

15

Establishment and Characterization of Monoclonal Antibodies Against SARS Coronavirus

Kazuo Ohnishi

Abstract

Immunological detection of viruses and their components by monoclonal antibodies is a powerful method for studying the structure and function of viral molecules. Here we describe detailed methods for establishing monoclonal antibodies against severe acute respiratory syndrome coronavirus (SARS-CoV). B cell hybridomas are generated from mice that are hyperimmunized with inactivated SARS-CoV virions. The hybridomas produce monoclonal antibodies that recognize viral component molecules, including the spike protein (S) and the nucleocapsid protein (N), enabling the immunological detection of SARS-CoV by immunofluorescence staining, immunoblot, or an antigen-capture ELISA system. In addition, several S protein-specific antibodies are shown to have *in vitro* neutralization activity. Thus the monoclonal antibody approach provides useful tools for rapid and specific diagnosis of SARS, as well as for possible antibody-based treatment of the disease.

Key words: monoclonal antibody; SARS-CoV; coronavirus; immunization; confocal immunofluorescence; Western blot; antigen-capture sandwich ELISA.

1. Introduction

The outbreak of severe acute respiratory syndrome (SARS) in 2003 ultimately led to 8000 people becoming infected, 916 of whom died. The causative agent was identified as SARS coronavirus (SARS-CoV) *(1,2)*. Even though the epidemic ended, the threat of reemergence persists, compounded by the absence of an established vaccine. One of the critical issues in controlling a pandemic is a system for early diagnosis that distinguishes SARS from other

From: *Methods in Molecular Biology, vol. 454: SARS- and Other Coronaviruses,*
Edited by: D. Cavanagh, DOI: 10.1007/978-1-59745-181-9_15, © Humana Press, New York, NY

types of pulmonary infections. Based on clinical experience, several options have been considered in the quest to develop the capacity to accurately diagnose SARS-CoV infection, including molecular biology techniques and serological tests such as antigen-capture ELISA assay and immunofluorescence assay to detect virus-infected cells in respiratory swabs *(3–7)*. The preparation of monoclonal antibodies (mAbs) is considered to be especially valuable for serological testing. Here we describe a method for the successful establishment and characterization of mAbs against SARS-CoV structural components. These mAbs enable the general immunological detection of SARS-CoV by methods such as immunofluorescent staining, immunoblotting, and immunohistology, in addition to the construction of a highly sensitive antigen-capture sandwich ELISA *(6)*.

2. Materials
2.1. Immunization of Mice with UV-Inactivated SARS-CoV

1. SARS-CoV, strain HKU-39849, is expanded, purified, and inactivated by the method described in Chapter 11.
2. BALB/c mice, 8- to 10-week-old females, (Japan SLC, Shizuoka, Japan).
3. Freund's complete adjuvant.
4. Freund's incomplete adjuvant.

2.2. Cell Fusion and Hybridoma Production

1. Cell-passage medium: RPMI1640 medium (Gibco/Invitrogen) supplemented with 10% fetal bovine serum (FBS), 5×10^{-5} M 2-mercaptoethanol, 1/100 volume of nonessential amino acid solution (NEAA, Gibco/Invitrogen), 1/100 volume of 200 mM GlutaMAX^{TM} solution (Gibco/Invitrogen) and 1/100 volume of penicillin (100 U/mL)/streptomycin (100 μg/ml) solution.
2. Fusion partner myeloma cell line: Sp2/O-Ag14 (ATCC CRL-1581) is maintained with cell-passage medium at 37°C in a 5% CO_2 incubator.
3. Polyethylene glycol reagent: PEG4000 (Sigma) is aliquoted to 2 g in glass tubes with lids and autoclaved. Aliquots can be stored at room temperature.
4. Hybridoma medium: RPMI1640 is supplemented with 15% FCS, 0.2 ng/ml recombinant mouse IL-6 (R & D Systems, MN, USA), 5×10^{-5} M 2-mercaptoethanol, 1/100 volume of nonessential amino acids, 1/100 volume of GlutaMAX^{TM}, and 1/100 volume of penicillin/streptomycin solution.
5. HAT-selection medium: HAT supplement solution (Gibco/Invitrogen) is added to the hybridoma medium at a ratio of 1/50.
6. HT-medium: HT supplement solution (Gibco/Invitrogen) is added to the hybridoma medium at a ratio of 1/100.
7. Serum-free hybridoma medium: Hybridoma-SFM medium (Gibco/Invitrogen) is supplemented with 0.2 ng/ml recombinant mouse IL-6 and 1/100 volume of penicillin/streptomycin solution.

2.3. Screening of Anti-SARS-CoV-Producing Hybridomas

1. NP-40 lysis buffer: 1% NP-40, 150 mM NaCl, 50 mM Tris-HCl (pH 7.5). Stored at room temperature.
2. SARS-CoV-infected Vero E6 cells: SARS-CoV-infected Vero E6 cells are cultured and UV-inactivated by the method described in Chapter 11.
3. ELISA-coating buffer: 50 mM sodium bicarbonate. The pH is about 9.5 without adjusting.
4. ELISA plates: Nunc-Immunoplate F96.Cert Maxisorp (Nunc, Denmark).
5. PBS-Tween: 10 mM phosphate buffer (pH 7.5), 140 mM NaCl, 0.05% Tween 20.
6. Blocking solution: 1% ovalbumin (OVA) in PBS-Tween.
7. Detecting (second) antibody: Alkaline phosphatase-conjugated anti-mouse IgG (1:2000 dilution, Zymed, South San Francisco, CA).
8. ELISA substrate: *p*-Nitrophenyl Phosphate Liquid Substrate System (PNPP, Sigma).
9. Cell-freezing solution: 8% dimethylsulfoxide/92%FCS.

2.4. Purification of Anti-SARS-CoV Monoclonal Antibody

1. Serum-free hybridoma medium (see Section 2.2, step 7).
2. Protein G resin: Protein G-Sepharose 6B column (GE Healthcare/Amersham, UK).
3. Glycine/HCl solution: 100 mM Glycine, pH 2.8 with HCl.
4. Saturated $(NH_4)_2SO_4$: Excess $(NH_4)_2SO_4$ is added to distilled water and stored at $4°C$. The supernatant is used.

2.5. Biotinylation of Monoclonal Antibodies

1. Biotinylation reagent: Sulfo-NHS-LC-biotin (Pierce, Rockford, IL).
2. Sodium bicarbonate buffer: 50 mM sodium bicarbonate buffer, pH should be about 8.5 without adjustment.

2.6. Immunofluorescence Assay

1. Chamber slides: Lab-Tek 8-well glass chamber slides (Nunc, Denmark).
2. IFA-staining buffer: 1% BSA (Sigma) in PBS. Sodium azide is added to 0.05% as preservative.
3. Paraformaldehyde solution: 4% paraformaldehyde is freshly dissolved in PBS by heating to $70°C$. Use cold.
4. PBS/TritonX-100 solution: 0.1% Triton X-100 in PBS.
5. DAPI solution: For the stock solution, 4,6-diamidino-2-phenylindole hydrochloride (DAPI, Invitrogen) is dissolved with DW at 5 mg/ml and stored at $-20°C$. Use with 300 nM (1/48,000 dilution).
6. Mounting solution: Fluoromount G (SouthernBiotech, Birmingham, Alabama USA).

2.7. Immunoblot with Monoclonal Antibodies

1. Blot membrane: PVDF membrane (Bio-Rad, CA, USA).
2. Blocking reagent: Starting Block™ (Pierce, USA).
3. Detecting (second) antibody: Peroxidase-conjugated F(ab')$_2$ fragment anti-mouse IgG (H + L) (use with 1:20,000 dilution, Jackson Immuno Research, West Grove, PA).
4. Chromatography paper: 3 MM CHR (Whatman, UK).
5. X-ray film (Kodak, Rochester, NY).
6. Chemiluminescent reagent: SuperSignal West Femto (Pierce, USA).

2.8. Antigen-Capture Sandwich ELISA

1. High-binding immunoassay microplate: Immulon-2 (Dynatech Labs, VA, USA) or the equivalent.
2. ELISA-coating buffer: (see Section 2.3, step 3).
3. Blocking solution: 1% OVA in PBS-Tween.
4. Detecting (second) reagent: β-D-galactosidase-labeled streptavidine (Zymed, San Francisco, CA).
5. Galactosidase substrate: 4-Methy-lumbelliferyl-β-D-galactoside (Sigma).
6. Reaction stop solution: 0.1 M Glycine-NaOH (pH 10.2).

3. Methods

The viruses generally elicit strong humoral immune response in mice and hence are good antigens for obtaining the mAbs. The SARS-CoV also raises a high titer of serum antibody in mice. However, the immunization protocol should be optimized to obtain the desired specificity and quality of mAbs of interest. The choice of parameters for immunization, such as the type of adjuvant, pretreatment of viral antigens (inactivation procedures), antigen dose, route of immunization, and the strain of mice, profoundly affect the properties of resulting mAbs. For example, we found that inactivation of SARS-CoV with UV or formaldehyde gave rise to different immunoglobulin isotypes in mice *(8)*. In addition, the choice of host animals for the immunization affects the mode of epitope recognition. With the common fusion-partner cell lines such as NS-1, P3U1, or Sp2/O, a variety of mouse strains, even different species such as rat and hamsters, are usable for the donor of antigen-specific B cells. However, BALB/c mice are the most commonly used immunization host because the most efficient fusion-partner myelomas are derived from BALB/c strain. The protocol described here is a general procedure to obtain mAbs against viral antigens, though the choice of parameters, i.e., the immunization protocol, hybridoma production and so on, should be optimized for each purpose (*see* **Note 1**).

3.1. Immunization of Mice with UV-Inactivated SARS-CoV

1. BALB/c mice are immunized subcutaneously with 20 µg of UV-inactivated purified SARS-CoV using Freund's complete adjuvant (FCA).
2. After 2 weeks, the mice are boosted with subcutaneous injection of 5 µg of UV-inactivated SARS-CoV using Freund's incomplete adjuvant (FIA).
3. On day 3 after the boost, sera from the mice are tested by ELISA (see Section 3.3) for the antibody titer against SARS-CoV.
4. The two mice showing the highest antibody titer are further boosted intravenously with 5 µg of the inactivated virus 14 days after the previous boost.
5. If the antibody titer is lower than expected, the booster injection can be repeated several times before the final boost.

3.2. Cell Fusion and Hybridoma Production

Three days after the final boost, spleen cells from immunized mice are fused with Sp2/O-Ag14 myeloma by the polyethylene glycol method of Kosbor et al. (*9*) (see Section 3.2, step 5. This protocol gave rise to more than 40 candidate hybridomas, some of which react with S and N protein of SARS-CoV (**Table 1**).

1. For 1 to 2 h before cell-fusion, 2 g of PEG in a sterile glass tube is melted in a microwave oven and dissolved in 2.4 ml of RPMI1640 (abbreviated as RPMI hereafter), prewarmed to 37°C, by repeated pipetting. The PEG solution is kept at 37°C for at least 1 h.
2. One or two spleens from mice are excised and the spleen cell suspension is prepared by passing through a sterile stainless mesh or by smashing with two slide glasses. The cells are washed by centrifugation (270 × g for 5 min) twice with 5% FCS/RPMI and once with RPMI.
3. The log-phase growing Sp2/O-Ag14 myeloma cells are washed twice with 5% FCS/RPMI and once with RPMI.

Table 1
Summary of the First Hybridoma Screening by ELISA[a]

Immobilized antigen	Experiment 1	Experiment 2	Total
(Total wells assayed)	1920	960	2880
SARS-CoV infected Vero cell-lysate	28	14	42
Recombinant-S	19	7	26
Recombinant-N	3	0	3

[a]Reproduced from (*6*) with permission.

4. Count the cells and mix them to a ratio of splenocyte:Sp2/O-Ag14 = 3:1. Spin the cells down and remove the supernatant thoroughly.
5. Add 0.5 ml PEG/RPMI solution to the cell pellet slowly (taking about 60 sec), loosening of cell pellet using the tip of a pipette. Stir the cell clumps gently with a pipette for additional 90 sec. The cells should be seen as small aggregates in this step.
6. Add 10 ml of prewarmed (37°C) RPMI slowly, taking 3–4 min for the first 1 ml drop by drop, and then taking about 6 min for the remaining 9 ml.
7. Incubate the tube at 37°C for at least 20 min.
8. Wash the cells twice with prewarmed RPMI.
9. Suspend the cells with HAT-medium to a concentration of $3-5 \times 10^4$ Sp2/O-Ag14 cells/ml and plate to 0.2 ml/well of the 96-well plates.
10. Feed the cells with HAT-medium by changing two-thirds of the volume of the wells on the days 3, 7, and 12.
11. The hybridoma colonies should be seen by microscope on days 3–7 and some fast-growing colonies should be recognizable by eye from day 7. Generally, if the immunization and cell fusion step was successful, more than 70% of the wells contain hybridoma colonies. When the sizes of the colonies becomes one-tenth to one-fifth of the 96-well bottom area, the first screening of positive clones should be taken by ELISA (see Section 3.3).
12. The hybridoma cells in the ELISA-positive wells are recovered and cloned by a limiting dilution method as follows. Cells are counted, diluted with HAT-medium, and plated into 96-well plates so that one well contains three or ten cells; the left-half of the plate contains three cells/well and the right-half of the plate contains ten cells/well.
13. The hybridoma colonies will be discernible after 7–10 days. Then the ELISA test should be performed again and the ELISA-positive wells containing single colonies are selected.
14. The cells from the selected wells are expanded in a 24-well plate and adapted to HT-medium for at least 1 week.
15. The cells are further adapted to the hybridoma medium for 1 week. If necessary, the cells are further adapted to a serum-free hybridoma medium.
16. The hybridomas are now ready for collecting the mAbs. The aliquots of the cells are resuspended in freezing buffer and stored at –135°C or in liquid nitrogen.

3.3. Screening of Anti-SARS-CoV-Producing Hybridomas by ELISA

The first screening is conducted by ELISA using SARS-CoV-infected Vero E6 cell lysate as the antigen. In this first screening, the uninfected Vero E6 cell lysate is used as the antigen for a negative control. In the case of the assay for the test bleed of immunized mice, the serial dilution of the sera with PBS-Tween (start from 1/20 dilution) is used in place of hybridoma culture supernatants.

1. The Vero cell lysate is prepared as follows. After the inactivation of virus by UV-irradiation, the cell lysate is prepared in NP-40 lysis buffer followed by

centrifugation at $15,000 \times g$ for 20 min to remove cell debris (*see* **Note 2**). The supernatant is used as SARS-CoV-infected Vero cell lysate and the aliquots are stored at $-20°C$. The uninfected Vero cells are treated in exactly the same way and used as the uninfected Vero cell lysate for the negative control plate.

2. The SARS-CoV-infected Vero cell lysate is diluted 100-fold using ELISA-coating buffer and the 96-well ELISA plates are antigen-coated by this solution as 0.1 ml/well at 4°C overnight.
3. The ELISA plates are washed with PBS-Tween twice and blocked with blocking solution (1% OVA in PBS-Tween) at room temperature for 1 h. Plates are washed once with PBS-Tween and 0.05 ml of PBS-Tween is added to each well.
4. The 0.05 ml of culture supernatants from HAT-selected hybridomas are transferred to each well and incubated at room temperature for 1 h (*see* **Note 3**).
5. After washing three times with PBS-Tween, 0.1 ml/well of the second antibody (alkaline phosphatase-conjugated anti-mouse IgG, diluted to 1/1000–1/5000 with PBS-Tween) is added and left at room temperature for 1 h.
6. After washing four times with PBS-Tween, 0.1 ml/well of the substrate solution (PNPP) is added. The color development (bright yellow) is stopped by adding 0.05 ml of 0.1 M EDTA and quantified by OD_{405}.

3.4. Purification of Anti-SARS-CoV Monoclonal Antibody

1. Hybridomas are grown in about 2 liters of serum-free hybridoma medium to the late-log phase/early stationary phase.
2. The culture supernatants are harvested by centrifugation at $1700 \times g$ for 20 min, and 1/100 volume of 1 M Tris-HCl (pH 7.4) and 1/500 volume of 10% NaN_3 are added.

The following steps are carried out at 4°C (in a cold room) or on ice.

3. The Protein G-sepharose 6B column (2–5 ml bed volume) is preequilibrated with PBS and the supernatant is loaded at a flow rate of about 1 drop/5 sec. This step takes 2–3 days.
4. The column is washed with PBS and the bound antibody is eluted with glycine/HCl solution. In this step, 1-ml fractions are collected into an Eppendorf tube that contains 0.1 ml of 1 M Tris-HCl (pH 8.0).
5 After measuring the OD_{280} of the fractions, the protein-containing fractions are pooled. An equal volume of saturated $(NH_4)_2SO_4$ is gradually added to the pooled fraction. The solution is kept on ice overnight.
6. Precipitated proteins are collected by centrifugation at $2300 \times g$ for 30 min and the pellet is dissolved with a small volume (1–2 ml) of PBS, dialyzed against PBS (three changes of 500 ml, more than 3 h each),
7. The dialyzed sample is centrifuged at $15,000 \times g$ for 20 min, aliquoted, and stored at $-20°C$.

This procedure generally gives 2–10 mg of purified mAb from a 2-liter culture of hybridomas.

3.5. Biotinylation of Monoclonal Antibodies

1. The protein concentration of the purified antibody is determined by OD_{280} using a molecular extinction coefficient equal to 1.4 (for immunoglobulins).
2. The protein concentration is adjusted to 2 mg/ml with PBS, and 1 ml is dialyzed against 50 mM sodium bicarbonate buffer (pH 8.5) at 4°C.
3. Dissolve 1 mg of sulfo-NHS-LC-biotin in 1 ml of distilled water, and quickly add 74 μl of this solution to the antibody solution. Mix well and stand in ice for 2 h.
4. Dialyze the solution against PBS at 4°C overnight with three changes of 500 ml PBS dialysis solution, each for more than 3 h.
5. The sample is transferred to an Eppendorf tube and centrifuged at 15,000 × *g*, 4°C for 20 min.
6. The supernatant is recovered and the protein concentration is determined by OD_{280} as above. The aliquots are stored at –20° to –80°C. Avoid repeated freeze-thawing. The working solution can be stored at 4°C for months to years. Preservatives such as sodium azide can be added to 0.05% if necessary.

3.6. Immunofluorescence Assay

1. Virus-infected cells such as Vero E6 cells are cultivated on Lab-Tek 8-well glass chamber slides.
2. The cells are washed with PBS twice and fixed with 4% paraformaldehyde for 10 min at room temperature.
3. The cells are washed three times for 5 min each time with PBS and quenched with 50 mM NH_4Cl for 10 min at room temperature.
4. The cells are then permeabilized with PBS/TritonX-100 for 5 min and washed twice with PBS-Tween.
5. The cells are equilibrated with IFA-staining buffer for 1 h at room temperature.
6. The biotinylated mAb is diluted with the IFA-staining buffer and cells are stained with the solution for 1 h at room temperature (*see* **Note 4**).
7. The cells are washed three times for 5 min each time with PBS-Tween.
8. The second reagent such as Texas red-streptavidin is diluted with the IFA-staining buffer (*see* **Note 4**) and cells are stained with the solution for 1 h at room temperature.
9. The cells are washed three times for 5 min each time with PBS-Tween. If double staining of DNA-containing structures (nuclei) is desired, DAPI can be added to the second wash.
10. An antifade mounting solution such as Fluoromount G is gently dropped onto the cells and a cover slip carefully placed over them. The edge of the cover slip is sealed by nail varnish.
11. The slides are left to stand for 1 h at room temperature or overnight at 4°C to support the antifade effect.
12. The slide is ready for confocal microscopy, with the excitation/emission wavelength for Texas red (596 nm/615 nm) and DAPI (358 nm/461 nm).

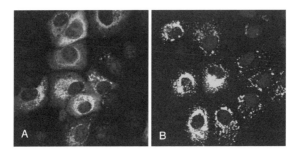

Fig. 1. Fluorescent immunostaining of SARS-CoV-infected Vero E6 cells with monoclonal antibodies (mAb). SARS-CoV-infected Vero E6 cells are paraformaldehyde-fixed, permeabilized with Trion X-100/PBS-Tween, and incubated with biotinylated mAbs. TexasRed-streptavidine is used for the detecting reagent and the nuclei are stained with DAPI: (A) Anti-S mAb, SKOT-3. (B) Anti-N mAb, SKOT-8.

Figure 1 shows an example of confocal microscopic observation of SARS-CoV-infected Vero E6 cells stained with anti-S mAb, SKOT-3 (A), and anti-N mAb, SKOT-8 (B). The nuclei are stained with DAPI.

3.7. Immunoblot with Monoclonal Antibodies

1. UV-inactivated purified SARS-CoV is electrophoresed and blotted to PVDF membrane by the standard procedure (*see* **Note 5**).
2. The blotted PVDF membrane is blocked with Starting Block^{TM} solution for 1 h at room temperature.
3. The membrane is reacted with the first mAbs diluted to 1 μg/ml (*see* **Note 4**) with 10% Starting Block^{TM}/PBS-Tween for 1 h at room temperature. This incubation is done by placing the PVDF membrane in the hybridization bag with 5 ml of antibody solution.
4. The membrane is washed with excess volume (about 20–30 ml) of PBS-Tween three times for 5 min each time in an appropriate container.
5. The membrane is reacted with peroxidase-conjugated F(ab')$_2$ fragment anti-mouse IgG diluted to 1:20,000 by 10% Starting Block^{TM}/PBS-Tween (*see* **Note 4**). The incubation is done in the hybridization bag for 1 h at room temperature.
6. The membrane is washed thoroughly with PBS-Tween four times for 5 min each time.
7. After washing, the membrane is placed on 3 MM filter paper and the liquid is removed but not dried. The membrane is then soaked with West Femto A:B = 1:1 mixture (in case of minigel size, i.e., 7 × 9 cm, 0.8–1.0 ml of A:B mixture is required). The membrane should be completely soaked with excess volume of A:B mixture.
8. The membrane is picked up by a flat-tip forceps and placed in between two transparent polyester sheets (*see* **Note 6**).

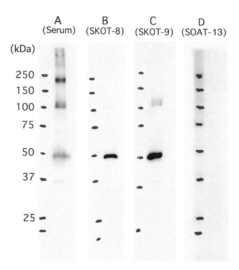

Fig. 2. Immunoblot of SARS-CoV proteins with monoclonal antibodies. Purified SARS-CoV proteins (0.5 μg/lane) are electrophoresed with SDS-PAGE (under reducing conditions), blotted to PVDF membrane, and detected by incubation with mAbs against SARS-CoV proteins. The detection is done with peroxidase-labeled-F(ab')$_2$ anti-mouse IgG followed by chemiluminescent reaction: (A) Mouse serum from SARS-CoV immunized mouse. (B) Anti-N mAb, SKOT-8. (C) Anti-N mAb, SKOT-9. (D) Anti-S mAb, SOAT13. The positions of molecular weight markers are shown on the left. (Reproduced from *(6)* with permission.)

9. The membrane/transparent sheet sandwich is placed in an X-ray film cassette with film and exposed for an appropriate time to obtain the best signals.

An example of the result is shown in **Fig. 2**, in which the purified SARS-CoV proteins (0.5 μg/lane) are electrophoresed with SDS-PAGE, blotted to PVDF membrane, and detected by anti-S and anti-N mAbs.

3.8. Antigen-Capture Sandwich ELISA

For the diagnosis of SARS-CoV infection, serological tests such as immunofluorescence assay (IFA, described above) and antigen-capture ELISA assay are two good options for detecting virus in, e.g., respiratory swabs or in virus-infected cells. By utilizing established mAbs against SARS-CoV, it is possible to construct a highly sensitive antigen-capture sandwich ELISA test system for detection of SARS-CoV. The sandwich ELISA consists of two mAbs, an antigen-capturing antibody and a detecting antibody, which recognize different epitopes of the target antigen. The antigen-capturing antibody is immobilized

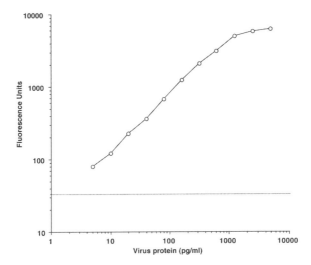

Fig. 3. Antigen-capture sandwich ELISA of SARS-CoV. Anti-N mAb, SKOT-8, is immobilized on the surface of a 96-well plate. Serially titrated purified SARS-CoV fractions are reacted for 1 h at room temperature and the bound virus proteins are detected by biotinylated SKOT-9 (anti-N) antibody followed by β-D-galactosidase-labeled streptavidin. They are then quantified by a chemiluminescent reaction using 4-methyl-umbelliferyl-β-D-galactoside as a substrate: (abscissa) concentration of purified SARS-CoV proteins; (ordinate) fluorescent unit. (Reproduced from (*6*) with permission.)

on the ELISA plate and captures the viral antigens in the test sample. The detecting antibody is normally labeled with a signal-producing enzyme such as a peroxidase or a phosphatase. The following method is the basic procedure for constructing a sandwich ELISA (*see* **Note 7**).

1. Purified mAb for the antigen-capture is immobilized on an Immulon-2 microplate by incubating 4 μg/ml antibody in the ELISA-coating buffer at 4°C overnight.
2. The microplate is blocked with 1% OVA for 1 h at room temperature or overnight at 4°C.
3. The plate is washed three times with PBS-Tween.
4. The UV-inactivated purified SARS-CoV samples (*see* **Note 1**), which are serially diluted with 1% OVA/PBS-Tween, are added to the wells and incubated for 1 h at room temperature.
5. After washing three times with PBS-Tween, wells are reacted with biotinylated detecting (second) mAb (0.1 μg/ml) for 1 h at room temperature.
6. After washing three times with PBS-Tween, wells are reacted with β-D-galactosidase-labeled streptavidine for 1 h at room temperature.

7. After washing four times with PBS-Tween, fluorescent substrate 4-methyl-umbelliferyl-β-D-galactoside is added and incubated for 2 h at 37°C.
8. The reaction is stopped by adding one-half volume of 0.1 M glycine-NaOH (pH 10.2).
9. The fluorescence of the reaction product, 4-methyl-umbelliferone (4-MU), is measured using FluoroScan II (Flow Laboratories Inc., Inglewood, CA) at excitation and emission wavelengths of 355 and 460 nm, respectively.

Figure 3 shows an example of an antigen-capture ELISA system for SARS-CoV, in which two anti-N mAbs, SKOT-8 (coating mAb) and biotinylated SKOT-9 (detecting mAb) are used. This sandwich ELISA detects SARS-CoV protein at a concentration as low as 40 pg/ml *(6)*.

4. Notes

1. For further understanding and detailed explanations of the hybridoma methodology, good textbooks are available *(10,11)*.
2. The validation for the complete inactivation of the virus is crucial. For a detailed account see Chapter 11 in this volume.
3. In this step the culture supernatants are diluted to one-half in order to reduce the background of the ELISA. If the background is still high, dilution of the culture supernatants can be one-fifth to one-tenth or more with PBS-Tween.
4. The titration of the mAbs and the second reagent is very important for obtaining the best results. It often varies from 1/100 to 1/100,000, and it should be determined for each mAb.
5. The SDS-polyacrylamide gel electrophoresis (PAGE) is carried out by the method of Laemmli *(12)*. The detailed conditions, such as the concentration of the gel or reducing/nonreducing and so on, should be changed case by case. Generally, loading 0.5–10 μg/lane of purified virus fraction would be enough for the detection of the viral antigen by the chemiluminescence method.
6. We conveniently utilize transparent sheets such as those used for an overhead projector (OHP-sheet).
7. In order to obtain good sensitivity and specificity of the test system, the optimization of the combination of two antibodies, i.e., coating mAb and detecting mAb, is required. This can be done by the ELISA with a matrix of candidate coating-mAbs and detecting-mAbs *(6)*.

Acknowledgments

The author would like to thank Drs. Fumihiro Taguchi and Shigeru Morikawa for their advice and discussion, Dr. Koji Ishii for providing recombinant SARS-CoV proteins, and Ms. Sayuri Yamaguchi for her technical assistance. This work was supported by grant from the Ministry of Public Health and Labor of Japan.

References

1. Drosten, C., Gunther, S., Preiser, W., van der Werf, S., Brodt, H. R., Becker, S., Rabenau, H., Panning, M., Kolesnikova, L., Fouchier, R. A., et al. (2003) Identification of a novel coronavirus in patients with severe acute respiratory syndrome. *N. Engl. J. Med.* **348**, 1967–1976.
2. Ksiazek, T. G., Erdman, D., Goldsmith, C. S., Zaki, S. R., Peret, T., Emery, S., Tong, S., Urbani, C., Comer, J. A., Lim, W., et al. (2003) A novel coronavirus associated with severe acute respiratory syndrome. *N. Engl. J. Med.* **348**, 1953–1966.
3. Peiris, J. S., Chu, C. M., Cheng, V. C., Chan, K. S., Hung, I. F., Poon, L. L., Law, K. I., Tang, B. S., Hon, T. Y., Chan, C. S., et al. (2003) Clinical progression and viral load in a community outbreak of coronavirus-associated SARS pneumonia: a prospective study. *Lancet* **361**, 1767–1772.
4. Chan, P. K., To, W .K., Ng, K. C., Lam, R. K., Ng, T. K., Chan, R. C., Wu, A., Yu, W. C., Lee, N., Hui, D. S., et al. (2004) Laboratory diagnosis of SARS. *Emerg. Infect. Dis.* **10**, 825–831.
5. Wang, W. K., Chen, S. Y., Liu, I. J., Chen, Y. C., Chen, H. L., Yang, C. F., Chen, P. J., Yeh, S. H., Kao, C. L., Huang, L. M., et al. (2004) Detection of SARS-associated coronavirus in throat wash and saliva in early diagnosis. *Emerg. Infect. Di.s* **10**, 1213–1219.
6. Ohnishi, K., Sakaguchi, M., Kaji, T., Akagawa, K., Taniyama, T., Kasai, M., Tsunetsugu-Yokota, Y., Oshima, M., Yamamoto, K., Takasuka, N., et al. (2005) Immunological detection of severe acute respiratory syndrome coronavirus by monoclonal antibodies. *Jpn. J. Infect. Dis.* **58**, 88–94.
7. Che, X. Y., Qiu, L. W., Pan, Y. X., Wen, K., Hao, W., Zhang, L. Y., Wang, Y. D., Liao, Z. Y., Hua, X., Cheng, V. C., et al. (2004) Sensitive and specific monoclonal antibody-based capture enzyme immunoassay for detection of nucleocapsid antigen in sera from patients with severe acute respiratory syndrome. *J. Clin. Microbiol.* **42**, 2629–2635.
8. Tsunetsugu-Yokota, Y., Ato, M., Takahashi, Y., Hashimoto, S.-I., Kaji, T., Kuraoka, M., Yamamoto, K.-I., Mitsuki, Y.-Y., Yamamoto, T., Ohshima, M., et al. (2007) Formalin-treated UV-inactivated SARS coronavirus vaccine retains its immunogenicity and promotes Th2-type immune responses. *Jpn. J. Infect. Dis.* In press.
9. Kozbor, D., and Roder, J. C. (1984) *In vitro* stimulated lymphocytes as a source of human hybridomas. *Eur. J. Immunol.* **14**, 23–27.
10. Harlow, E., and Lane, D. (1998) *Using Antibodies: A Laboratory Manual.* Cold Spring Harbor Laboratory, New York.
11. Shepherd, P., and Dean, C. (2000) *Monoclonal Antibodies: A Practical Approach*: Oxford University Press, New York.
12. Laemmli, U. K. (1970) Cleavage of structural proteins during the assembly of the head of bacteriophage T4. *Nature* **227**, 680–685.

16

Production of Monospecific Rabbit Antisera Recognizing Nidovirus Proteins

Jessika C. Zevenhoven-Dobbe, Alfred L. M. Wassenaar, Yvonne van der Meer, and Eric J. Snijder

Abstract

The importance of monospecific antisera for the experimental analysis of viral proteins is undisputed. They make it possible to identify and analyze the target protein against a background of a large number of other proteins, either in whole fixed cells or in cell lysates. This chapter describes our experience with the production of such rabbit antisera directed against proteins of coronaviruses and other nidoviruses. The use as antigens of either synthetic peptides (coupled to a carrier protein) or proteins expressed in *Escherichia coli* is described, and detailed protocols for immunization and preparation of test bleeds are provided.

For screening of the immune response following immunization, detailed protocols for three commonly used techniques are described, all of which are based on the use of infected cells or cells expressing the protein of interest, side by side with appropriate controls. The in situ immunodetection of the target in fixed cells by immunofluorescence microscopy is described, as are protocols for techniques that can be applied to cell lysates containing the target protein (Western blotting and immunoprecipitation). The latter techniques are performed in combination with polyacrylamide gel electrophoresis, thus allowing confirmation of the molecular weight of the target that is recognized by the antiserum.

Key words: antigen; peptide; immunization; rabbits; replicase; structural proteins; immunofluorescence assay; Western blotting analysis; immunoprecipitation; coronavirus; nidovirus

From: *Methods in Molecular Biology, vol. 454: SARS- and Other Coronaviruses,*
Edited by: D. Cavanagh, DOI: 10.1007/978-1-59745-181-9_16, © Humana Press, New York, NY

1. Introduction

1.1. Coronavirus/Nidovirus Targets for Antibody Production

Coronaviruses, and other nidoviruses such as arteriviruses, produce one of the largest sets of viral polypeptide species among the RNA viruses. This feature is related to the polycistronic nature of their genome, which contains up to 12 open reading frames and, in particular, to the polyprotein strategy used to produce the large viral replicase/transcriptase, the complex set of nonstructural proteins that is responsible for replication and transcription of the nidovirus genome [see *(1)* and the references therein]. In coronaviruses (**Fig. 1**), the larger of the two replicase polyproteins (pp1ab) can be more than 7000 amino acids long and is processed into 15 or 16 cleavage products, now commonly referred to as nonstructural protein (nsp) 1 to 16 *(2,3)*. In arteriviruses, despite their much smaller replicase size (pp1ab is "only" ~3100 to 3900 amino acids), the complexity is not very different and replicase processing yields 13 or 14 cleavage products. The regulated autoproteolysis of nidovirus replicase pp1a and pp1ab is driven by two to four proteinases that reside in the ORF1a-encoded part of the replicase. Cleavage sites for coronavirus and arterivirus proteinases have been identified for several prototypic nidoviruses through a combination of theoretical and experimental methods and can now be confidently predicted for other members of these virus groups.

Since an additional 4 to 12 proteins are expressed from subgenomic mRNAs produced from the 3'-end of the viral genome (**Fig. 1**), including the structural polypeptides responsible for virion formation, the proteome of the average nidovirus consists of between 20 and 30 protein species. It should be noted that intermediates of replicase polyprotein processing, which have been described for various nidoviruses, are not taken into consideration in this count and are likely to even further increase the repertoire of viral polypeptides produced in infected cells. Replicase precursors and processing intermediates in fact complicate certain types of analyses, such as those that rely on microscopy, since antibodies will usually not discriminate among precursors, intermediates, and processing end products. In the infected cell, many nidovirus proteins are targeted to specific locations (**Fig. 2**), some of them even to multiple specific locations.

Over a period of about 15 years, our laboratory has been involved in the production and characterization of polyclonal rabbit antisera directed against specific structural and, in particular, nonstructural proteins of nidoviruses *(4–8)*. Our experience in this area is summarized in this chapter.

1.2. Antibodies and Their Applications

The usefulness and importance of monospecific antibodies for the experimental analysis of almost any protein in biology is undisputed. In particular, when

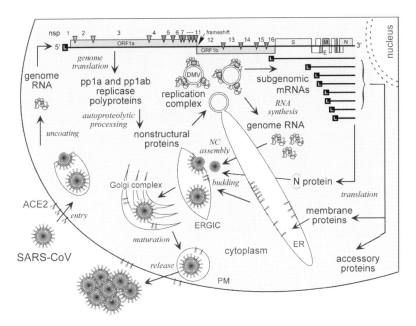

Fig. 1. Overview of the coronavirus/nidovirus life cycle, using SARS-CoV as an example *(1)*. Depicted is the cytoplasmic replication cycle, which starts with virus entry and release of the genome into the cytoplasm. Subsequently, the genome is translated to produce replicase polyproteins pp1a and pp1ab, which are cleaved to yield 16 nonstructural proteins *(2)*. A complex for viral RNA synthesis is assembled on cytoplasmic double-membrane vesicles (DMVs). This complex is involved in genome replication and the synthesis of a nested set of subgenomic mRNAs, which are required to express the genes in the 3′-proximal third of the genome. Translation of the smallest subgenomic mRNA yields the viral nucleocapsid protein (N) that assembles into new nucleocapsid (NC) structures together with newly generated genome RNA. Other subgenomic mRNAs encode viral envelope proteins that are (largely co-translationally) inserted into membranes of the host cell's exocytic pathway and migrate toward the site of virus assembly [the ER-Golgi intermediate compartment (ERGIC)]. Following budding of the NC through membranes that contain viral envelope proteins, the new virions leave the cell via the exocytic pathway. [Adapted from *(1)*.]

the target protein has to be studied against a background of a large number of other proteins, either in whole fixed cells or in cell lysates, specific antibodies can provide a rapid and reliable method for discriminating signal from noise. Three commonly used antibody-based detection techniques, which are also important for screening of the immune response during antiserum production, are discussed below:

- In situ immunodetection of the target in fixed cells, either by immunofluorescence (IF) microscopy or immunoelectron microscopy (IEM).

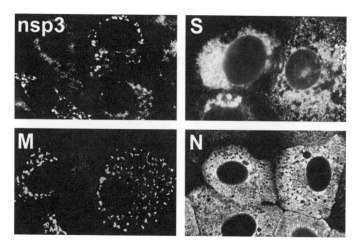

Fig. 2. Immunofluorescence microscopy images of Vero-E6 cells infected with SARS-coronavirus illustrating the expression of four important viral genes and the different subcellular localization of their products. Nonstructural protein 3 (nsp3) is one of the replicase subunits expressed from the genomic RNA, revealing the localization of the membrane-bound viral replication complex. Spike protein S localizes to different compartments of the exocytic pathway. Membrane protein M accumulates in the Golgi complex, in particular early in infection. Nucleocapsid protein N is a largely cytoplasmic protein.

- Detection of the target by Western blotting (WB), i.e., electrophoresis of a cell lysate containing the target and transfer of the (denatured) proteins to a solid carrier such as nitrocellulose or polyvinylidene fluoride (PVDF) membranes.
- Immunoprecipitation (IP) of the target from a cell lysate, often used in combination with radioactive labeling of protein synthesis in the cells for a specific time prior to cell lysis.

Although in several techniques experimental data obtained with monoclonal antibodies can be superior, in particular because of reduced background signal, the generation and screening of hybridoma cell lines requires a considerable investment, in terms of both labor and capital. Furthermore, in the context of specific research questions, polyclonal antisera may sometimes even be preferred over monoclonal antibodies since they contain a mixture of immunoglobulin molecules, derived from different B-cell lines in the immunized animal, often recognizing multiple epitopes of the target protein. Moreover, if desired, the chances of cross-reactivity of the antiserum with related proteins, e.g., from different strains of the same virus or from closely related other virus species, are considerably better for polyclonal antisera, particularly when they have been raised using a larger and/or conserved part of the target protein.

Obviously, the primary goal during polyclonal antiserum production in laboratory animals is to obtain a reasonable volume of a high-affinity antiserum. For rabbits, if necessary, bleeds of ~15–20 ml can be obtained at 4-week intervals, and the final bleed can yield up to 50 ml of serum from about 80 ml of whole blood. Since monoclonal antibodies are routinely produced in mice, the use of other species such as rabbits also creates the possibility of convenient in situ double-labeling experiments (see below). In such experiments, using species-specific secondary antibodies, a polyclonal rabbit antiserum against the target of choice and, for instance, a mouse monoclonal antibody recognizing a cellular protein can be used in combination to detect both proteins in the same specimen simultaneously.

2. Materials

2.1. Coupling of Synthetic Peptides to BSA

1. Freeze-dried synthetic peptide.
2. Phosphate-buffered saline (PBS): 10X PBS is prepared using 80.0 g/liter NaCl (1369 mM), 2.0 g/liter KCl (26.8 mM), 14.4 g/liter $Na_2HPO_4.2H_2O$ (80.8 mM), and 2.0 g/liter KH_2PO_4 (14.6 mM). The pH is 6.8, but will change to 7.4 upon dilution to 1X PBS.
3. Bovine serum albumin (BSA) solution: 50 mg/ml.
4. Glutaraldehyde solution 25% (w/v); note that glutaraldehyde is toxic! Prepare a 0.3% glutaraldehyde solution in PBS (120 µl of the 25% stock plus 9.9 ml of PBS).
5. Glycine stock: 1 M.
6. Dialysis membrane with a molecular weight cut-off value of 12–14 kDa (e.g., Slide-A-Lyzer Dialysis Cassettes, cat. no. 66330, Pierce).

2.2. Preparation for Immunization or Boosts

1. Freund's complete adjuvant (FCA).
2. Freund's incomplete adjuvant (FICA).
3. Phosphate-buffered saline (PBS): See above.
4. Syringe: 3-ml with luer-lock (e.g., Becton Dickinson #300910).
5. Three-way stopcock with luer-lock (e.g., Codan #445852).

2.3. Immunofluorescence Assay

1. Coverslips with infected cells or cells expressing the protein of interest.
2. Fixative: 3% paraformaldehyde in PBS.
3. PBS: see Section 2.1, step2.
4. PBS with 5% fetal calf serum (FCS).
5. PBS with 10 mM glycine.
6. PBS with 0.1% Triton X-100.

7. Bisbenzimide dye Hoechst 33258 (Sigma #14530) for nuclear DNA staining. Prepare a 100 μg/ml stock in H_2O (dissolving it in PBS will give precipitates).
8. Anti-rabbit IgG fluorescent conjugate (e.g., Jackson Immunoresearch; donkey-anti-rabbit IgG Cy3 conjugate).
9. 24-well cluster.
10. Microscopy glass slides.
11. Mowiol (embedding medium): Mix 2.4 g Mowiol 4-88 (e.g., Calbiochem #475904) with 6 g of glycerol and 2 ml of H_2O at room temperature for 2 h; add 12 ml 0.2 M Tris-HCl (pH 8.5) and stir at 50°C until all Mowiol has dissolved. Centrifuge for 15 min at 5000 × g. Then add 2.5% w/v Dabco (1,4-diazabicyclo(2.2.2)octane), e.g., Sigma #D2522, which improves the life span of fluorescent dyes), mix, fill out in small aliquots, and store at −20°C.

2.4. Western Blotting

1. Protein lysis buffer (20 mM Tris-HCl, pH 7.6; 150 mM NaCl; 0.5% Deoxycholine; 1% Nonidet P-40; 0.1% SDS).
2. 0.5 M EDTA.
3. Transblot SD semidry transfer cell (e.g., Biorad #170-3940).
4. PVDF membrane (e.g., Amersham Biosciences, Hybond P, #RPN303F).
5. Whatman paper.
6. Methanol.
7. 10X Western blot transfer buffer (WTB): 250 mM Tris, 1.92 M glycine.
8. PBS: see Section 2.1, step 2.
9. PBS-T (PBS with 0.5% Tween 20).
10. PBS-TM (PBS-T with 5% nonfat dry milk).
11. PBS-TMB (PBS-TM with 1% BSA).
12. Anti-rabbit IgG horseradish peroxidase conjugate, e.g., swine-anti-rabbit IgG HRPO (e.g., DakoCytomation #P02-17)
13. Chemiluminescence kit (e.g., Amersham Biosciences, ECL plus Western blotting detection kit, # RPN2132)

2.5. Immunoprecipitation

1. Protein lysis buffer, see Section 2.4, step 1.
2. IP buffer (20 mM Tris-HCl, pH 7.6; 150 mM NaCl; 5 mM EDTA; 0.5% Nonidet P-40; 0.1% Deoxycholine; 0 to 1% SDS, concentration to be varied depending on the antiserum).
3. Weak wash buffer A (20 mM Tris-HCl pH 7.6; 150 mM NaCl; 5 mM EDTA; 0.1% NP40).
4. Weak wash buffer B (20 mM Tris-HCl pH 7.6; 0.1% NP40).
5. Laemmli sample buffer (LSB): 50 mM Tris-HCl, pH 6.8; 100 mM DTT; 10% glycerol; 2% SDS; 0.03% bromo phenol blue.
6. Pansorbin (heat-killed, formalin-fixed *Staphylococcus aureus* cells; Calbiochem #507858) or Protein A/G sepharose beads (e.g., Amersham Biosciences #17-5280-04 or #17-0618-02).

2.6. IgG Purification and Direct Coupling to Alexa Fluor-488

1. PURE1A Protein A Antibody Purification Kit (Sigma #PURE1A-1KT).
2. Alexa Fluor 488 Protein Labeling Kit (Invitrogen/Molecular Probes #A10235).

3. Methods

3.1. Preparation of the Antiserum

The (obvious) standard prerequisites for the production of a polyclonal rabbit antiserum are: (i) an antigen, (ii) a (pathogen-free) rabbit to be immunized, (iii) a test method to detect the response, and (iv) a permit for animal experiments. The last relates to institutional and/or governmental regulations controlling animal use, which differ from country to country and are not discussed here in detail. Important considerations are the choice of adjuvant, method and site of administration of the antigen, sedation of animals, maximum volume of test bleeds, and various safety precautions regarding animals and personnel.

3.1.1. Antigens: "Natural" Proteins versus Synthetic Peptides

Two types of antigens have been used in our studies: synthetic peptides and proteins expressed in and purified from *Escherichia coli*. The production of antigens in *E. coli* are not covered in any detail in this chapter and the reader is referred to the extensive literature on this topic published elsewhere (and in Chapter 13 in this volume). *E. coli* allows for the cheap production of large amounts of antigen, but clearly such antigens have to be purified prior to use for immunization. This can be facilitated, e.g., by the use of a variety of tags (such as fusion to glutathione-S-transferase, maltose-binding protein, or the commonly used hexahistidine tag) for which convenient affinity resins are available. Reasonably pure antigens (>80% pure) are required and in the case of larger tags one should consider removing the tag proteolytically to ensure that the immune response will be directed against the target rather than against its tag. If the *E. coli* expression product turns out to be insoluble, which will usually interfere with its straightforward affinity purification, it may still be possible to purify a reasonably pure protein sample from inclusion bodies by repeated extraction of this material with increasing concentrations of urea (or guanidium isothiocyanate). As long as small amounts of this material can be used (i.e., the protein concentration is sufficiently high), after washing of the protein pellets with PBS, such urea-containing samples can be used for mixing with the adjuvant and immunization without the need for dialysis.

Synthetic peptides offer the advantage of being pure, and with certainty they contain exclusively the amino acid sequence of the selected target. Since they normally cover only 10 to 25 residues of this target, their sequence has to be selected with care (see below).

3.1.2. Design and Synthesis of Peptides for Immunization Purposes

It is not straightforward to confidently predict antigenic peptides from an amino acid sequence. Several programs to support this activity can be found on the Internet and algorithms for this purpose are included in most DNA/protein analysis software *(9–11)*. If the structure of the target is known, peptides located on its surface are to be preferred. Entirely polar and helical peptides should be avoided. In our experience, peptides located at (or close to) the N- or C-terminus of the target also have a high probability of being immunogenic. This was true in particular for peptides derived from the nidovirus replicase polyproteins, which were usually selected on the basis of known or predicted cleavage sites (although based on relatively small numbers; success rate with terminal peptides was around 75% and with internal peptides only around 25%).

The peptides to be synthesized usually are 10–25 amino acids long. Peptides that are very hydrophobic may be more difficult to handle (e.g., poor solubility prior to coupling). To protect peptides against host proteases and thus increase their stability, the C-terminal carboxyl group can be replaced with an amide group during synthesis. (This may not be advisable when using peptides that normally form the C-terminus of a protein.) To facilitate coupling to BSA used as the carrier protein (see below), one or two lysine residues can be added to one side of the peptide (to the N-terminus when targeting the C-terminus of a protein, and vice versa).

Synthetic peptides are available from a variety of commercial or in-house sources. In our institute, peptides are synthesized by solid-phase strategies on an automated multiple peptide synthesizer (SyroII, MultiSynTech, Witten, Germany). The purity of the peptides is determined by analytical reversed-phase HPLC and should be at least 80%. The identity and homogeneity of the peptides is confirmed by matrix-assisted laser desorption ionization time-of-flight mass spectrometry and analytical reversed-phase chromatography. Freeze-dried peptides, dissolved peptides, or coupled peptides in solution are best stored at –85°C.

3.1.3. Coupling of Synthetic Peptides to BSA

Synthetic peptides have to be coupled to a carrier protein to enhance their immunogenicity. Bovine serum albumin (BSA) and keyhole limpet hemocyanin (KLH) are the two most commonly used carrier proteins. We have always relied on BSA (a very soluble and stable plasma protein) and a coupling reaction using glutaraldehyde, which links the amino groups of carrier and synthetic peptide (in lysine residues and at the amino terminus of the peptide). For example, cross-linking at lysine residues can be represented as:

$$BSA(Lys)-NH_2 + O{=}CH(CH2)_3CH{=}O + NH_2-(Lys)peptide \rightarrow$$
$$BSA(Lys)-N-CH(CH2)_3CH-N-(Lys)peptide + 2H_2O$$

1. In a microfuge tube, prepare a solution of ~5 mg/ml of the peptide in PBS. (Routinely, our peptide synthesis yields about 5 mg of product, of which ~4 mg is the peptide and the remainder consists of salts). Weigh the dry peptide on a piece of paper and carefully split it with a scalpel (wear gloves). For a solution of ~5 mg/ml, dissolve about half of the peptide in a final volume of 0.4 ml of PBS. Solubility in PBS is usually good if the freeze-dried product looks "large and fluffy." If it looks small and granular, it may be better to first dissolve the peptide in a small volume of DMSO and then dilute this solution in PBS. The final concentration of DMSO should be less than 5%.
2. Add BSA (the carrier protein) to the peptide solution at the correct molar ratio. The aim is to couple one peptide per about 50 amino acids of carrier protein, for about 12 peptides per BSA molecule. The amount of peptide can be estimated with the following formula, which assumes an average molecular mass (Mr) of 110 Da per amino acid: (volume)*(concentration)/(x-mer*110), or— in a 21 mer example— (0.4 ml)*(5 mg/ml)/(21*110 μg/μmol) = 0.86 μmole. Calculate the required amount of BSA, which is 618 amino acids long (including 59 lysine residues) and thus weighs ~68 mg/μmol. Mixing peptide and BSA at a molar ratio of 12:1 means that in our example using 0.86 μmol of peptide, we would use 0.86/12 = 0.072 μmol of BSA (or 4.9 mg, or 98 μl of a 50-mg/ml solution).

Note that the above calculations are based on a number of assumptions and should be considered as a rule of thumb. Clearly, since the sequence is known, the Mr of the peptide can be calculated more precisely. However, since we usually add extra lysines to the peptide there will be different ways in which it can be coupled to the carrier and a variety of complexes will be formed (BSA "trees" with a varying number of peptide "branches" scattered all over the backbone), presenting the epitope in many different ways.

3. To the peptide-BSA mixture, add PBS to give a final volume of 670 μl; then (carefully, slowly, and while mixing) add 330 μl of a freshly prepared 0.3% glutaraldehyde solution (equaling approximately 10 μmol).
4. Incubate the coupling reaction on a roller device at room temperature for 1 h to allow cross-linking of peptide and carrier protein.
5. Add 200 μl of 1-M glycine in order to quench the remaining free glutaraldehyde and continue rolling for 1 h.
6. Dialyze the coupling reaction overnight against PBS to remove small molecules such as uncoupled peptide, glycine, and glycine-glutaraldehyde. Dialyze two or three times against at least 500 volumes of PBS at 4°C. The final step can be

overnight. Usually, after dialysis, the concentration of peptide-BSA complex is around 4 mg/ml and the solution is ready for direct use as an antigen in immunization.

3.1.4. Preparation of Primary and Booster Immunizations

When using synthetic peptides or proteins purified from *E. coli*, adjuvants containing immunostimulatory molecules are applied to enhance the immune response to the antigen. Freund's adjuvants have been used in our studies; Freund's complete adjuvant (FCA) for the initial immunization and Freund's incomplete adjuvant (FICA) for subsequent booster immunizations. FCA (but not FICA) contains heat-inactivated *Mycobacterium tuberculosis* or *Mycobacterium butyricum* (or extracts thereof), which stimulate both cellular and humoral immunity. The water-in-oil emulsion that is the basis for FCA guarantees the slow release of the antigen from the site of immunization, but it should be noted that the mineral oil component is quite toxic and induces granulomatous reactions. Mix FCA/FICA with the antigen-containing solution to form a "toothpaste-like" stable emulsion. See below for details. In our standard protocol, ~200 µg of antigen (peptide-BSA complex) is used for the primary immunization, and the same amount for subsequent booster reactions.

1. Before use, mix the FCA, e.g., gently on a roller device to get the bacteria into suspension. For the primary immunization, the antigen (200–250 µg) is diluted with PBS to give a final volume of 0.8 ml PBS, which is mixed with 0.4 ml of FCA. Transfer this mixture to a 3-ml syringe (luer-lock). Use a P1000 micropipette; put the tip in the opening of the syringe, and slowly pull the plunger back. In the same manner, fill a second syringe with 0.4 ml of FCA and 0.4 ml of air (to promote the formation of the emulsion).
2. Connect the two syringes with a three-way stopcock (**Fig. 3**). Mix the contents of the two syringes until a thick, white emulsion is obtained. This means that the suspension is converted into an oily mass that includes the water. This emulsion, when properly prepared, is stable and will not disperse when a droplet is put into PBS. The emulsion will ensure the slow release of the antigen into the animal.
3. Disconnect the syringes and use the antigen suspension the same day. Optionally, the syringes can be stored for later use, overnight at 4°C or for longer periods at –80°C.
4. For booster immunizations FICA is used instead of FCA. Material for two boosts can be prepared in one action by doubling the volumes above. Filled syringes can be stored at –80°C for prolonged periods of time.

3.1.5. Rabbits and the Schedule for Immunization and Bleeding

Young adult New Zealand white rabbits (preferably females, weighing between 2 and 3 kg) have been used routinely in our studies (**Fig. 3**). The young age

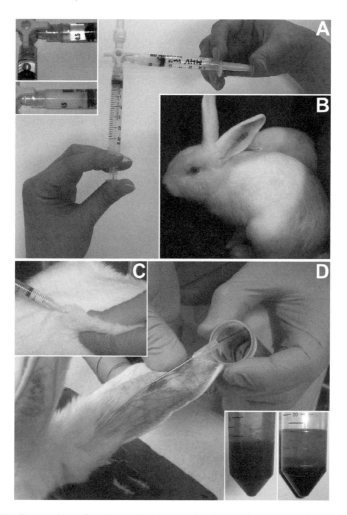

Fig. 3. (A) Prepration of syringes for immunization using two syringes and a three-way stopcock. Insets show syringes with antigen and Freund's adjuvants before and after mixing. (B) Young adult New Zealand white rabbit with ear tattoo. (C) Subcutaneous immunization of a rabbit. (D) Collection of a test bleed of 15–20 ml from an ear vein of a rabbit. Insets show a fresh bleed and a bleed after clotting and centrifugation the next day.

of the animals is thought to be particularly important to ensure a strong primary response to the antigen. Although it is often advised to use two animals per antigen, in our experience—with just a few exceptions—both animals generally give the same (positive or negative) result, although titers of the antiserum and background signal(s) may vary between such duplicates. Still, if the number of animals available or housing capacity is limited, and also for financial and ethical reasons, it may be wiser to use a single animal per antigen and—in the event of failure—return to such an antigen in a subsequent round of immunizations.

The rabbits are injected subcutaneously at three or four places on their back (**Fig. 3**). Given the strong inflammatory response induced by FCA, the animals are monitored closely following the primary immunization. The interval between the primary immunization and the first booster, and between subsequent boosters, is about 4 weeks. It is essential to obtain a preimmunization bleed on (or before) the day of the primary immunization, which will be used to assess the response against the antigen (see below). The standard schedule for immunization and bleeding would look like this:

- Day 1: Preimmunization bleed and primary immunization.
- Day 28: First booster immunization.
- Day 38: First test bleed.
- Day 56: Second booster immunization.
- Day 66: Second test bleed.
- Day 70: Final bleed (~50 ml of serum can be obtained).

Bleeds should be limited to 15% of the animal's total blood volume and a 4-week recovery period should be allowed. As a "rule of thumb," the blood volume of rabbits is about 56 ml/kg of body weight—assuming the animal is mature, healthy, and adequately nursed.

In our experience, a (good) response is unlikely (but not impossible. . .) if there is no positive signal after two or three boosts. Although there have been some exceptions to this rule, it is advisable—for financial reasons as well as ethical considerations—to terminate the experiment after ~70 days and, if necessary, try again with an additional animal or an alternative antigen later on.

3.1.6. Preparation of Serum from Bleeds

1. Collect 15–20 ml of blood (usually from an ear vein of the rabbit; **Fig. 3**) in a 50-ml centrifuge tube and allow complete clotting of the blood in a water bath at 37°C for 1 h.
2. Use a Pasteur pipette with a melted tip to gently liberate the clot from the wall of the tube. Be sure to cause minimal lysis of red blood cells, to ensure a clear serum sample later on. Subsequently, leave the clot to shrink overnight at 4°C (**Fig. 3**).
3. Centrifuge the tube at 1300 × g for 10 min to further reduce the volume of the clot.
4. Carefully, with a pipette, remove as much serum as possible from the tube without touching/damaging the clot; in case of doubt, use a second collection tube for the last few ml (also an additional spin might help), which may be less clear and might be saved as a backup sample only.
5. Prepare some smaller and some larger aliquots of the serum and store at –20°C or –80°C. Sera can be kept for many years, but repeated freezing and thawing should

be avoided. Also, it is advisable to use tubes with a screw cap lid containing a rubber seal, which will minimize evaporation during prolonged storage. Store a small sample (0.5 to 1 ml) at 4°C for testing.

6. Working stocks of antisera (in screw cap tubes) can be kept at 4°C for many months; if desirable, 0.05% (final concentration) of sodium azide (highly toxic!) can be added as a preservative, but in our experience this is not necessary if sera are kept cool and handled with care. It is advisable to spin antisera (in particular the less clear ones) for 1 min at full speed in a microcentrifuge prior to use.

3.2. Testing of the Antiserum

The native proteins that are the targets for antiserum production are usually highly structured molecules and obviously the reactivity of the antiserum will be influenced by the conformation of the antigen used for immunization. In particular, when synthetic peptides are used the structural resemblance between antigen and target may be limited. This probably explains why—in our experience— peptides derived from the termini of the target protein, which are often less structured than the protein core, generally work better than internal peptides.

Although an ELISA approach, based on the antigen used for immunization, can be employed for an initial screening of the immune response in the animal, this result may be relatively meaningless when it comes to the question of whether the antiserum will ultimately react with the viral protein target. One will in fact be measuring the response against the combination of peptide and carrier protein or, in the case of immunization with proteins expressed in and purified from *E. coli*, against the target and contaminants present in the antigen. All the immunizations we carried out with BSA-coupled peptides produced sera that reacted positively in an ELISA using plates coated with the uncoupled peptide, but less than half of those turned out to be useful in IF, WB, or IP. Therefore, it is advisable to screen immediately using samples containing the native, full-length viral protein target.

Moreover, during this screening process, the conformation of the target is an issue. When screening formaldehyde-fixed whole cells by microscopy techniques the target is fixed but structured. During WB analysis proteins are subjected to denaturing conditions and fixed to the membrane used for blotting, whereas IPs are done in solution and can be performed under both native and denaturing conditions. In fact, the level of denaturation in IP experiments can be easily influenced by varying the percentage of SDS in the IP buffer. Often, IPs with antisera raised using synthetic peptides work better in the presence of relatively high (0.25–0.5%) concentrations of SDS, probably because the epitope in the denatured target resembles the antigen used for immunization more closely under these conditions. An additional significant advantage is the fact that the higher SDS concentrations will strongly reduce the background IP signal and result in a much cleaner analysis (see below).

In our experience, antisera that work well in IF assays usually also work well in WB and IP. Conversely, sera that are negative in IF may still be highly reactive in WB and/or denaturing IP. Consequently, it is important to test new antisera in at least two of these three assays. In our daily practice, we have usually relied on IF for preliminary screening and WB for confirmation. There is an additional reason to confirm the IF results by subsequent WB or IP analysis: in IF assays antisera can sometimes show a strong reaction with cellular proteins (**Fig. 4**), either—at low dilutions—because of the lack of a specific response against the target or—at higher dilutions—because of an unexpected cross-reactivity

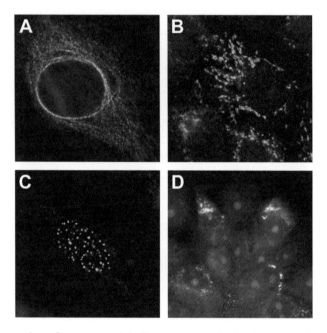

Fig. 4. Examples of erroneous labeling patterns obtained with various rabbit antisera in immunofluorescence microscopy: (A) Antiserum raised against nsp11 of the arterivirus EAV recognizing a cytoskeletal component. (B) Antiserum raised against SARS-CoV nsp1 producing a mitochondrion-like labeling pattern. (C) Antiserum raised against SARS-CoV nsp3 *(4)*, which was very specific in Vero-E6 cells (*see* **Fig. 5**), resulting in a strong, punctate nuclear labeling in BHK-21 cells that expressed the ACE-2 receptor for SARS-CoV. (D) SARS-CoV-infected cells labeled several weeks after fixation and embedded using ProLong mounting fluid. An aspecific nucleolar background labeling was observed, especially in the green range of the fluorescence spectrum. The specific, much brighter signal is derived from an anti-nsp3 labeling using a rabbit antiserum that was directly coupled to Alexa Fluor-488 *(4)*.

with cellular proteins. WB or IP analysis provides important information about the molecular mass of the target that is recognized, and may thus prevent premature conclusions about the specificity of the antiserum. The correct assessment of specificity is also aided by the inclusion of two important controls: the preimmunization serum (which should obviously not show the same signal) and mock-infected control cells (which might reveal that it is in fact a cellular target that is being recognized). When performing initial screening with IF assays, it is practical to use cell cultures that have been infected at an MOI of 0.5–1 and therefore contain a mixture of virus-infected and mock-infected cells in the same specimen.

3.2.1. Immunofluorescence Assay

Semiconfluent cells seeded on glass cover slips (10-mm-diameter) are the preferred specimens for initial testing. The cells can either be infected with the target virus or transfected with a vector expressing the target protein. Obviously, cells should be fixed at a time point when the target protein is convincingly expressed. The most reactive rabbit antisera that we have produced can be used in IF assays at dilutions of between 1:1000 and 1:5000. For initial testing, however, dilutions on the order of 1:100 to 1:500 are advised.

1. Wash the cells once with PBS and fix the cells at room temperature with 3% paraformaldehyde in PBS for at least 30 min (or overnight).
2. Wash the cells once with PBS containing 10 mM glycine (coverslips in sealed dishes can be stored at 4°C in PBS for many weeks).
3. Using sharp tweezers, carefully transfer coverslips to the wells of a prelabeled 24-well cluster containing PBS-glycine. Be sure to remember, every time you handle the coverslips, on which side there are cells. While in the cluster, cells should always be facing upward.
4. Permeabilize the cells at room temperature for 10 min in PBS containing 0.1% Triton X-100.
5. In 5 min, wash the cells three times with PBS-glycine and leave them in the last wash step.
6. Dilute the primary antiserum (initial dilutions, e.g., 1:100 and 1:500) in PBS containing 5% FCS; 50 µl per coverslip (10-mm-diameter) will be needed.
7. Cut a large piece of parafilm, place it in a larger dish, label the position of the various samples, and place 50-µl drops of the antiserum dilutions on the parafilm. One by one, take the coverslips from the 24-well cluster, remove excess PBS by touching a tissue to the side of the coverslips, and place the coverslips on the drops with the cells facing the antiserum.
8. Incubate for 30–60 min at room temperature (or 37°C); make sure the samples do not dry out during this incubation (e.g., by placing a wet tissue in the dish).

9. Return the coverslips to the 24-well cluster; in 20 min wash three times with PBS-glycine and leave them in the last wash step.

10. Prepare a dilution of the fluorescently labeled secondary antibody. Optimal dilutions of conjugates should be determined separately using a well-defined primary antiserum. Again 50 µl per coverslip will be needed. Optionally, 1 µg/ml of Hoechst 33258 can be added to this dilution for staining of nuclear DNA.

11. Incubate and wash as described under steps 7 to 9.

12. Take the coverslips from the 24-well cluster. Embed the specimens onto glass microscopy slides in a mounting fluid [e.g., Mowiol or ProLong (Molecular Probes)]. Avoid air bubbles in the mounting fluid by slowly and carefully sliding the coverslip on a small drop (~5 µl) of mounting fluid.

13. Store the specimens in the dark at 4°C. The mounting fluid should harden at least overnight before high magnification lenses and immersion oil are used.

14. Analyze the specimens in a fluorescence microscope using the filter sets required for the label attached to the secondary antibody.

3.2.2. *Immunofluorescence Double-Labeling Experiments*

Double-labeling experiments can be done to compare the localization of two (or more) proteins of interest in the same cell by combining a rabbit antiserum recognizing one protein and, e.g., a mouse monoclonal antibody recognizing a second target. Obviously, the two primary antisera have to be detected with suitable conjugates carrying different fluorescent labels, preferably with well-separated emission spectra. Following initial optimization (testing of different dilutions, balancing of the two signals, and controls for specificity of primary and secondary antibodies), one can usually carry out the labeling using a two-step approach. First, specimens are simultaneously incubated in a combined dilution of the two primary antibodies and, subsequently, after extensive washing, in a combined dilution of the two conjugates. In case of background and/or specificity problems, however, it may be necessary to perform the labeling in four consecutive steps.

In IF assays, double labeling with two antisera from the same species is only possible when one of the two is directly coupled to a fluorescent group. We have recently been successful (**Fig. 5**) (*4*) in purifying the Ig fraction from small volumes (2–3 ml) of rabbit antisera using a commercially available protein A column and have subsequently conjugated these antibodies to Alexa Fluor-488. The IF assay then consisted of three incubation steps: (i) incubation with the uncoupled rabbit antiserum, (ii) incubation with the anti-rabbit conjugate recognizing the first antibody, and (iii) incubation with the Alexa Fluor-488-coupled second rabbit antiserum. The order of these steps is very important to avoid binding of the anti-rabbit conjugate to the directly labeled second rabbit antiserum.

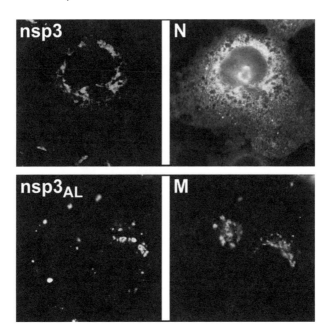

Fig. 5. Examples of immunofluorescence microscopy double-labeling experiments. Top panel: double labeling of EAV-infected Vero cells with an anti-nsp3 rabbit antiserum (left) and a mouse monoclonal antibody recognizing the N protein (right). Bottom panel: double labeling of SARS-CoV-infected Vero cells with two rabbit antisera, an anti-nsp3 serum that was directly coupled to Alexa Fluor-488 (left) and an anti-M rabbit antiserum that was visualized with a Cy3-conjugated donkey-anti-rabbit IgG secondary antibody (right). See text for details.

3.2.3. Western Blotting Analysis

During Western blotting (WB) analysis samples containing the protein of interest are separated according to size in acrylamide gels and transferred to a membrane, which is subsequently incubated with the antiserum. Protein bands reacting with the antiserum are detected using a secondary antibody and accompanying assay (a variety of enzyme-linked or fluorescent conjugates is available; here we use a peroxidase-coupled secondary antibody). As in the case of IF assays, lysates to be tested can be derived from cells that are either infected or transfected with an appropriate expression vector. For infection lysates, we use a high multiplicity of infection to avoid a mixture of infected and uninfected cells in the sample. Cells are lysed in protein lysis buffer, which leave the nuclei intact and allow their removal by centrifugation. For initial testing, an amount of lysate equaling $\sim 5 \times 10^5$ cells per 5 cm gel width (or 10–15 slots) can be used. Alternatively, purified protein (e.g., expressed in *E. coli*) can be used (100–500 ng is

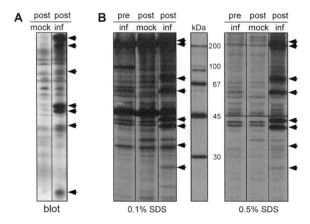

Fig. 6. Example of a standard test of a rabbit antiserum in Western blot and immuno-precipitation. The antiserum used here was raised against a domain in the C-terminal region of pp1a of the arterivirus EAV *(8)* and recognizes a large set of processing intermediates and end products, which are indicated by arrowheads: (A) Western blot analysis: EAV- and mock-infected cell lysates were prepared at 8 h postinfection, run on an SDS-polyacrylamide gel, blotted to PVDF membrane, and incubated with the postimmune serum at a 1:1000 dilution. Detection was with an anti-rabbit IgG HRPO conjugate and a chemiluminescence assay. [Reproduced from *(13)*.] (B) Immunopre-cipitation: EAV-infected cells were labeled with ^{35}S-methionine/cysteine for 3 h, from 5–8 h postinfection and lysates were used for IP using 5 μl of antiserum per sample. Fol-lowing the IP, samples were run on an SDS-polyacrylamide gel and signal was detected using autoradiography. As specificity controls, the preimmune serum was tested on EAV-infected cells and the postimmune serum was tested on mock-infected cells. Fur-thermore, for the left panel the binding of the antiserum to the antigen was done in the presence of 0.1% SDS, whereas 0.5% SDS was used for the right panel. The comparison illustrates how higher SDS percentages can reduce the background signal and improve the overall picture.

usually sufficient). Most antisera can be used at a 1000-fold dilution or higher, but we advise starting the testing with a 1:500 dilution (**Fig. 6A**).

1. Wash the cells with cold PBS and—to a 10-cm^2 dish with 1–2 × 10^6 cells—add 300 μl of cold protein lysis buffer.
2. Incubate for 5 min while monitoring lysis with a microscope. Transfer the lysate to a labeled microfuge tube and spin down the nuclei for 2 min at full speed in a microcentrifuge.
3. Transfer the supernatant to the new tube, leave the pellet (nuclei, often barely visi-ble) behind and add 1/100 volume of 0.5 M EDTA.
4. The lysate can now be used for WB or stored in the –20°C/–80°C freezer.

5. Prepare an SDS-PAGE gel (or minigel) of a suitable acrylamide percentage (depending on the size of the protein of interest) and run: (i) the lysate containing the protein of interest, (ii) a control lysate (mock-infected or untransfected cells), and (iii) a molecular weight marker. Several sets of these samples can be run on one gel to try different antiserum dilutions, or wider slots can be used and strips cut with the right samples after blotting.

6. With a pencil mark one side of the membrane; this is the side of the membrane that will face the gel during blotting and, subsequently, the solutions during incubation. Prewet the PVDF membrane in methanol for 5 min; never touch the PVDF membrane with bare fingers; always use gloves! Dilute 10X WTB to give 1X WTB and incubate the membrane in 1X WTB for 10 min.

7. Take the gel from between the glass plates and briefly wash it in 1X WTB (not longer than 5 min so that the gel does not swell).

8. Build the electroblot stack—from cathode to anode—with the following layers: three sheets of Whatman paper presoaked in WTB (and optionally one sheet of Whatman paper soaked in WTB with 0.1% SDS), the equilibrated gel, the equilibrated and marked membrane, three sheets of presoaked Whatman paper. Avoid air bubbles between the layers!

9. Blot at 15 V (fixed) for 20 min.

10. Take out the membrane and rinse it in H_2O for 5 min.

11. Dilute the antiserum (1:500) and the preimmune serum as a control (1:500) in PBS-TMB and incubate the blots for 1 h at room temperature while swirling.

12. Wash the blot three times in PBS-T for 5 min.

13. Dilute the peroxidase-conjugated secondary antibody (swine-anti-rabbit IgG HRPO) in PBS-TMB and incubate the blots for 1 h at room temperature while swirling.

14. Wash the blot three times in PBS-T for 5 min.

15. Prepare the solutions for the chemiluminescence assay according to the manufacturer's instructions.

16. Gently "semidry" the blot with some Whatman paper (let the fluid run off; do not really dry it!).

17. Spread a piece of plastic foil on the bench and tape the edges to make a smooth surface.

18. Put the chemiluminescence solution on the plastic foil and incubate the blot with the "protein side down" for 5 min.

19. Remove the excess of fluid with some Whatman paper and wrap the blot in plastic foil.

20. Expose an X-ray film to the blot for 1–2 min, develop the film, and check if a longer exposure is required. To avoid overexposure of the film, it may be wise to wait about 30 min before making the first exposure.

3.2.4. Immunoprecipitation

Proteins of interest can be "fished out" of cell lysates by immunoprecipitation. Briefly, antibodies are allowed to bind to the protein and the protein-antibody complexes are then purified by having them bind to beads carrying the Ig-binding proteins A or G on their surface, which are subsequently spun down and washed repeatedly to remove unbound proteins. The simplest form of protein A carrying beads are formalin-fixed *Staphylococcus aureus* cells (e.g., pansorbin; Calbiochem), but alternatively protein A or G coupled to sepharose can be used. Usually, metabolically labeled protein lysates (e.g., ^{35}S-methionine-labeled) are used for immunoprecipitation studies and samples are run on SDS-PAGE gels that can then be used for autoradiography.

In contrast to WB analysis, immunoprecipitation relies on recognition of the target in solution. Depending on the conditions used, proteins can either be close to their native confirmation, partially denatured, or completely denatured (e.g., in the presence of high concentrations of SDS). Thus, the conformation of the antigen used for immunization may influence the results. In our experience, linear antigens such as synthetic peptides may yield antisera that work considerably better in immunoprecipitation assays carried out under stringent denaturing conditions (e.g., 0.5–1% SDS), which will often also reduce the background in the assay (**Fig. 6B**). Nevertheless, the optimal concentration of SDS should be determined for every antigen-antiserum pair since the optimum may be as low as 0.1% SDS, e.g., when using antibodies recognizing mainly (or exclusively) conformational epitopes. Furthermore, the amount of antiserum in the immunoprecipitation assay is important. For initial screening of rabbit antisera we usually test 1 and 5 μl of the antiserum in a 300 μl total IP volume.

1. Take 50 μl of ^{35}S-labeled cell lysate in protein lysis buffer *(12)* (equaling about 2–4 × 10^5 cells; obviously the expression level of the target protein is to be considered as well); add 250 μl of IP buffer and the antiserum (1 or 5 μl). Incubate these ~300-μl samples overnight (or for at least 3 h) at 4°C while swirling.
2. Wash the pansorbin cells (25 μl for each sample) by mixing with an equal volume of IP buffer, spin down for 20 sec at full speed in a microcentrifuge, and resuspend the pansorbin cell pellet in the original volume (again, 25 μl for each sample).
3. Add 25 μl of washed pansorbin cells to each immunoprecipitation and incubate at 4°C for more than 1 h while swirling.
4. Spin down for 1 min at full speed in a microcentrifuge, remove the supernatant, and add 500 μl of weak wash buffer A; vortex until the pellet is completely resuspended. (Optionally, this step can be repeated once to get a cleaner result.)
5. Spin down for 1 min at full speed in a microcentrifuge again and repeat the same washing procedure with weak wash buffer B.
6. Resuspend the pellet in 30 μl of LSB by pipetting up and down.
7. Before loading the samples on an SDS-PAGE gel, denature at 96°C for 3 min. (For some proteins, e.g., very hydrophobic ones, it may be better not to heat the samples and just leave them in LSB for 15 min or longer.)
8. Spin the pansorbin cells down for 2 min at full speed in a microcentrifuge; do not resuspend the pellet! Transfer the supernatant to a new tube and load part of this sample onto an SDS-PAGE gel.

9. Following SDS-PAGE, process the gel for autoradiography or phosphorimager analysis according to standard protocols and expose for 1–7 days.

4. Notes

1. Not all bleeds of the same rabbit have the same concentration of antibody and thus the dilutions to be used can vary.
2. Before embedding, blot the edge of the coverslip as well as the side without cells on tissue paper. Remnants of PBS on the coverslip can cause background when the dried PBS forms crystals.
3. Mounting solutions can sometimes have strange side effects, especially when the cells are not freshly fixed. For example, the use of ProLong Gold mounting fluid can generate a green nonspecific nucleolar signal.
4. When microscopy slides used for embedding are not clean, wipe them first with a tissue with 70% ethanol, next with water, and then with a dry paper towel/tissue. When ethanol is not removed, a (sometimes bright) orange background may result.
5. Occasionally, antisera judged to be negative in IF using paraformaldehyde-fixed cells can become positive using methanol-fixed cells (or the other way round). However, methanol fixation is less gentle and generally the morphology of the cell is less well preserved. Fixation is performed for 10 min at $-20°C$ (using ice-cold 100% methanol or 95% methanol/5% acetic acid) and cells are subsequently transferred to and kept in PBS-glycine. Methanol immediately dissolves all cellular membranes, so there is no need for permeabilization with Triton-X100 during the IF assay.
6. Western blotting results can be improved by using longer washing steps.

Acknowledgments

The authors would like to thank Jan Wouter Drijfhout and Willemien Benckhuijsen (LUMC Department of Immunohaematology) for advice and assistance on peptide design and synthesis. We are grateful to the staff of the LUMC animal facility for almost 15 years of pleasant collaboration and reliable housing and handling of the rabbits used for antiserum production. This work was supported (in part) by the European Commission int the context of the activities of the Euro-Asian SARS-DTV Network (SP22-CT-2004-511064).

References

1. Snijder, E. J., Siddell, S. G., and Gorbalenya A. E. (2005) The order Nidovirales. In: Mahy, B. W., and ter Meulen, V. (eds.) *Topley and Wilson's Microbiology and Microbial Infections*: Virology volume. Hodder Arnold, London, pp. 2390–2404.
2. Ziebuhr, J., Snijder, E. J., and Gorbalenya, A. E. (2000) Virus-encoded proteinases and proteolytic processing in the *Nidovirales*. *J. Gen. Virol.* **81**, 853–879.

3. Snijder, E. J., Bredenbeek, P. J., Dobbe, J. C., Thiel, V., Ziebuhr, J., Poon, L. L. M., et al. (2003) Unique and conserved features of genome and proteome of SARS-coronavirus, an early split-off from the coronavirus group 2 lineage. *J. Mol. Biol.* **331**, 991–1004.

4. Snijder, E. J., van der Meer, Y., Zevenhoven-Dobbe, J., Onderwater, J. J. M., van der Meulen, J., Koerten, H. K., et al. (2006) Ultrastructure and origin of membrane vesicles associated with the severe acute respiratory syndrome coronavirus replication complex. *J. Virol.* **80**, 5927–5940.

5. Tijms, M. A., van der Meer, Y., Snijder, E. J. (2002) Nuclear localization of non-structural protein 1 and nucleocapsid protein of equine arteritis virus. *J. Gen. Virol.* **83**, 795–800.

6. Denison, M. R., Spaan, W. J. M., van der Meer, Y., Gibson, C. A., Sims A. C., Prentice, E., et al. (1999) The putative helicase of the coronavirus mouse hepatitis virus is processed from the replicase gene polyprotein and localizes in complexes that are active in viral RNA synthesis. *J. Virol.* **73**, 6862–6871.

7. van Dinten, L. C., Wassenaar, A. L. M., Gorbalenya, A. E., Spaan, W. J. M., and Snijder, E. J. (1996) Processing of the equine arteritis virus replicase ORF1b protein: identification of cleavage products containing the putative viral polymerase and helicase domains. *J. Virol.* **70**, 6625–6633.

8. Snijder, E. J., Wassenaar, A. L. M., and Spaan, W. J. M. (1994) Proteolytic processing of the replicase ORF1a protein of equine arteritis virus. *J. Virol.* **68**, 5755–5764.

9. Hopp, T. P., and Woods, K. R. (1981) Prediction of protein antigenic determinants from amino-acid-sequences. *Proc. Natl. Acad. Sci. USA–Biol. Sci.* **78**, 3824–3828.

10. Jameson, B. A., and Wolf H. (1988) The antigenic index—a novel algorithm for predicting antigenic determinants. *Comp. Appl Biosci.* **4**, 181–186.

11. Kolaskar, A. S., and Tongaonkar, P. C. A. (1990) Semiempirical method for prediction of antigenic determinants on protein antigens. *FEBS Lett.* **276**, 172–174.

12. deVries, A. A. F, Chirnside, E. D., Horzinek, M. C., and Rottier P. J. M. (1992) Structural proteins of equine arteritis virus. *J. Virol.* **66**, 6294–6303.

13. van Aken, D., Zevenhoven-Dobbe, J. C., Gorbalenya, A. E., and Snijder, E. J. (2006) Proteolytic maturation of replicase polyprotein pp1a by the nsp4 main proteinase is essential for equine arteritis virus replication and includes internal cleavage of nsp7. *J. Gen. Virol.* **87**, 3473–3482.

VI

MANIPULATING THE GENOMES OF CORONAVIRUSES

17

Manipulation of the Coronavirus Genome Using Targeted RNA Recombination with Interspecies Chimeric Coronaviruses

Cornelis A.M. de Haan, Bert Jan Haijema, Paul S. Masters, and Peter J.M. Rottier

Abstract

Targeted RNA recombination has proven to be a powerful tool for the genetic engineering of the coronavirus genome, particularly in its 3' part. Here we describe procedures for the generation of recombinant and mutant mouse hepatitis virus and feline infectious peritonitis virus. Key to the two-step method is the efficient selection of recombinant viruses based on host cell switching. The first step consists of the preparation—using this selection principle—of an interspecies chimeric coronavirus. In this virus the ectodomain of the spike glycoprotein is replaced by that of a coronavirus with a different species tropism. In the second step this chimeric virus is used as the recipient for recombination with synthetic donor RNA carrying the original spike gene. Recombinant viruses are then isolated on the basis of their regained natural (e.g., murine or feline) cell tropism. Additional mutations created in the donor RNA can be co-incorporated into the recombinant virus in order to generate mutant viruses.

Key words: coronavirus; mouse hepatitis virus; feline infectious peritonitis virus; reverse genetics; targeted RNA recombination; host cell switching; tropism

1. Introduction

Targeted RNA recombination was the first reverse genetics method developed to introduce mutations into the coronavirus genome [for a recent review, see (*1*)]. Actually, its first demonstration involved the repair of an 87-nucleotide deletion

From: *Methods in Molecular Biology, vol. 454: SARS- and Other Coronaviruses,*
Edited by: D. Cavanagh, DOI: 10.1007/978-1-59745-181-9_17, © Humana Press, New York, NY

in the N gene of a temperature-sensitive mutant of the mouse hepatitis virus (MHV) by recombination with a synthetic donor RNA equivalent to the subgenomic mRNA for the wild-type N gene *(2)*. Recombinant virus was selected by its increased temperature stability and efficient growth at the nonpermissive temperature, a tedious process. The frequency of RNA recombination and hence the ease with which recombinant viruses could be isolated was subsequently improved by the use of defective interfering (DI) RNAs as donors, i.e., RNA constructs consisting of sequences derived from the 5' and 3' terminal parts of the MHV genome *(3,4)*. Yet, selection of mutant viruses remained the Achilles' heel of the method, particularly when one was seeking to construct temperature-sensitive or otherwise crippled mutants. Hence, the approach did not become common practice.

A decisive improvement emerged from studies on the assembly of the spike into the coronavirion. With the use of virus-like particles it was shown that this process is governed by the spike (S) protein's carboxy-terminal region, comprising the transmembrane and cytoplasmic domains *(5)*. Spikes were still incorporated into virus-like particles after replacement of their ectodomain by that of another coronavirus. Such chimeric spikes also retained their biological functions—receptor binding and membrane fusion—as judged by their ability to cause cell-cell fusion. Given the generally strict species-specific nature of coronaviruses in tissue culture these features enabled the design of a powerful positive selection strategy based on interspecies chimeric coronaviruses, as was first described for MHV *(6)*.

Here we describe the detailed procedures for the method of targeted RNA recombination using this strategy of host cell switching. Protocols are presented for MHV and FIPV (feline infectious peritonitis virus) as examples. For any given virus the procedure starts with the generation of an interspecies chimeric coronavirus. Thus, a donor DI RNA is prepared containing a chimeric S gene. Cells are infected with the virus in question and additionally transfected with the donor RNA. In such cells homologous recombination of the donor RNA with viral genomic RNA can give rise to the formation and release of chimeric virus. This chimeric virus is subsequently selected by passing the culture supernatant onto cells that are infectable only by the virus from which the chimeric S protein ectodomain was derived. Felinized MHV (fMHV) and murinized FIPV (mFIPV) are thus the chimeric viruses described in the examples; they are isolated and grown in feline and murine cells, respectively. The chimeric viruses subsequently function as the recipients for the second round of homologous recombination designed to regenerate the original virus or to obtain mutants thereof. Donor DI RNA is again prepared that now contains the original wild-type S gene, plus any additional intended mutations. In cells infected with the chimeric virus, recombination with this RNA can lead to formation of (mutants

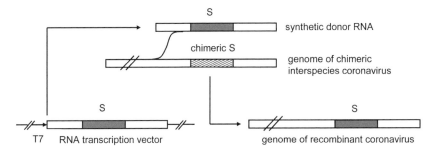

Fig. 1. Second round of homologous recombination to regenerate the original virus or to obtain mutants thereof. Donor DI RNA with the original wild-type S gene plus any additional intended mutations are introduced into cells that have been infected with the chimeric virus. Recombination with this RNA can lead to formation of (mutants of) the original virus. These are isolated by passing the collection of progeny viruses onto its natural cells.

of) the original virus, which can now, in turn, be isolated by passing the collection of progeny viruses onto its natural cells. A schematic outline of the latter recombination process is depicted in **Fig. 1**.

Owing to restrictions inherent in the selection principle, the targeted RNA recombination system described here will have its main value in the study and manipulation of functions specified by the genomic regions downstream of the polymerase gene. For a survey of important contributions that the targeted recombinant approach has made in the different areas of its application, the reader is referred to the review by Masters and Rottier *(1)*.

2. Materials

2.1. Plasmids and RNA Synthesis

1. Plasmids pMH54 *(6)* and pBRDI1 *(7)* are used for the generation of MHV and FIPV recombinants, respectively.
2. Restriction enzymes *Pac* I and *Not* I (New England Biolabs) are used in combination with the supplied reaction buffer.
3. 3 M ammonium acetate solution (*see* **Note 1**).
4. Ethanol 70% (v/v).
5. Ethanol 96% (v/v).
6. T7 RNA polymerase kit (Ambion).
7. Agarose (Invitrogen).
8. 1X Tris-borate electrophoresis buffer (TBE): 0.09 M Tris-borate, 0.002 M EDTA. 0.5X Tris-acetate electrophoresis buffer (TAE): 0.02 M Tris-acetate, 0.0005 EDTA.

2.2. Cell Culture and Plaque Assay

1. Dulbecco's Modified Eagle's Medium (DMEM; Biowhittaker) supplemented with 10% fetal bovine serum (FBS; HyClone), 100 IU of penicillin/ml and 100 μg of streptomycin/ml (p/s) (DMEM++).
2. Both murine LR7 cells (5) and feline FCWF cells (ATCC) are used for the generation of MHV and FIPV recombinants (*see* **Note 2**).
3. Solutions of trypsin (500 mg/liter trypsin 1:250) and EDTA (200 mg/liter) (Trypsin EDTA solution; Biowhittaker).
4. Dulbecco's phosphate buffered saline (DPBS) without Ca and Mg.
5. Select agar (Invitrogen).
6. 2X Minimal Essential Medium Eagle (EMEM; Biowhittaker) supplemented with 20% FBS and 2X p/s.

2.3. Infection and Electroporation

1. Recipient viruses fMHV and mFIPV are used for the generation of MHV and 2. FIPV recombinants, respectively.
2. 0.4-cm Gene Pulser cuvette (Biorad).
3. Biorad Gene Pulser II apparatus with Capacitance extender plus.

3. Methods

The procedures for the two steps of the method—preparation of the chimeric recipient virus and re-creation of the original virus (with or without additional mutations)—are identical, except for the reciprocal way the recombinant viruses are selected. Hence, we have chosen to describe here the protocols for the second step only, i.e., for generating recombinant forms of wild-type MHV and FIPV; the chimeric viruses fMHV and mFIPV are assumed to have been prepared similarly.

For the preparation of recombinant MHV and FIPV, we used, respectively, RNA transcription vectors pMH54 and pBRDI1. The generation of these plasmids has been described in detail by Kuo et al. (6) and by Haijema and co-workers (7), respectively. pMH54 specifies a defective MHV-A59 RNA transcript consisting of the 5'-end of the genome (467 nt) fused to codon 28 of the HE gene and running to the 3'-end of the genome; pBRDI1 specifies a similar FIPV transcript.

Both viral cDNA constructs are cloned behind a bacteriophage T7 RNA polymerase promoter. Mutations to be created in the MHV or FIPV genomes are introduced into these cDNA constructs in order to be carried through the transcribed RNA into the recombinant virus.

For the generation of recombinant MHV, feline FCWF cells are infected with fMHV, after which donor RNA transcribed from pMH54 is transfected by electroporation. Next, the recombinant MHV is selected on murine LR7 cells. For the generation of recombinant FIPV, the LR7 cells are infected with mFIPV, after which the donor RNA transcribed from pBRDI1 is transfected by electroporation. Subsequently, the recombinant FIPV is selected on FCWF cells.

3.1. Preparation of Recipient Virus Stocks

1. Maintain the LR7 and FCWF cells by regular passaging, using Trypsin EDTA solution when the cells approach confluency, to provide new maintenance cultures as well as cultures for the generation of the recipient virus stocks.
2. A 1:15 split of LR7 cells and a 1:3 split of FCWF cells will usually provide cultures that approach confluence after 72 h.
3. Wash LR7 or FCWF cells once with DMEM before inoculating with mFIPV or fMHV, respectively, in DMEM at a multiplicity of infection (MOI) of 0.05–0.1 $TCID_{50}$ (50% tissue culture infectious doses) per cell (*see* **Note 2**).
4. After 1 h incubation at 37°C, remove the inoculum and wash the cells once with DMEM, and then overlay them with DMEM++.
5. Harvest the culture medium containing the viruses typically after 24 h incubation at 37°C, when the monolayers exhibit extensive cytopathic effects and detach from the plastic.
6. Transfer the medium to 50-ml polypropylene tubes and clarify by centrifugation for 10 min at 2500 rpm at room temperature using a Labofuge GL (Heraeus) or a comparable centrifuge.
7. Store the supernatant in aliquots at –80°C until use. Typical titers obtained for fMHV and mFIPV are around 10^6 $TCID_{50}$/ml.

3.2. Preparation of Synthetic Donor RNA

1. Linearize 10 μg of pMH54 or pBRDI1 with 30 U *Pac* I or *Not* I restriction enzyme, respectively, using the reaction buffers supplied by New England Biolabs in a reaction volume of 100 μl for 2 h at 37°C.
2. Check the linearization of the plasmids by analyzing an aliquot of the reaction mixture using 1% (w/v) agarose gel electrophoresis in 0.5X TAE.
3. Precipitate the linearized plasmid DNA by addition of one-tenth volume of 3 M ammonium acetate and 2.5 volumes of 96% (v/v) ethanol, followed by incubation at –20°C for 30 min.
4. Pellet the DNA precipitates using an Eppendorf centrifuge (13,000 rpm, 15 min, 4°C).
5. Wash the pellet with 70% (v/v) ethanol, dry it, and take it up in 10 μl of water (*see* **Note 3**).
6. Transcribe synthetic RNA in a 20-μl reaction volume, using the T7 Message Machine kit (Ambion) according to the manufacturer's instructions, with the following modifications.

    a. Instead of 1 µg of DNA, 3 µl of the concentrated, linearized plasmid DNA is used.

    b. In addition, 3 µl of 30 mM GTP (present in the kit) is added, to enhance transcription of large RNAs (*see* **Note 4**).

7. Check RNA synthesis by analyzing one-tenth of the reaction volume by electrophoresis in a 1% (w/v) agarose gel in 1X TBE.
8. Store the RNA at −80°C until use.

3.3. Infection and Electroporation

1. Typically, one 80-cm^2 flask containing a culture of cells reaching confluence is sufficient for two RNA electroporations.
2. Infect FCWF or LR7 cells with the recipient viruses fMHV or mFIPV, respectively, at an MOI of 0.5–1 TCID$_{50}$/cell in DMEM++.
3. After 4 h of incubation at 37°C, detach the cells from the plate using Trypsin EDTA solution (1.5 ml per 80 cm^2).
4. Add approximately 10 ml DMEM++ to the cell suspension.
5. Pellet the cells by centrifugation in 50-ml polypropylene tubes for 10 min at 500 rpm at room temperature.
6. Remove the supernatant and gently take up the pellet in DPBS without Ca and Mg.
7. Pellet the cells by centrifugation in 50-ml polypropylene tubes for 10 min at 500 rpm at room temperature.
8. Remove the supernatant and gently take up the cell pellet in DPBS without Ca and Mg (1 ml/80 cm^2 original monolayer). Gently pipette the cell suspension carefully up and down until all cell clumps have disappeared.
9. Transfer 800 µl of the cell suspension to a 0.4-cm Gene Pulser cuvette containing 9 µl of the synthetic donor RNA.
10. Mix the cells and RNA by pipetting gently.
11. Subject the RNA/cell mixture to two consecutive pulses using the Biorad Gene Pulser II with capacitance extender plus (0.3 kV, 950 µF).
12. Add the electroporated cell suspension to 8 ml DMEM++ and mix gently.
13. Transfer the DMEM++ to two 25-cm^2 flasks containing subconfluent monolayer cultures of LR7 or FCWF cells, in order to allow propagation of the respective recombinant viruses.

3.4. Selection and Purification of Recombinant Viruses

    Recombinant MHV or FIPV are selected by growth on LR7 or FCWF cells, respectively. After 24 h incubation of these cells—overlaid with the infected/transfected cells—at 37°C, cytopathic effects are visible when recombinant wild-type MHV or FIPV is generated. The appearance of cytopathic effects may take more time when preparing viruses with crippling mutations.

1. When cytopathic effects are clearly visible, collect the culture medium and clarify by centrifugation as described above.
2. Store the supernatants at –80°C until the recombinant viruses are purified by two consecutive plaque assays.
3. To this end, autoclave a 3% (w/v) agar solution in water and cool it to 43°C.
4. At the same time, warm the 2X EMEM concentrate containing FBS and p/s at 37°C.
5. Inoculate cells reaching confluence (LR7 or FCWF cells for MHV or FIPV, respectively) with increasing dilutions in DMEM of the cleared cell culture supernatant.
6. After 1 h at 37°C, remove the inoculum and overlay the cells with a 1:1 mixture of the agar solution and the 2X EMEM concentrate (2 ml/10 cm^2 monolayer) (*see* **Note 5**).
7. After 5 min incubation at room temperature, transfer the cells to 37°C. Typically, plaques are visible after a 24- h incubation.
8. Pick plaques using disposable tips of a 200-µl pipette. First, remove the extreme end of the tip using scissors.
9. Subsequently, push the tip through the agar overlay, at the position of a free plaque.
10. Transfer the agar in the tip to 0.5 ml DMEM.
11. Store this at –80°C until it is used in a second round of plaque purification.
12. After the second round of plaque purification, generate a stock of recombinant virus by inoculating a 25-cm^2 flask containing either LR7 or FCWF cells, using one-half of the purified plaque.
13. Harvest stocks (designated as passage 1) when extensive cytopathic effects are visible.
14. Clear the harvest by centrifugation and store in aliquots at –80°C.

3.5. Verification of Recombinant Phenotype

1. In general, generate at least two independent recombinants to verify that the observed phenotypic characteristics are the result of the intended genomic modifications.
2. Purify the genomic RNA from the passage 1 stock using the QIAamp Viral RNA mini kit (Qiagen).
3. Genetically analyze the recombinant viruses by reverse transcription-PCR using genomic RNA as a template.
4. Analyze the RT-PCR products using size or restriction fragment analysis and direct sequencing.

4. Notes

1. Unless otherwise stated, prepare all solutions in MilliQ water.
2. Instead of LR7 cells, 17 Cl1, L2, DBT, or Sac(–) cells can be used.
3. Wear gloves and use RNase-free water and materials.
4. Capped synthetic RNA transcripts are used. It is recommended that the synthetic RNA be used within 48 h.
5. Mix by pipetting once up and down; then use immediately. If the temperature of the agar solution used is too high, the cells will be killed. If the temperature is not high enough, the solution will solidify prematurely.

References

1. Masters, P. S., and Rottier, P. J. (2005) Coronavirus reverse genetics by targeted RNA recombination. *Curr. Top. Microbiol. Immunol.* **287**, 133–159.
2. Koetzner, C. A., et al. (1992) Repair and mutagenesis of the genome of a deletion mutant of the coronavirus mouse hepatitis virus by targeted RNA recombination. *J. Virol.* **66**, 1841–1848.
3. van der Most, R. G., et al. (1992) Homologous RNA recombination allows efficient introduction of site-specific mutations into the genome of coronavirus MHV-A59 via synthetic co-replicating RNAs. *Nucleic Acids Res.* **20**, 3375–3381.
4. Masters, P. S., et al. (1994) Optimization of targeted RNA recombination and mapping of a novel nucleocapsid gene mutation in the coronavirus mouse hepatitis virus. *J. Virol.* **68**, 328–337.
5. Godeke, G. J., et al. (2000) Assembly of spikes into coronavirus particles is mediated by the carboxy-terminal domain of the spike protein. *J. Virol.* **74**, 1566–1571.
6. Kuo, L., et al. (2000) Retargeting of coronavirus by substitution of the spike glycoprotein ectodomain: crossing the host cell species barrier. *J. Virol.* **74**, 1393–1406.
7. Haijema, B. J., Volders, H., and Rottier, P. J. (2003). Switching species tropism: an effective way to manipulate the feline coronavirus genome. *J. Virol.* **77**, 4528–38.

18

Generation of Recombinant Coronaviruses Using Vaccinia Virus as the Cloning Vector and Stable Cell Lines Containing Coronaviral Replicon RNAs

Klara Kristin Eriksson, Divine Makia, and Volker Thiel

Abstract

Coronavirus reverse genetic systems have become valuable tools for studying the molecular biology of coronavirus infections. They have been applied to the generation of recombinant coronaviruses, selectable replicon RNAs, and coronavirus-based vectors for heterologous gene expression. Here we provide a collection of protocols for the generation, cloning, and modification of full-length coronavirus cDNA using vaccinia virus as a cloning vector. Based on cloned coronaviral cDNA, we describe the generation of recombinant coronaviruses and stable cell lines containing coronaviral replicon RNAs. Initially, the vaccinia virus-based reverse genetic system was established for the generation of recombinant human coronavirus 229E. However, it is also applicable to the generation of other coronaviruses, such as the avian infectious bronchitis virus, mouse hepatitis virus, and SARS coronavirus.

Key words: Coronavirus; RNA virus; reverse genetics; vaccinia virus; full-length cDNA; *in vitro* transcription; recombinant coronaviruses

1. Introduction

The extraordinarily large size of the positive-stranded coronavirus RNA genome posed a significant obstacle for the establishment of coronavirus reverse genetic systems based on cloned full-length cDNA. Conventional cloning techniques using plasmid DNA cloning vectors were not suitable to stably accommodate large coronaviral cDNAs. Moreover, in numerous cases, specific coronaviral cDNA sequences turned out to be resistant to cloning in conventional plasmid DNAs or

From: *Methods in Molecular Biology, vol. 454: SARS- and Other Coronaviruses,*
Edited by: D. Cavanagh, DOI: 10.1007/978-1-59745-181-9_18, © Humana Press, New York, NY

were unstable upon propagation in prokaryotic hosts. Finally, however, several laboratories succeeded in establishing coronavirus reverse genetic systems based on full-length cDNA *(1–3)*. Not surprisingly, those researchers had to solve the problem of "cDNA instability" and, accordingly, the solutions they provided are all based on nonconventional approaches.

The reverse genetic system described here is based on the use of a vaccinia virus as a cloning vector that replicates in eukaryotic cells. Vaccinia virus is a large DNA virus with a genome size of approximately 200 kb that is able to stably accommodate foreign DNA sequences of coronavirus genome size (27–31 bp). The basic techniques required to clone (Section 3.1), modify (Section 3.2), and rescue (Section 3.3) recombinant coronaviruses are described. One application of the system, namely, the generation of coronavirus replicon RNAs and cell lines, is described in Section 3.4.

2. Materials

1. QiaexII Gel elution kit (Qiagen).
2. Buffer A (10 mM Tris-Cl pH 9.0, 1 mM EDTA).
3. MagNA Lyser Instrument, MagNa Lyser Green Beads (Roche).
4. Phosphate-buffered saline (PBS).
5. 0.25% (w/v) trypsin.
6. 36% sucrose.
7. Sorvall or Beckman ultracentrifuge, AH-629 or SW-28 rotors.
8. RNase-free DNase.
9. Proteinase K, PCR grade (Roche).
10. Proteinase K digestion buffer (1X concentration: 100 mM Tris-Cl pH 7.5, 5 mM EDTA, 0.2% (w/v) SDS, 200 mM NaCl).
11. RNase-free water.
12. T4 DNA Ligase (high-concentrate; Fermentas).
13. Pulse field gel instrument and equipment.
14. Lipofectin, Lipofectamine2000 (Invitrogen).
15. Sonication water bath (Branson 3210).
16. RiboMax large-scale RNA production system—T7 (Promega).
17. $m^7G(5')ppp(5')G$ cap analog (30 mM).
18. LiCl solution (7.5 M LiCl, 50 mM EDTA pH 7.5).
19. Sodium dodecylsulfate (SDS).
20. Electroporation instrument (e.g., BioRad Gene Pulser, 0.4-cm electroporation cuvettes).
21. Cell culture medium chemicals for GPT⁺ selection: (a) mycophenolic acid (MPA), 10 mg/ml in 0.1 M NaOH (400X stock); (b) xanthine, 10 mg/ml in 0.1 M NaoH (40X stock) and (c) hypoxanthine, 10 mg/ml in 0.1 M NaOH (667X stock).
22. Cell culture medium chemicals for GPT⁻ selection: 6-thioguanine (6-TG), 1 mg/ml (1000X stock).

23. Cells: BHK-21, CV-1, D980R *(4)*.
24. Viruses: Vaccinia virus vNotI/tk *(5)*, fowlpox virus.

3. Methods

3.1. Cloning of Coronavirus cDNA in Vaccinia Virus

This section describes the steps involved in the cloning of a full-length coronavirus cDNA. Starting from viral RNA, a set of plasmid DNAs should be generated together covering the full-length coronavirus genomic sequence. The plasmid insert cDNAs are then assembled by *in vitro* ligation to obtain a full-length coronavirus cDNA fragment. This fragment will be inserted into a vaccinia virus genome, again by *in vitro* ligation. The cloned full-length coronavirus cDNA in vaccinia virus is then amenable to mutagenesis by vaccinia virus-mediated homologous recombination.

3.1.1. Generation of Plasmid DNAs Covering a Coronavirus Full-Length cDNA

1. Analyze the coronavirus genome for useful naturally encoded endonuclease restriction sites that can later be used to ligate cloned cDNA inserts. Preferably use restriction enzymes that produce nonpalindromic sticky ends with at least three nucleotide (nt) overhangs. Avoid the use of restriction enzymes that generate blunt ends, since ligation efficiencies of blunt end fragments are low. If there are no useful sites available at particular regions of the cDNA sequence, restriction sites may be generated that introduce silent nucleotide changes. Alternatively, introduce sites at the border of the coronavirus cDNA fragments for restriction enzymes that cleave outside of their recognition sequence and orientate the sites so that cleavage occurs in the coronavirus cDNA region **(Fig. 1A)**.

2. Generate a set of plasmid DNAs covering the entire coronavirus cDNA using standard plasmid DNA cloning techniques. cDNA insert fragments should have a size of approximately 5 kb. Make sure that the cDNA fragment borders are flanked by appropriate endonuclease restriction sites in order to release the cloned cDNA fragments by endonuclease digestion (see above). If particular plasmid clones appear unstable upon propagation in *Escherichia coli*, change the plasmid backbone, preferably to a low copy plasmid. If plasmid DNAs remain unstable upon propagation proceed with the cloning of the remaining part of the coronavirus cDNA and insert the respective unstable cDNA sequence on the vaccinia virus level by vaccinia virus-mediated recombination using RT-PCR cDNA fragments (see Section 3.2).

3. The cDNA fragments corresponding to the 5'- and 3'-end of the genome should contain additional sequences as follows **(Fig. 2)**. To facilitate cloning into the vaccinia virus genome by *in vitro* ligation with *Not*I-cleaved vaccinia virus DNA (see Section 3.1.4) both end fragments should contain an *Eag*I or *Bsp*120I site.

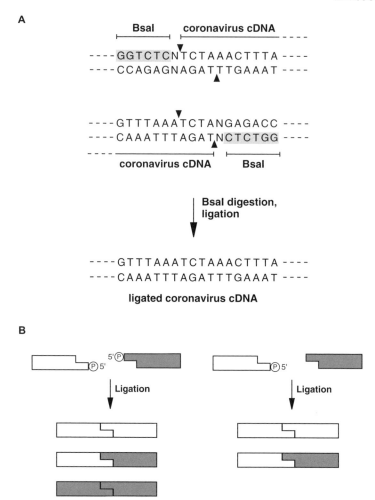

Fig. 1. Strategies to minimize the number of possible ligation products: (A) Ligation of two coronavirus cDNA fragments using *Bsa*I restriction endonuclease. *Bsa*I recognition sequences can be engineered adjacent to the coronavirus cDNA to obtain *Bsa*I-cleaved cDNA ends without heterologous sequences. The sticky ends are not palindromic and are comprised of a coronavirus-encoded sequence. The subsequent ligation reaction is directional and gives rise to only one possible reaction product. (B) The use of alkaline phosphatase to reduce the number of possible ligation products is illustrated. The left panel shows a conventional ligation using cDNA fragments with palindromic sticky ends. In this case three different ligation products are possible. The right panel shows a ligation reaction if one cDNA fragment has been dephosphorylated with alkaline phosphatase prior to the ligation reaction. In this case only two ligation products are possible.

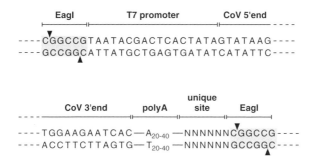

Fig. 2. Structure of 5'- and 3'-ends of cloned full-length coronavirus cDNA. Upstream of the coronavirus 5'-end there should be an *Eag*I or *Bsp*120I restriction endonuclease site to allow insertion of the cDNA into the *Not*I site of the vaccinia virus genomic DNA by *in vitro* ligation. Between the *Eag*I or *Bsp*120I site and the 5'-end of the coronavirus cDNA sequence there should be a bacteriophage T7 RNA polymerase promoter and one G nucleotide (if not yet present at the 5'-end of the coronavirus genome) for the initiation of the *in vitro* transcription reaction. Downstream of the 3'-end of the coronavirus genome, a stretch of 20–40 A nucleotides (synthetic poly(A) tail) and a unique (i.e., not present in the coronavirus genomic sequence) restriction endonuclease site should be cloned. The unique restriction site is needed for the generation of runoff *in vitro* transcripts. Furthermore, an *Eag*I or *Bsp*120I site is needed to insert the cDNA into the *Not*I site of the vaccinia virus genome by *in vitro* ligation. If the coronavirus sequence does not encode an *Eag*I or *Bsp*120I site, the unique site is not needed since *Eag*I or *Bsp*120I cleavage can be done to produce templates for the generation of runoff *in vitro* transcripts.

(DNA cleaved with *Eag*I or *Bsp*120I can be ligated with *Not*I cleaved DNA.) Furthermore, the 5'-end fragment should contain a promoter for the bacteriophage T7 RNA polymerase and one G nucleotide (if not yet present at the genomic 5'-end of the coronavirus genomic RNA) for the proper initiation of the *in vitro* transcription. The 3'-end fragment should encode a stretch of A nucleotides (approximately 20 nt) followed by a restriction site that is not present in the coronavirus cDNA sequence.

3.1.2. Assembly of Long cDNA Fragments by in vitro Ligation

1. Design a strategy for the sequential assembly of cloned cDNA inserts. Examples have been described for the construction of full-length human coronavirus 229E (HCoV-229E), avian infectious bronchitis virus (IBV), and mouse hepatitis virus, strain A59 (MHV-A59) cDNAs (*3,6,7*).
2. Liberate cloned insert cDNA fragments from plasmid DNAs by restriction endonuclease cleavage. Start with 50–100 µg plasmid DNA. Isolate the cDNA fragments by gel purification using standard agarose gels. Avoid exposure of cDNA fragments to UV light during the isolation process (*see* **Note 1**). If palindromic sticky

ends are present at the cDNA fragment termini, the number of possible ligation products can be reduced by dephosporylation of one ligation partner by alkaline phosphatase treatment (**Fig. 1B**). Note that the ends of cDNA fragments corresponding to the 5'- and 3'-genomic end should be cleaved with *Eag*I or *Bsp*120I and dephosphorylated by alkaline phosphatase treatment.

3. Ligate cDNA fragments in analytical scale using high-concentrate T4 DNA ligase. Leave ligation reaction overnight at room temperature. Analyze ligation products on standard agarose gels. The samples should be heated to 65°C for 5 min before loading on the gel.

4. If the ligation reaction worked efficiently in the analytical scale, use the same conditions for a preparative scale ligation. Analyze an aliquot of the reaction on a standard agarose gel. If inefficient ligation is encountered repeatedly, revise the assembly strategy by using alternative restriction sites.

5. If the preparative ligation reaction was efficient, purify the desired ligation fragments by gel purification (avoid UV exposure; *see* **Note 1**) using the QiaexII gel elution procedure (Qiagen).

6. Use the ligated and purified cDNA fragments from step 5 for further ligation reactions with further cDNA fragments until a full-length cDNA fragment has been obtained. Alternatively assemble fragments to obtain a set of not more than two or three cDNA fragments together encompassing the entire coronavirus cDNA sequence (*see* **Note 2**).

3.1.3. Preparation of Vaccinia Virus DNA

This section describes the preparation of purified vaccinia virus DNA that can be used: (i) for the *in vitro* ligation with the assembled coronavirus full-length cDNA (see Section 3.1.4), and (ii) as template for *in vitro* transcription reactions (see Section 3.3.1). The protocol describes the vaccinia virus purification and subsequent DNA preparation in a preparative scale (virus derived from 10 to 20 150-cm^2 flasks of infected BHK-21 cells). However, the protocol can also be down-scaled.

1. Grow 10 to 20 150-cm^2 flasks of BHK-21 cells to 80% confluency and infect with vaccinia virus. Vaccinia virus infection should be done with an appropriate multiplicity of infection (MOI) to obtain complete a cytopathic effect (CPE) 3 days postinfection (p.i.).

2. Three days p.i. freeze cells by putting the flasks into a freezer for at least 2 h. Thaw, collect, and pellet cells (1000 rpm, 5 min, 4°C). Wash cell pellet with PBS. Resuspend concentrated infected cells in 1 ml Buffer A per 150-cm^2 tissue culture flask.

3. Fill MagNA Lyser green bead tubes to maximum filling level of 1 ml.

4. Homogenize using MagNA Lyser machine (1 × 20 sec; speed 5000) (*see* **Note 3**).

5. Centrifuge at 1000 rpm 4°C for 2 min.

6. Pipette supernatant into a clean fresh tube. Treat supernatant with 0.1 volume 0.25% (w/v) trypsin and incubate at 37°C for 20 min.
7. Adjust the trypsin-treated cell homogenate with buffer A to a volume of 18 ml and carefully overlay an 18-ml 36% (w/v) sucrose cushion in a 36-ml ultracentrifugation tube and centrifuge (13,500 rpm at 4°C for 80 min; Sorvall or Beckman ultracentrifuge, Rotor AH 629 or SW 28).
8. Discard the supernatant and resuspend the pellet in 0.4 ml Buffer A.
9. Digest with RNase-free DNase (1–5 U) for 20 min (*see* **Note 4**) and then stop DNase treatment by adjusting the solution to 10 mM EDTA and incubate for 10 min at 65°C.
10. Add 1 vol of 2X Proteinase K digestion buffer (final concentration is 1X proteinase K digestion buffer) and 4 μl proteinase K; incubate at 50°C for 2 h.
11. Extract DNA with 1 vol phenol/chloroform/isoamylalcohol (25:24:1), mix gently (do not vortex! *See* **Note 5**), and centrifuge (14,000 rpm, 5 min, room temperature, Eppendorf centrifuge). Take the water phase and perform a second round of DNA extraction with 1 vol chloroform/isoamylalcohol (24:1), mix gently (do not vortex!) and centrifuge (14,000 rpm, 5 min, room temperature, Eppendorf centrifuge).
12. Take the water phase and add 2.5 vol 100% ethanol, mix gently (do not vortex!), and pellet the DNA by centrifugation (14,000 rpm, 5 min, room temperature, Eppendorf centrifuge) (*see* **Note 5**).
13. Discard the supernatant and wash the DNA with 70% ethanol; centrifuge again (14,000 rpm, 5 min, room temperature, Eppendorf centrifuge).
14. Discard the supernatant completely and resolve the DNA in 50–200 μl RNase-free water.

3.1.4. Ligation of Insert cDNA with Vaccinia Virus DNA

This section describes the integration of the assembled coronaviral cDNA fragments (Section 3.1.2) into the vaccinia virus genome by *in vitro* ligation. Inserted DNA fragments can be a full-length coronavirus cDNA or two or three cDNA fragments that can be ligated just prior to the ligation to the vaccinia virus genomic DNA. As parental virus we recommend the vaccinia virus v*Not*I/tk that encodes a unique *Not*I cloning site *(5)*.

1. Set up a standard *in vitro* ligation reaction if the insert DNA consists of more than one fragment. This procedure is only recommended if the number of possible ligation products is limited (e.g., by avoiding palindromic sticky ends or by limiting the number of possible ligation products by using one dephosporylated DNA end per ligation reaction). The reaction should include a standard ligation buffer (including ATP) and can have a volume of up to 100 μl. Let the reaction go for 1–2 h at room temperature while preparing the *Not*I-cleaved vaccinia virus DNA (step 2). We recommend a molar ratio of insert:vaccinia virus DNA of 1:1 and a prior test of the reaction on an analytical scale.

2. Cleave vaccinia virus v*Not*I/tk DNA with *Not*I for 1–2 h at 37°C. The volume of the reaction can be up to 50 μl.

3. Mix the ligation reaction and the *Not*I restriction reaction and adjust buffers to 1X concentrations (1X ligation buffer and 1X *Not*I restriction buffer). Add fresh T4 ligase and *Not*I enzymes and incubate overnight at room temperature (*see* **Note 6**).

4. Heat the reaction to 65°C for 5 min, centrifuge (14,000 rpm, room temperature, Eppendorf centrifuge). Take the supernatant, add fresh *Not*I enzyme, and incubate at 37°C for 1–2 h (see **Note 7**).

5. Analyze the reaction products (or an aliquot thereof) on a pulse field gel. Prior to loading the sample(s) heat to 65°C for 5 min to achieve appropriate separation of DNA fragments in the pulse field gel.

6. Store ligation products at −20°C.

3.1.5. Rescue of Recombinant Vaccinia Viruses Containing Full-Length Coronavirus cDNA Insert

1. Seed CV-1 cells in a six-well dish 1 day before transfection. Cells should be 80% confluent for optimal transfection efficiency.

2. Infect 80% confluent CV-1 cells with fowlpox virus (MOI 1–10) for 1–2 h (*see* **Note 8**).

3. Transfect ligation reaction from Section 3.1.4 (without any further purification) into fowlpox virus-infected CV-1 cells using Lipofectin as described by the manufacturer (Invitrogen). Do not vortex at any time and use cut pipette tips when handling vaccinia virus DNA (*see* **Note 5**).

4. After 3–4 h trypsinize cells and seed them together with fresh (uninfected) CV-1 cells (4:1 excess of fresh CV-1 cells) into a 96-well plate.

5. At 5–10 days posttransfection collect cells and medium from wells displaying CPE (this is the first vaccinia virus stock) (*see* **Note 9**).

6. Transfer half of the first vaccinia virus stock to fresh CV-1 cells plated in a six-well dish. Wait until full CPE becomes apparent and collect the second vaccinia virus stock.

7. To analyze the obtained recombinant vaccinia viruses take half of the second vaccinia virus stock, pellet cells, and prepare DNA from the cell pellet according to Section 3.1.3, steps 10–14 (use 1X Proteinase K buffer).

8. To confirm the identity of recombinant vaccinia viruses perform Southern blot, PCR, and/or sequencing analyses.

3.2. Modification of Coronavirus cDNA by Vaccinia Virus-Mediated Homologous Recombination

The cloned coronavirus cDNA is amenable to mutagenesis by vaccinia virus-mediated homologous recombination. Two steps of homologous recombination

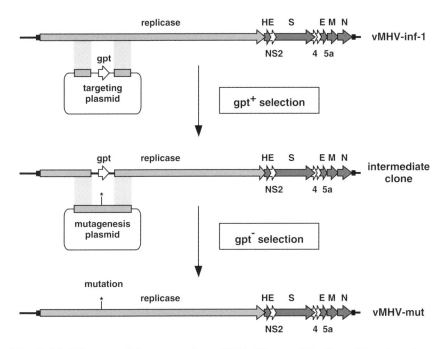

Fig. 3. Modification of the coronavirus cDNA. The modification of the cloned coron-avirus cDNA by vaccinia virus-mediated homologous recombination is illustrated. The parental vaccinia virus vMHV-inf-1 is used in combination with a targeting plasmid to target the region of interest in the cloned cDNA. Upon gpt⁺ selection an interme-diate clone is obtained that is subjected to a second round of recombination with a mutagenesis plasmid encoding the mutation of choice. The final mutant cDNA clone can be obtained after gpt⁻ selection. Note that the intermediate clone can also be used to introduce other mutations in the targeted region by using a different mutagenesis plasmid.

are performed using appropriate plasmid DNAs **(Fig. 3)**. The first "targeting" plasmid DNA contains a sequence of approximately 500 bp corresponding to a region encoded upstream of the region that should get targeted for mutagenesis (left flank), the *E. coli* guanosin-phosphoribosyltransferase (gpt) gene located downstream of a vaccinia virus promoter and a sequence of approximately 500 bp corresponding to a region encoded downstream of the region that should be targeted for mutagenesis (right flank).

The targeting plasmid is used to target the region of interest for a second round of recombination using a "mutagenesis" plasmid that contains the left and right flank of the targeting plasmid and, between the flanks, the region of interest encoding the desired mutation. Both plasmids can be constructed by standard cloning techniques. We routinely use the plasmid pGPT-1 *(4)* that is based on pBluescriptKS$^+$ (Stratagene) and contains a fragment encoding the *E. coli* gpt gene downstream of a vaccinia virus promoter cloned into the pBluescriptKS$^+$ multiple cloning site. The plasmid pGPT-1 is available from the authors upon request. Note that vaccinia virus-mediated homologous recombination also works when a linear DNA fragment, such as an RT-PCR product, is used instead of a plasmid DNA *(7)*.

3.2.1. Vaccinia Virus-Mediated Homologous Recombination

This section describes the procedure of vaccinia virus-mediated homologous recombination. The protocol is used to generate a "transfection" stock containing recombined viruses at a ratio of approximately 1:1000 (recombinant viruses:parental viruses). The transfection stock is subsequently subjected to plaque purification under gpt$^+$ or gpt$^-$ selection (see Sections 3.2.2. and 3.2.3) to obtain stocks of recombinant vaccinia virus clones.

1. Seed CV-1 cells in a six-well dish so that they are 80–95% confluent on the day of infection.
2. Infect with parental vaccinia virus at an MOI of 1. Incubate for 1–2 h at 37°C.
3. At 1–2 h postinfection transfect 5 µg of the targeting or mutagenesis plasmid using Lipofectin or Lipofectamine 2000 according to the manufacturer's (Invitrogen) instructions.
4. Wash cells at 3–6 h posttransfection and culture cells for 2–3 days until full CPE becomes apparent.
5. Prepare a virus stock (which is the "transfection stock") from the infected/transfected culture by scraping the cells off the plate in 0.5 ml of the culture medium. Store at –20°C.

3.2.2. Targeting the Region of Interest: Selection of Recombinant gpt$^+$ Vaccinia Viruses

1. Seed CV-1 cells in a six-well dish so that they are 80–95% confluent on the day of infection.
2. Replace culture medium with gpt$^+$ selection medium (MEM containing 5% FCS, antibiotics and 25 µg/ml MPA, 250 µg/ml xanthine, and 15 µg/ml hypoxanthine) at least 6 h prior to infection.
3. Freeze-thaw the vaccinia virus transfection stock (see Section 3.2.1.) three times using dry ice on ethanol and sonicate for 1–5 min immediately prior to infection (*see* **Note 10**).
4. Infect CV-1 cells with different dilutions (10^{-2}, 10^{-3}, and 10^{-4}) of the transfection stock (*see* **Note 11**).
5. Culture cells for 2–3 days in the gpt$^+$ selection medium. Pick plaques as soon as they are easily detectable (usually on day 2 p.i.) by marking them at the bottom of

the dish, followed by scraping the cells of the plaque off the dish, and aspirate in 100 μl of medium using a standard pipette (*see* **Note 12**).

6. Perform another two rounds of gpt⁺ plaque selection (steps 4 and 5) by infecting the CV-1 cells with 5–20 μl of a picked plaque. Always perform freeze-thaw cycles and sonication prior to infection.

7. After the third round of plaque selection infect CV-1 cells in a six-well dish with half of a picked plaque and culture cells until full CPE. Store half of this stock for further use and use the other half for DNA preparation and analysis of the recombinant vaccinia virus clone (i.e., PCR, Southern blot, sequencing).

3.2.3. Inserting the Mutation of Choice: gpt Negative Selection

1. Seed D980R cells in a six-well dish so that they are 60–80% confluent on the day of infection (*see* **Note 13**). Cells can be seeded in gpt⁻ selection medium containing 0.5–1 μg/ml 6-TG. The cells should be cultured in gpt⁻ selection for at least 6 h prior to infection.

2. Perform plaque selection as described in Section 3.2.2. steps 4–7. The only differences are the cells (D980R cells) and the gpt⁻ selection medium.

3.3. Rescue of Recombinant Coronaviruses from Cloned Full-Length cDNA

The rescue of recombinant coronaviruses is based on two steps. First, a full-length coronavirus RNA is produced using the genomic DNA of a vaccinia virus containing the full-length coronavirus cDNA insert as a template for *in vitro* transcription. Second, the recombinant full-length RNA is transfected into eukaryotic cells. Within these cells the coronavirus replication cycle will be initiated by translation of replicase gene products from the transfected RNA and, finally, recombinant coronaviruses are released into the tissue culture supernatant.

3.3.1. Generation of Infectious Full-Length Coronavirus RNA by in vitro Transcription

1. Prepare vaccinia virus DNA from purified virus stocks as described in Section 3.1.3.

2. Cleave the vaccinia virus DNA (1–10 μg) with the restriction enzyme for which a unique recognition site downstream of the synthetic poly(A) tail has been introduced (**Fig. 2**).

3. Extract DNA with 1 vol phenol/chloroform/isoamylalcohol (25:24:1), mix gently, and centrifuge (14,000 rpm, 5 min, room temperature, Eppendorf centrifuge). Take the water phase and perform a second round of DNA extraction with 1 vol

chloroform/isoamylalcohol (24:1), mix gently, and centrifuge (14,000 rpm, 5 min, room temperature, Eppendorf centrifuge).

4. Take water phase and precipitate the cleaved vaccinia virus DNA by adding 1/20 vol of 5 M NaCl and 2.5 vol of 100% ethanol and centrifuge (14,000 rpm, 5 min, Eppendorf centrifuge). Do not overdry vaccinia virus DNA.
5. Wash DNA pellet with 70% ethanol; centrifuge again.
6. Completely remove the supernatant and resolve the DNA in 10–20 µl RNase-free water.
7. Set up the *in vitro* transcription reaction using the RiboMax Kit (Promega) as follows (*see* **Note 14**):

5X transcription buffer	10 µl
m^{7}G(5')ppp(5')G cap analog (30 mM)	5 µl
GTP (100 mM)	0.7 µl
ATP, CTP, UTP (100 mM), each	3.75 µl
Template DNA (1–10 µg)	x µl
RNase-free water	y µl
Enzyme mix (RNasin, T7 RNA pol.)	5 µl
Total	50 µl

7. Incubate at 30°C for 2 h.
8. Add 2 µl of RNase-free DNase, incubate at 37 °C for 20 min. Either store the reaction at –80 °C until transfection or (optional) precipitate the RNA (steps 9–11).
9. Add half the volume of LiCl solution and freeze the sample for at least 30 min.
10. Pellet RNA by centrifugation (14,000, 15 min, 4 °C, Eppendorf centrifuge)
11. Wash RNA pellet (should appear yellowish) with 70% ethanol and resolve in RNase-free water. Store at –80 °C.
12. Analyze the RNA on an agarose gel containing 1% SDS. Stain the RNA after gel electrophoresis with ethidium bromide.

3.3.2. Rescue of Recombinant Coronaviruses

1. One day before RNA transfection seed BHK-21 (*see* **Note 15**) cells so that there are 5 × 10^6 to 1 × 10^7 BHK-21 cells for each transfection. RNA transfection will be performed by electroporation.
2. Trypsinize, collect, and pellet 5 × 10^6 to 1 × 10^7 BHK-21 cells (centrifuge 1000 rpm, 5 min, 4 °C). Perform all further steps on ice.
3. Wash cells with 20 ml ice-cold PBS. Make sure that cells are well separated. Take one drop to count the cells and pellet the rest again (1000 rpm, 5 min, 4 °C).
4. Resolve 5 × 10^6 to 1 × 10^7 BHK-21 cells in 0.8 ml ice-cold PBS and fill into a 0.4-cm electroporation cuvette.

5. Add RNA and electroporate with two pulses (settings on BioRad Gene Pulser: Resistance = ∞, 230 V, high-capacity 1000 μF) (*see* **Note 15**).
6. Transfer the electroporated cells from into a 10-cm culture dish with 10 ml warm culture medium and add 1×10^6 fresh cells that are susceptible for the coronavirus that should be rescued (e.g., murine 17Cl1 cells for MHV rescue, human MRC-5 cells for the rescue of HCoV-229E).
7. Change the medium after 3–6 h when cells have attached to the bottom of the culture dish.
8. Recombinant coronaviruses should be released into the tissue culture medium between days 1 and 3 postelectroporation. Check for released virus on days 1–3 by transferring part of the supernatant onto susceptible fresh cells. Store culture supernatant for further analysis at –80 °C.

3.4. Coronavirus Replicon RNAs

Replicon RNAs are autonomously replicating RNAs encoding: (i) all replicative proteins required for the expression of a functional replication complex, and (ii) *cis*-acting elements required for the recognition of the replicon RNA by the replicase complex. Usually replicon RNAs are devoid of sequences leading to production of progeny particles. Coronavirus replicon RNAs differ from those of other positive-stranded RNA viruses in that they have to encode the nucleocapsid protein, which has been shown to be important for efficient coronavirus RNA replication (*8,9*).

It has been shown for HCoV-229E and SARS-CoV replicons that stable cell lines can be generated if the replicon RNA mediates the expression of a selection marker (*4,10*). Two selection markers, conferring neomycin/G418 (*4*) or blasticidin (*10*) resistance, have been used successfully for establishing stable coronavirus replicon cell lines. The HCoV-229E replicon RNA encodes the neomycin resistance gene inserted downstream of the nonstructural protein (Nsp) 1 and a sequence encoding a "2A-like" autoprocessing peptide. The 2A-like autoprocessing peptide mediates a co-translational liberation of a slightly modified Nsp1 carboxyterminus and subsequent translation of the neomycin resistance gene. In order to ensure translation of the remaining Nsps of the replicase gene (Nsps 2–16), an internal ribosomal entry site (IRES) derived from the encephalomyocarditis virus (EMCV) has been placed upstream of the Nsp2-coding sequence.

The SARS-CoV replicon RNA contains a gene encoding a fusion protein comprising the green fluorescent protein (GFP) and the blasticidin deaminase (GFP-BlaR) that has been cloned downstream of the replicase gene as a separate transcription unit under the control of the transcription regulatory sequence (TRS) of the SARS-CoV spike gene. In both cases, transfection of *in vitro* synthesized replicon RNA into eukaryotic cells and subsequent selection using G418 or blasticidin resulted in the establishment of stable cell lines containing actively replicating coronavirus replicon RNAs. To facilitate the detection of replicon-containing cell lines, green fluorescence resulting from replicon-mediated GFP expression has been used. For the SARS-CoV replicon RNA this has been achieved by the use of the GFP-BlaR fusion protein *(10)*. To achieve GFP expression by the HCoV-229E replicon RNA, the GFP gene has been inserted as a separate transcription unit downstream from the replicase gene, driven by the TRS of the HCoV-229E spike gene *(4)*.

Coronavirus replicon cell lines can be used as a noninfectious system to analyze coronavirus replication and transcription or to identify and evaluate replicase inhibitors. The following protocols describe the generation of coronavirus replicon cell lines and their use in the evaluation of coronavirus replicase inhibitors.

3.4.1. Generation of Coronavirus Replicon Cell Lines

1. Based on a full-length coronavirus cDNA cloned in vaccinia virus, a replicon RNA-encoding cDNA can be generated using vaccinia virus-mediated homologous recombination as described in Section 3.2.
2. Generate replicon RNA by *in vitro* transcription as described in Section 3.3.1.
3. Introduce the replicon RNA into a host cell line of choice (*see* **Note 16**) by electroporation as described in Section 3.3.2 (steps 1–5).
4. Plate the transfected cells in normal growth medium. Change the medium after 3–6 h when cells have attached to the bottom of the culture dish and continue to culture the cells for 1–2 days in growth medium without selection pressure. Split if necessary.
5. Start the selection of stable lines at antibiotic concentrations only slightly above the level at which nontransfected cells die (*see* **Note 17**).
6. Increase the antibiotic concentration gradually during the following 2–3 weeks until resistant colonies appear.
7. Pick colonies for subculture in separate wells and test them for maintenance of replicon RNA. Expression of a reporter protein, such as GFP, by the replicon RNA facilitates the screening of replicon RNA-containing resistant colonies.
8. When stable clones have been obtained, further culturing can be done under low selection pressure (*see* **note 18**). Replicon cells can be stored in liquid nitrogen.

3.4.2. Identification and Evaluation of Coronavirus Replicase Inhibitors Using Replicon Cell Lines

1. Seed the replicon cells in selection medium so that they are 50–70% confluent on the next day. You can use 96-, 24-, or 6-well dishes.
2. Prior to adding antiviral compounds, wash the cells and culture them in standard medium without selection drugs.

3. Add graded doses of antiviral compound(s) to the cells and culture them for 1–3 days (*see* **Note 19**). For comparison include nontreated cells and culture them under identical conditions.

4. In order to assay for cytotoxicity of candidate inhibitors and to determine the selectivity index, include a cytotoxicity/cell viability test. This can be done with replicon cells or the respective parental cell line.

5. Determine GFP expression on days 1, 2, and 3 posttreatment by fluorescence microscopy and flow cytometry (*see* **Note 20**).

4. Notes

1. We found that DNA that has been exposed to UV light is difficult and sometimes impossible to clone into vaccinia virus DNA. When purifying DNA fragments from agarose gels, cut small slices at the edges of the fragment band out of the gel and stain them with ethidium bromide. Use UV light to visualize the borders of the DNA band in the slices and mark the position. Insert the slices back into the gel and cut the piece of agarose between the two marked positions out of the gel. The DNA recovered from those agarose pieces have not been exposed to UV light and are easily clonable in vaccinia virus.

2. It is possible to insert more than one DNA fragment into the vaccina virus genome by *in vitro* ligation *(37)*. Up to three DNA fragments can be ligated prior to adding *Not*I-cleaved vaccinia virus DNA (see Section 3.1.4, step 1). However, we recommend this procedure only if the fragments can be efficiently ligated (a full-length cDNA fragment should be visible in agarose gels). Furthermore, the number of possible ligation products should be minimized using strategies illustrated in **Fig. 1**.

3. In order to prepare vaccinia virus, DNA the virus particles have to be liberated from cells and cell debris. This can be achieved by using a tight Dounce homogenisator or, as described here, by using the MagNA Lyser protocol. To establish appropriate conditions we recommend doing a pilot experiment in which several conditions are compared. After homogenization check for virus titers and decide for the most vigorous homogenization conditions that still leave the virus particles intact.

4. This step results in DNase digestion of free DNA (mostly of cellular origin) and will leave the DNA in virus particles intact.

5. Vaccinia virus genomic DNA has a size of approximately 200 kb. Standard pipette tips are usually too narrow and pipetting will result in shearing the DNA. To avoid this, cut the pipette tips to generate an opening of about 2–3 mm and avoid vigorous pipetting. Avoid drying the DNA. If the large vaccinia virus DNA is overdried it will no longer be possible to dissolve it in water.

6. The ligation of insert DNA fragment(s) and the *Not*I-cleaved vaccinia virus arms is facilitated by adding *Not*I into the ligation reaction *(11)*. Religated vaccinia virus arms are recleaved by the *Not*I enzyme allowing a new round of ligation. The 5'- and 3'-ends of the insert fragment are cleaved with *Eag*I or *Bsp*120I

and dephosphorylated. Therefore: (i) insert-insert ligation is not possible (owing to dephophorylated ends), and (ii) *Eag* or *Bsp*120I-*Not*I ligation products (i.e., insert-vector ligation) are not recleavable with *Not*I. As a result the ligation-restriction reaction will drive the overall reaction toward an accumulation of insert-vector ligation products.

7. The ligation-restriction reaction may still contain a small proportion of unre-cleaved vector-vector ligation products. These products may lead to functional vaccinia virus genomes without the insert cDNA fragment and may cause a high background in the rescue of recombinant vaccinia viruses (Section 3.1.5). We therefore recommend an additional *Not*I cleavage reaction after the ligation-restriction reaction.

8. Since the vaccinia virus genomic DNA is not infectious, a helper virus has to be provided to rescue recombinant vaccinia viruses from DNA. We recommend fowlpox virus as a helper virus, since a fowlpox virus infection is abortive in mammalian cells, but can still serve to rescue vaccinia virus from DNA. Therefore, the recovered viruses will be vaccinia virus only (and no fowlpox virus).

9. On days 2–4 p.i. the cells may look heavily infected, most likely owing to fowlpox virus infection. However, recombinant vaccinia virus cannot yet be expected in the cell culture. Just continue to cultivate the cells; most cell layers will recover. Usually, the first vaccinia virus-mediated CPE can be expected on day 5 p.i. and a peak is observed around day 7 p.i.. If cells get too confluent the medium can be changed.

10. Freeze-thawing and sonication is needed to release and separate vaccinia virus particles from the cells and cell debris. For selection of pure recombinant vac-cinia virus this procedure is critical.

11. A ratio of 1:1000 of recombinant vaccinia viruses:parental vaccinia viruses can be expected. Thus, under selection pressure, single plaques should appear on CV-1 cell layers that have been infected with a 10^{-3} dilution of the transfection stock. At 2 h p.i., an overlay of 1% low-melting agarose in selection medium can be made. This is done to reduce the risk of contamination of recombinant plaques with parental virus. Because most vaccinia virus is contained within an infected cell within the time frame (2–3 days) of the selection, agarose overlays are usually not necessary.

12. Six plaques is a reasonable number to pick. There is a limited risk of picking "false" plaques or plaques contaminated by parental virus that necessitates selec-tion of a few plaques in parallel.

13. D980R cells grow fast and vaccinia virus plaques are not as easily recogniz-able as on CV-1 cells. During the selection it is necessary to keep cells in good condition to facilitate the formation of easily detectable plaques. The risk of overgrowth can be reduced by seeding cells at a lower density, and cell death from starvation can be reduced by replacing the medium with fresh selection medium on day 2 p.i..

14. It is possible, although not recommended by the manufacturer, to use the Promega RiboMax Kit to generate capped *in vitro* transcripts. One simply has to add a cap structure analog to the reaction. In the given *in vitro* transcription protocol the ratio of $m^7G(5')ppp(5')G$ cap analog to GTP is 2:1.

15. The optimal conditions for the electoporation of long RNA molecules are dependent on the cells and the electroporation device. BHK-21 cells are known to be suitable for efficient RNA electroporation and should be the first choice. We recommend doing a pilot optimization to determine optimal conditions for RNA transfection. It is now well established that the coronavirus nucleocapsid (N) protein facilitates the rescue of recombinant coronaviruses in several systems *(2,6,7)*. We recommend co-electroporating an mRNA encoding the coronavirus N protein (5–10 µg N mRNA produced by *in vitro* transcription). It is even more efficient to generate and use a BHK-21-derived cell line stably expressing the N protein *(7)*.

16. We observed replicon RNA replication in a wide variety of eukaryotic host cells. Although coronaviruses are usually species specific, coronavirus replicon RNAs are able to replicate in many cell lines once introduced into the host cell cytoplasm by transfection. Cell lines tested in our laboratory (using the HCoV-229E replicon) include cells of human (e.g., MRC-5, HeLa cells) and animal (e.g., BHK-21, 17clone1, L929 cells) origin.

17. We recommend determining the lowest concentration of the selection drug where nontransfected cells die for the cell line of choice.

18. Replicon cell lines based on commonly used cells such as baby hamster kidney (BHK) or Chinese hamster ovary (CHO) cells are generally easy to culture. To increase the number of cells expressing a high level of replicon-derived transcripts it is important to split the lines often enough to maintain them constantly subconfluent. GFP is a convenient marker to determine the percentage of GFP-expressing cells by flow cytometry.

19. Depending on the cell density and the stability of the compound it might be necessary to change the medium daily.

20. GFP is a valuable reporter protein to determine the percentage of green fluorescent cells as a marker for the percentage of cells with actively replicating RNA or to determine the mean fluorescence as a value that indicates GFP expression levels. Some inhibitors may lead to a reduced overall number of green fluorescent cells, whereas some inhibitors may just reduce the mean fluorescence. To generate more quantitative data on the inhibitory effect of a compound and to gain some insight into the kinetics of inhibition, other reporter proteins, such as luciferase proteins or alkaline phosphatase, may be used.

Acknowledgment

This work was supported by the Swiss National Science Foundation and the European Commission (SARS-DTV SP22-CT-2004-511064).

References

1. Almazan, F., Gonzalez, J.M., Penzes, Z., et al. (2000) Engineering the largest RNA virus genome as an infectious bacterial artificial chromosome. *Proc. Natl. Aca.d Sci. USA* **97**(10), 5516–5521.
2. Yount, B., Curtis, K. M., and Baric, R. S. (2000) Strategy for systematic assembly of large RNA and DNA genomes: transmissible gastroenteritis virus model. *J. Virol.* **74**(22), 10600–10611.
3. Thiel, V., Herold, J., Schelle, B., and Siddell, S. G. (2001) Infectious RNA transcribed *in vitro* from a cDNA copy of the human coronavirus genome cloned in vaccinia virus. *J. Gen. Virol.* **82**(Pt 6), 1273–1281.
4. Hertzig, T., Scandella, E., Schelle, B., et al. (2004) Rapid identification of coronavirus replicase inhibitors using a selectable replicon RNA. *J. Gen. Virol.* **85**(Pt 6), 1717–1725.
5. Merchlinsky, M., and Moss, B. (1992) Introduction of foreign DNA into the vaccinia virus genome by *in vitro* ligation: recombination-independent selectable cloning vectors. *Virology* **190**(1), 522–526.
6. Casais, R., Thiel, V., Siddell, S.G., Cavanagh, D., and Britton, P. (2001) Reverse genetics system for the avian coronavirus infectious bronchitis virus. *J. Virol.* **75**(24), 12359–12369.
7. Coley, S. E., Lavi, E., Sawicki, S. G., et al. (2005) Recombinant mouse hepatitis virus strain A59 from cloned, full-length cDNA replicates to high titers *in vitro* and is fully pathogenic in vivo. *J. Virol.* **79**(5), 3097–3106.
8. Almazan, F., Galan, C., and Enjuanes, L. (2004) The nucleoprotein is required for efficient coronavirus genome replication. *J. Virol.* **78**(22), 12683–12688.
9. Schelle, B., Karl, N., Ludewig, B., Siddell, S. G., and Thiel, V. (2005) Selective replication of coronavirus genomes that express nucleocapsid protein. *J. Virol.* **79**(11), 6620–6630.
10. Ge, F., Luo, Y., Liew, P. X., and Hung, E. (2007) Derivation of a novel SARS-coronavirus replicon cell line and its application for anti-SARS drug screening. *Virology* **360**(1), 150–158.
11. Thiel, V., and Siddell, S. G. (2005) Reverse genetics of coronaviruses using vaccinia virus vectors. *Curr. Top. Microbiol. Immunol.* **287**, 199–227.

19

Transient Dominant Selection for the Modification and Generation of Recombinant Infectious Bronchitis Coronaviruses

Maria Armesto, Rosa Casais, Dave Cavanagh, and Paul Britton

Abstract

We have developed a reverse genetics system for the avian coronavirus infectious bronchitis virus (IBV) in which a full-length cDNA corresponding to the IBV genome is inserted into the vaccinia virus genome under the control of a T7 promoter sequence. Vaccinia virus as a vector for the full-length IBV cDNA has the advantage that modifications can be introduced into the IBV cDNA using homologous recombination, a method frequently used to insert and delete sequences from the vaccinia virus genome. Here we describe the use of transient dominant selection as a method for introducing modifications into the IBV cDNA. We have used it successfully for the substitution of specific nucleotides, deletion of genomic regions, and the exchange of complete genes. Infectious recombinant IBVs are generated in situ following the transfection of vaccinia virus DNA containing the modified IBV cDNA into cells infected with a recombinant fowlpox virus expressing T7 DNA-dependent RNA polymerase.

Key words: transient dominant selection (TDS); vaccinia virus; infectious bronchitis virus (IBV); coronavirus; avian; reverse genetics; nidovirus; fowlpox virus; T7 RNA polymerase.

1. Introduction

The avian coronavirus, infectious bronchitis virus (IBV), is a highly infectious pathogen of domestic fowl and like other coronaviruses is an enveloped virus that replicates in the cell cytoplasm and contains a single-stranded,

From: *Methods in Molecular Biology, vol. 454: SARS- and Other Coronaviruses,*
Edited by: D. Cavanagh, DOI: 10.1007/978-1-59745-181-9_19, © Humana Press, New York, NY

positive-sense RNA genome of 28 kb for IBV. Molecular analysis of the role of individual genes in pathogenesis of RNA viruses has been advanced by the availability of full-length cDNAs, for the generation of infectious RNA transcripts that can replicate and result in infectious viruses. The assembly of full-length coronavirus cDNAs was hampered owing to regions from the replicase gene being unstable in bacteria. We therefore devised a reverse genetics strategy for IBV involving insertion of the full-length cDNA, under the control of a T7 RNA promoter, into the vaccinia virus genome. This is followed by the in situ recovery of infectious IBV in cells transfected with the vaccinia virus DNA and infected with a recombinant fowlpox virus expressing T7 RNA polymerase *(1)*.

An advantage of using vaccinia virus, in addition to the stability of the IBV cDNA, is the ability to generate modified IBV cDNAs by homologous recombination for the subsequent rescue of recombinant IBVs (rIBVs). We use the vaccinia virus-based transient dominant selection (TDS) recombination method *(2)* for modifying the IBV cDNA sequence within the vaccinia virus genome *(3–5)*. The method relies on a three-step procedure. In the first step, the modified IBV cDNA is inserted into a plasmid containing a selective marker under the control of a vaccinia virus promoter. In our case we use a plasmid, pGPTNEB193 [**Fig. 1**; *(6)*], which contains a dominant selective marker gene, *Escherichia coli* guanine phosphoribosyltransferase (*Ecogpt*; *(7)*), under the control of the vaccinia virus $P_{7.5K}$ early/late promoter. In the second step, the complete plasmid sequence, containing a region of the IBV cDNA to be modified, is integrated into the IBV sequence in the vaccinia virus genome (**Fig. 2**). This occurs as a result of a single crossover event involving homologous recombination between the IBV cDNA in the plasmid and the IBV cDNA sequence in the vaccinia virus genome. Recombinant vaccinia viruses (rVV) expressing the *Ecogpt* gene are selected for resistance against mycophenolic acid (MPA) in the presence of xanthine and hypoxanthine. In the third step, the MPA-resistant rVVs are grown in the absence of MPA selection, resulting in loss of the *Ecogpt* gene owing to a single homologous recombination event between duplicated sequences, present in the vaccinia virus genome resulting from integration of the plasmid sequence (**Fig. 3**). During the third step two recombination events can occur, each of them with equal frequency. One event will result in the generation of the original (unmodified) IBV sequence and the other in the generation of an IBV cDNA containing the desired modification.

Infectious rIBVs are generated from the rVV DNA transfected into primary chick kidney (CK) cells previously infected with a recombinant fowlpox virus expressing T7 RNA polymerase [rFPV-T7; *(8)*]. In addition, a plasmid, pCi-Nuc *(1,9)*, expressing the IBV nucleoprotein (N), under the control of both the cytomegalovirus (CMV) RNA polymerase II promoter and the T7 RNA promoter, is co-transfected into the CK cells. Expression of T7 RNA polymerase in the presence of the IBV N protein and the rVV DNA, containing the

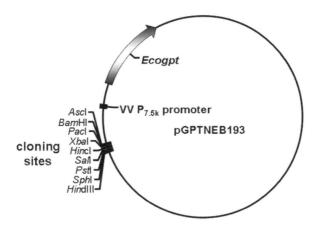

Fig. 1. Schematic diagram of the recombination vector for insertion of genes into a vaccinia virus genome using TDS. Plasmid pGPTNEB193 contains the *Ecogpt* selection gene under the control of the vaccinia virus early/late $P_{7.5k}$ promoter, a multiple cloning region for the insertion of the sequence to be incorporated into the vaccinia virus genome and the *bla* gene (not shown) for ampicillin selection of the plasmid in *E. coli*. For modification of the IBV genome, a sequence corresponding to the region being modified, plus flanking regions of 500 to 800 nucleotides, for recombination purposes is inserted into the multiple cloning sites using an appropriate restriction endonuclease. The plasmid is purified from *E. coli* and transfected into Vero cells previously infected with a recombinant vaccinia virus containing a full-length cDNA copy of the IBV genome.

full-length IBV cDNA under the control of a T7 promoter, results in the generation of infectious IBV RNA, which in turn results in the production of infectious rIBVs **(Fig. 4)**.

The overall procedure can be divided into two parts, modification of the IBV cDNA by TDS in recombinant vaccinia viruses and the recovery of infectious rIBV. The generation of the *Ecogpt* plasmids, based on pGPTNEB193, containing the modified IBV cDNA, is by standard *Escherichia coli* cloning methods *(10,11)* and is not described here. General methods for growing vaccinia virus and for using the TDS method for modifying the vaccinia virus genome have been published *(12,13)*.

2. Materials

2.1. Production of Vaccinia Virus Stocks

1. BHK-21 maintenance medium: Glasgow-Modified Eagle's Medium (G-MEM; Sigma) containing 2 mM L-glutamine (Gibco), 0.275% sodium bicarbonate, 1%

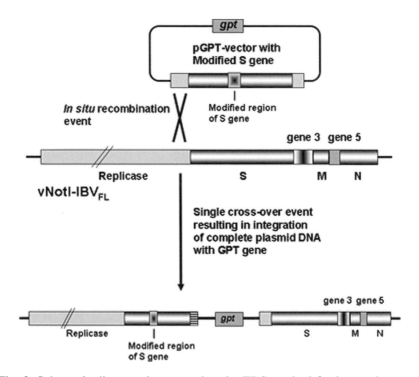

Fig. 2. Schematic diagram demonstrating the TDS method for integrating a modified IBV sequence into the full-length IBV cDNA within the genome of a recombinant vaccinia virus (vNotI-IBV$_{FL}$). The diagram shows a potential first single-step recombination event between the modified IBV sequence within pGPTNEB193 and the IBV cDNA within vNotI-IBV$_{FL}$. In order to guarantee a single-step recombination event any potential recombinant vaccinia viruses are selected in the presence of MPA; only vaccinia viruses expressing the *Ecogpt* gene are selected. The main IBV genes are indicated: the replicase, spike (S), membrane (M), and nucleocapsid (N) genes. The IBV gene 3 and 5 gene clusters that express three and two gene products, respectively, are also indicated. In the example shown a modified region of the S gene is being introduced into the IBV genome.

    fetal calf serum (Autogen Bioclear), 0.3% tryptose phosphate broth (TPB; BDH), 500 U/ml nystatin (Sigma) and 100 U/ml penicillin and streptomycin (Sigma).

2. TE buffer: 10 mM Tris-HCl pH 9, 1 mM EDTA.
3. BHK-21 cells.
4. T150 (150-cm^2) flasks.
5. 50-ml Falcon tubes.
6. Screw-top Eppendorf microfuge tubes.
7. Large Tupperware boxes, Biohazard tape, masking tape.
8. A Beckman CS-15R centrifuge or equivalent.

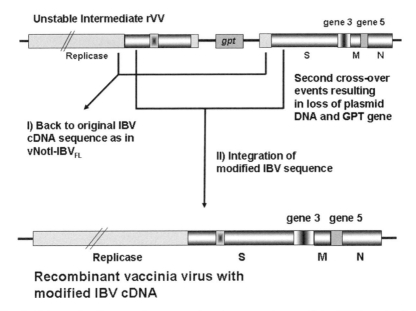

Fig. 3. Schematic diagram demonstrating the second step of the TDS method. Integration of the complete pGPTNEB193 plasmid into the vaccinia virus genome results in an unstable intermediate because of the presence of tandem repeat sequences, in this example the 3′-end of the replicase gene, the S gene, and the 5′-end of gene 3. The second single-step recombination event is induced in the absence of MPA; loss of selection allows the unstable intermediate to lose one of the tandem repeat sequences including the *Ecogpt* gene. The second step recombination event can result in either: (I) the original sequence of the input vaccinia virus IBV cDNA sequence, in this case shown as a recombination event between the two copies of the 3′-end of the replicase gene, which results in loss of the modified S gene sequence along with *Ecogpt* gene; or (II) retention of the modified S gene sequence and loss of the original S gene sequence and *Ecogpt* gene as a result of a potential recombination event between the two copies of the 5′-end of the S gene sequence. This event results in a modified S gene sequence within the IBV cDNA in a recombinant vaccinia virus.

2.2. Infection/Transfection of Vero Cells

1. Vero cells.
2. Modified Eagle's Medium: 10X (E-MEM; Sigma).
3. BES (N,N-Bis(2-hydroxyethyl)-2-aminoethanesulfonic acid) cell maintenance medium: 1X E-MEM, 0.3% TPB, 0.2% bovine serum albumin (BSA; Sigma), 20 mM BES, 0.21% sodium bicarbonate, 2 mM L-glutamine, 250 U/ml nystatin and 100 U/ml penicillin and streptomycin.

A

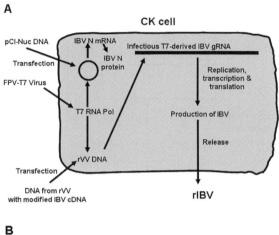

B

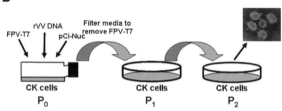

Fig. 4. A schematic representation of the recovery process for obtaining rIBV from DNA isolated from a recombinant vaccinia virus containing a full-length IBV cDNA under the control of a T7 promoter: (A) In addition to the vaccinia virus DNA containing the full-length IBV cDNA under the control of a T7 promoter, a plasmid, pCI-Nuc, expressing the IBV nucleoprotein, required for successful rescue of IBV, is transfected into CK cells previously infected with a recombinant fowlpox virus, FPV-T7, expressing T7 RNA polymerase. The T7 RNA polymerase results in the synthesis of an infectious RNA from the vaccinia virus DNA that consequently leads to the generation of infectious IBV being released from the cell. (B) Any recovered rIBV present in the media of P_0 CK cells is used to infect P_1 CK cells. The medium is filtered though a 0.22-μm filter to remove any FPV-T7 virus. IBV-induced CPE is normally observed in the P_1 CK cells following a successful recovery experiment. Any rIBV is passaged a further two times, P_2 and P_3, in CK cells. Total RNA is extracted from the P_1 to P_3 CK cells and the IBV-derived RNA analyzed by RT-PCR for the presence of the required modification.

   4. 2X E-MEM medium: E-MEM (2X), 10% fetal calf serum, 0.35% sodium bicarbonate, 4 mM L-glutamine, 1000 U/ml nystatin, and 200 U/ml penicillin and streptomycin.
   5. Six-well plates.
   6. OPTIMEM medium (Invitrogen).
   7. Lipofectin (Invitrogen).

8. MXH solution: separate stocks of: (1) mycophenolic acid (MPA; Sigma) 10 mg/ml in 0.1 N NaOH, (2) xanthine (Sigma) 10 mg/ml in 0.1 N NaOH (heated at 37°C to dissolve), and (3) hypoxanthine (Sigma) 10 mg/ml in 0.1 N NaOH.
9. *Ecogpt* selection medium: Add 250 µl of MPA, 2.5 ml of xanthine and 149 µl of hypoxanthine to 50 ml of prewarmed 2X EMEM medium. Add 50 ml of 2% low-melting-point agarose (keep in a water bath at 42°C) to the 2X EMEM containing MPA, xanthine, and hypoxanthine and mix well before adding it to vaccinia virus infected cells (*see* **Note 7**).
10. Screw-top Eppendorf microfuge tubes.
11. A Beckman CS-15R centrifuge or equivalent.

2.3. Selection of MPA Resistant Recombinant Vaccinia Viruses (*gpt*⁺ Phenotype)

1. Vero cell monolayers in six-well plates.
2. PBSa solution: 172 mM NaCl, 3 mM KCl, 10 mM Na_2HPO_4, and 2 mM KH_2PO_4 pH 7.2.
3. 1X E-MEM.
4. *Ecogpt* selection medium containing 1% low-melting agarose (*see* **Note 7**).
5. 1X E-MEM containing 1% low-melting agarose and 0.01% neutral red.

2.4. Selection of MPA-Sensitive Recombinant Vaccinia Viruses (Loss of *gpt*⁺ Phenotype)

Two or three of the plaque purified gpt⁺ rVVs are used for generation of new rVVs in the absence of MPA selection medium to generate viruses with a gpt⁻ phenotype.

1. Vero cell monolayers in six-well plates.
2. PBSa.
3. 1X E-MEM.
4. 1X E-MEM containing 1% low-melting agarose (*see* **Note 7**).
5. 1X E-MEM containing 1% low-melting agarose and 0.01% neutral red (*see* **Note 7**).
6. *Ecogpt* selection medium containing 1% low-melting agarose (*see* **Note 7**).

2.5. Screening of Recombinant Vaccinia Viruses

Small stocks of the plaque-derived rVVs have to be produced for extraction of DNA for screening purposes. The vaccinia DNA is used as a template for PCR and sequence analysis to check for the presence of the modified sequence and confirmation that the *Ecogpt* gene has been lost.

1. Vero cell monolayers in six-well plates.
2. 1X BES medium.
3. PBSa.
4. Freshly prepared proteinase K (Sigma; 10 mg/ml) in H_2O.
5. 2X proteinase K buffer: 200 mM Tris/HCl pH 7.5, 10 mM EDTA, 0.4% SDS, 400 mM NaCl.
6. Phenol-chloroform (Amresco), chloroform (BDH), and absolute ethanol (BDH).
7. 70% ethanol.
8. Bench top microfuge.

2.6. Vaccinia Virus Purification

Vaccinia virus DNA for a recovery of IBV requires partial purification of the rVV through a sucrose cushion.

1. 50-ml Falcon tubes.
2. TE buffer.
3. Filtered 30% sucrose (w/v) in 1 mM Tris/HCl pH 9.
4. A Beckman CS-15R centrifuge or equivalent.
5. Superspin 630 rotor and Sorvall OTD65B ultracentrifuge or equivalent.

2.7. Extraction of Vaccinia Virus DNA

1. Freshly prepared proteinase K (Sigma; 10 mg/ml) in H_2O.
2. 2X proteinase K buffer: 200 mM Tris/HCl pH 7.5, 10 mM EDTA, 0.4% SDS, 400 mM NaCl.
3. 15-ml Falcon tubes.
4. Sodium acetate 3 M.
5. Phenol-chloroform, chloroform, absolute ethanol, and 70% ethanol.
6. A Beckman CS-15R centrifuge or equivalent.

2.8. Analysis of Vaccinia Virus DNA by Pulse Field Agarose Gel Electrophoresis

1. 10X TBE buffer: 1 M Tris, 0.9 M boric acid pH 8, and 10 mM EDTA.
2. Agarose (Biorad, pulsed field certified ultrapure DNA-grade agarose).
3. DNA markers (e.g., 8–48 kb markers, Biorad).
4. Ethanol, ethidium bromide (0.5 mg/ml), MQ water.
5. CHEF-DR® II pulse field gel electrophoresis (PFGE) apparatus (Biorad).
6. 6X sample loading buffer: 62.5% glycerol, 62.5 mM Tris-HCl pH 8, 125 mM EDTA and 0.06% bromophenol blue (BDH).
7. Microwave oven, sealed plastic container to hold gel, orbital shaker, and water bath

2.9. Preparation of rFP-T7 Stock Virus

1. CEF growth medium: 1X 199 Medium with Earle's Salts, 0.3% TPB, 8% new born calf serum (NBCS), 0.225% sodium bicarbonate, 2 mM L-glutamine, 100 U/ml penicillin, 100 U/ml streptomycin and 500 U/ml nystatin.
2. CEF maintenance medium: as above but containing 2% NBCS.
3. CEF cells.
4. T75 (75-cm^2) flasks.
6. A Beckman CS-15R centrifuge or an equivalent centrifuge.

2.10. Infection CK Cells with rFPV-T7

1. CK cell maintenance medium: 1X BES medium: BES (N,N-Bis(2-hydroxyethyl)-2-aminoethanesulfonic acid) cell maintenance medium: 1X E-MEM, 0.3% TPB, 0.2% bovine serum albumin (BSA; Sigma), 20 mM BES, 0.21% sodium bicarbonate, 2 mM L-glutamine, 250 U/ml nystatin, and 100 U/ml penicillin and streptomycin.
2. CK cells.
3. 60-mm dishes.
4. PBSa.
5. Stock rFPV-T7 virus.

2.11. Transfection of Vaccinia Virus DNA into CK Cells

1. OPTIMEM 1 with GLUTAMAX-1 (Invitrogen).
2. Lipofectin reagent (Invitrogen).
3. Stock rFPV-T7 virus.
4. The rVV DNA as prepared in Section 3.7, step 2.
5. Plasmid pCi-Nuc, which contains IBV nucleoprotein under the control of the CMV and T7 promoters.
6. Millex® GP 0.22-μm syringe-driven filters (Millipore).

2.12. Serial Passage of rIBVs

1. CK cell maintenance medium: 1X BES medium (as in Section 2.2.1, step 3).
2. CK cells.
3. 60-mm dishes.
4. PBSa.

3. Methods
3.1. Production of Vaccinia Virus Stocks

1. Freeze-thaw the vaccinia virus stocks three times (37°C/dry ice) and sonicate for 2 min using a cup form sonicator (Heat Systems Ultrasonic Inc., Model W-375), continuous pulse at 70% duty cycle, seven-output control (*see* **Notes 1–4**).

2. Add G-MEM to the sonicated virus and infect twenty T150 flasks of confluent monolayers of BHK-21 cells using 2 ml of the diluted vaccinia virus per flask at a MOI of 0.1–1. Incubate the infected cells for 1 h at 37°C and 5% CO_2.

3. Add 20 ml of prewarmed (37°C) G-MEM and incubate the infected cells at 37°C and 5% CO_2 until the cells show an advanced CPE (normally about 2–3 days). At this stage the cells should easily detach from the plastic.

5. Either continue to step 6 or freeze the flasks at –20°C until use.

6. If prepared from the frozen state the flasks have to be defrosted by leaving them at room temperature for 15 min and then at 37° C until the medium over the cells has thawed.

7. Tap the flasks to detach the cells from the plastic, using a cell scraper if necessary.

8. Transfer the medium containing the cells to 50-ml Falcon tubes and centrifuge at $750 \times g$ for 15 min at 4°C to pellet the cells.

9. Discard the supernatant (99% of vaccinia virus is cell-associated) and resuspend the cells in 2 ml of TE buffer. This preparation can be used for a working stock of virus (follow point 10) or for further purification of the virus for extraction of DNA.

10. Aliquot the resuspended cells in 1-ml aliquots in screw top microfuge tubes and store at –70°C until required.

11. Determine the titer of the virus stock using Vero cells before use. The titer should be within the order of 10^{8-9} PFU/ml.

3.2. Infection/Transfection of Vero Cells

1. Infect six-well plates of 70% confluent monolayers of Vero cells with the rVV at an MOI of 0.2. Use two independent wells per recombination. (*see* **Notes 1–4**)

2. Incubate at 37°C 5% CO_2 for 2 h in to allow the virus to infect the cells.

3. After 1 h of incubation, prepare the following solutions for transfection: (a) Solution A: For each transfection: Dilute 5 µg of modified pGPTNEB193 (containing the modified IBV cDNA) in 1.5 ml of OPTIMEM medium. (b) Solution B: Dilute 12 µl of Lipofectin in 1.5 ml of OPTIMEM for each transfection.

4. Incubate solutions A and B separately for 30 min at room temperature; then mix the two solutions and incubate the mixture at room temperature for 15 min.

5. During the 15 min, remove the inoculum from the infected cells and wash the cells twice with OPTIMEM.

6. Add 3 ml of the transfection mixture (solutions A + B) to each well.

7. Incubate for 60–90 min at 37°C, 5% CO_2.

8. Remove the transfection mixture from the cells and replace it with 3 ml of BES medium.

9. Incubate the transfected cells overnight at 37°C, 5% CO_2.

10. The following morning add the MXH components, MPA 12.5 µl, xanthine 125 µl, and hypoxanthine 7.4 µl, directly to each well.

11. Incubate the cells at 37°C, 5% CO_2 until they display advanced vaccinia virus induced CPE (normally 2 days).

12. Harvest the infected/transfected cells into the cell medium of the wells and centrifuge for 3–4 min at 300 × g. Discard supernatant and resuspend the pellet in 0.4 ml of 1X E-MEM.

13. Freeze-thaw the vaccinia virus stocks (*see* **Notes 1–4**) three times (37°C/dry ice) and sonicate for 2 min using a cup form sonicator (Heat Systems Ultrasonic Inc., Model W-375), continuous pulse at 70% duty cycle, seven-output control. This will be the stock virus for selection of a rVV containing the intended modification. The virus can be stored at –20° or –70°C.

3.3. Selection of MPA-Resistant Recombinant Vaccinia Viruses (GPT+ Phenotype)

Isolation of gpt$^+$ rVVs is by plaque assay on Vero cells.

1. Freeze-thaw the vaccinia virus three times (37°C/dry ice) and sonicate for 2 min using a cup form sonicator (Heat Systems Ultrasonic Inc., Model W-375), continuous pulse at 70% duty cycle, seven-output control (*see* **Notes 1–4**).

2. Remove the medium from Vero cells in six-well plates and wash the cells twice with PBSa.

3. Prepare 10^{-1} and 10^{-2} dilutions of the recombinant vaccinia virus in 1X E-MEM (normally, dilute 150 µl of virus in 1350 µl of medium).

4. Remove the PBSa from the Vero cells and add 500 µl of the diluted virus per well (assay each dilution in duplicate).

5. Incubate for 1– 2 h at 37°C, 5% CO_2.

6. Remove the inoculum and add 3 ml of the *Ecogpt* selection medium in 1% low-melting agarose overlay (*see* **Note 7**).

7. Incubate for 4 days at 37°C, 5% CO_2 and stain the cells by adding 2 ml of 1X E-MEM containing 1% agarose and 0.01% neutral red.

8. Incubate the cells at 37°C, 5% CO_2 for 6 h and pick ten-well isolated plaques for each recombinant by taking a plug of agarose directly above the plaque. Place the plug of agarose in 400 µl of 1X E-MEM.

9. Perform two further rounds of plaque purification for each selected recombinant vaccinia virus (two or three of the picked plaques from step 8) in the presence of selection medium, as described in steps 1–8, using a dilution of 10^{-1} for each virus.

3.4. Selection of MPA-Sensitive Recombinant Vaccinia Viruses (Loss of gpt+ Phenotype)

1. Take the MPA resistant plaque-purified rVVs and freeze-thaw the virus three times (37°C/dry ice) and sonicate for 2 min using a cup form sonicator (Heat Systems Ultrasonic Inc., Model W-375), continuous pulse at 70% duty cycle, seven-output control (*see* **Notes 1–4**).

2. Remove the medium from Vero cells in six-well plates and wash the cells with PBSa.
3. Prepare 10^{-1} and 10^{-2} dilutions of the gpt^{+} plaque-purified recombinant vaccinia viruses in 1X E-MEM
4. Remove the PBSa from the Vero cells and add 500 μl of the diluted gpt^{+} plaque-purified recombinant vaccinia viruses to each well (assay each dilution in duplicate).
5. Incubate the infected Vero cells for 1–2 h at 37°C, 5% CO_2.
6. Remove the inoculum and add 3 ml of overlay containing 1X E-MEM and 1% agarose.
7. Incubate the infected Vero cells for 4 days at 37°C, 5% CO_2 and stain the cells by adding 2 ml of 1X E-MEM containing 1% agarose and 0.01% neutral red. At the end of the day or the following morning, choose approximately ten isolated plaques for each recombinant and resuspend in 400 μl of 1X E-MEM.
8. Plaque purify each recombinant vaccinia virus three times in the absence of selection medium following the same procedure in Section 3.3, as described for plaque purification in presence of selection medium. However, dilutions of 10^{-1}, 10^{-2}, and 10^{-3} are required. Dilution 10^{-1} is plated in the presence of *Ecogpt* selection medium, to identify the presence of any MPA-resistant rVVs. Dilutions 10^{-2} and 10^{-3} are carried out in the absence of selection medium. Once there is no evidence of MPA-resistant rVVs in the MPA selection controls, it can be assumed that the *Ecogpt* gene has been lost and the recombinant vaccinia viruses can be screened for the presence of the required modifications and the presence/absence of the *Ecogpt* gene confirmed.
9. Select several plaques and place the plug of agarose in 400 μl of 1X EMEM.

3.5. Screening of Recombinant Vaccinia Viruses

1. Take the plaque-purified rVVs and freeze-thaw three times (37°C/dry ice) and sonicate for 2 min using a cup form sonicator (Heat Systems Ultrasonic Inc., Model W-375), continuous pulse at 70% duty cycle, seven-output control. (*see* **Notes 1–4**).
2. Wash the Vero cells with PBSa.
3. Dilute 150 μl of the sonicated rVVs in 350 μl of 1X BES medium.
4. Remove the PBSa from the Vero cells and add 500 μl of the diluted rVVs.
5. Incubate at 37°C, 5% CO_2 for 1–2 h.
6. Remove the virus inoculum and add 2.5 ml of 1X BES medium.
7. Incubate the infected Vero cells at 37°C, 5% CO_2 until the cells show signs of vaccinia virus-induced CPE in about 70–80% of the Vero cell monolayer (approx. 4 days).
8. Scrape the Vero cells into the medium and centrifuge for 1 min at 13,000 rpm $(16,000 \times g)$.
9. Discard the supernatants and resuspend the cells in 800 μl of 1X BES medium.

10. Take 700 μl of the resuspended cells as virus stocks and store at –20°C.

11. To the remaining 100 μl of the resuspended cells add 100 μl 2X proteinase K buffer and 2 μl of the 10 mg/ml proteinase K stock to give a final concentration of 0.1 mg/ml. Gently mix to prevent shearing of the vaccinia virus DNA and incubate at 50°C for 2 h (*see* **Note 5**).

12. Add 200 μl of phenol-chloroform to the proteinase K-treated samples and mix by inverting the tube five to ten times and centrifuge at 13,000 rpm (16,000 × *g*) for 5 min.

13. Take the upper phase (aqueous phase) and repeat step 12 twice more.

14. Add 200 μl of chloroform to the upper phase from the final step of 13. Mix well and centrifuge at 13,000 rpm (16,000 × *g*) for 5 min.

15. Take the upper phase and precipitate the vaccinia virus DNA by adding 2.5 volumes of absolute ethanol; the precipitated DNA should be visible. Centrifuge the precipitated DNA at 13,000 (16,000 × *g*) for 20 min. Discard the supernatant and wash the pelleted DNA with 400 μl 70% ethanol.

16. Centrifuge at 13,000 rpm (16,000 × *g*) for 10 min, carefully discard the supernatant and remove the last drops of 70% ethanol using a capillary tip.

17. Resuspend the DNA in 30 μl of water, briefly heat the DNA at 50°C (with the lid of the Eppendorf tube opened) to remove any remaining ethanol, and store at 4°C.

18. At this stage the rVV DNA from step 17 is analyzed by PCR and/or sequence analysis for the presence/absence of the *Ecogpt* gene and for the modifications within the IBV cDNA sequence. The rVVs that have lost the *Ecogpt* gene and contain the desired IBV modifications are used to produce larger stocks of virus, as described in Section 3.1 (but using smaller amounts) for further analysis and for the preparation of larger stocks of vaccinia virus DNA for recovery of rIBV.

3.6. Vaccinia Virus Purification

1. Prepare large batches of vaccinia virus as described in Section 3.1. Ten T150 flasks are normally sufficient (*see* **Notes 1–4**).

2. Freeze-thaw the 2-ml aliquots, from Section 3.1, step 9, three times (37°C/dry ice) and sonicate for 2 min using a cup form sonicator (Heat Systems Ultrasonic Inc., Model W-375), continuous pulse at 70% duty cycle, seven-output control.

3. Place the aliquots on ice and then pool identical aliquots in 50-ml Falcon tubes and centrifuge (Beckman CS-15R) at 750 × *g* for 10 min at 4°C to remove the cell nuclei.

4. Add TE buffer to the supernatants to give a final volume of 13 ml.

5. Add 16 ml of the 30% sucrose solution into a Beckman ultra-clear (25 × 89 mm) ultracentrifuge tube and carefully layer 13 ml of the cell lysate from step 4 onto the sucrose cushion. Place the tubes in a superspin 630 rotor.

6. Centrifuge the samples using a Sorvall OTD65B ultracentrifuge with a superspin 630 rotor at 14,000 rpm (36,000 × *g*) at 4°C for 60 min.

7. The partially purified vaccinia virus particles form a pellet under the sucrose cushion. After centrifugation carefully remove the top layer (usually pink) and the sucrose layer with a pipette. Wipe the sides of the tube carefully with a tissue to remove any sucrose solution.

8. Resuspend each pellet using 5 ml of TE buffer and store at –70°C.

3.7. Extraction of Vaccinia Virus DNA

1. Defrost the partially purified vaccinia virus from step 3.6.8 at 37°C and add 5 ml of prewarmed 2X proteinase K buffer and 100 μl of 10 mg/ml proteinase K. Incubate at 50°C for 2 h (*see* **Notes 1–4**).

2. Add 5 ml of the proteinase K treated vaccinia virus DNA into two 15-ml Falcon tubes.

3. Add 5 ml of phenol-chloroform, mix by inverting the tube five to ten times, and centrifuge at 1100 × *g* for 15 min at 4°C. Cut the end of the pipette tips and transfer the upper phase to a clean 15-ml Falcon tube. Repeat this step once more, placing the upper phase in a clean 15-ml Falcon tube. (*see* **Note 5**).

4. Add 5 ml of chloroform, mix by inverting the tube five to ten times, and centrifuge at 1100 × *g* for 15 min at 4°C. Transfer 2.5 ml of the upper phase to two clean 15-ml Falcon tubes.

5. Precipitate the vaccinia virus DNA by adding 2.5 volumes of absolute ethanol and 0.1 volumes of 3 M sodium acetate. Centrifuge at 1200 × *g*, at 4°C for 30 min. In order to visualize the DNA pellet 2 μl of pellet paint (Novagen) per sample can be added before the 3 M sodium acetate, mix, add the ethanol, mix again, and incubate for 2 min at room temperature before centrifugation.

6. Discard the supernatant and wash the DNA using 10 ml of 70% ethanol. Leave on ice for 5 min and centrifuge at 1200 × *g*, 4°C for 30 min. Discard the supernatant and remove the last drops of ethanol using a capillary tip. Dry the inside of the tube using a tissue to remove any ethanol.

7. Resuspend the vaccinia DNA in 300 μl of water and briefly heat at 50°C to remove any remaining ethanol. Gently flick the tube until the DNA dissolves. Note: more water may have to be added, depending on the viscosity of the DNA solution.

8. Leave the tubes at 4°C overnight. If the pellet has not totally dissolved, add more water.

9. Keep the vaccinia virus DNA at 4°C. DO NOT FREEZE (*see* **Note 6**).

10. Digest 1 μg of the DNA with a suitable restriction enzyme in a 20-μl volume to check the quality of the DNA by pulse field agarose gel electrophoresis.

3.8. Analysis of Vaccinia Virus DNA by Pulse Field Agarose Gel Electrophoresis

1. Prepare 2.3 liters of 0.5X TBE buffer for preparation of the agarose gel and as an electrophoresis running buffer. 100 ml is required for a 12.7 × 14-cm agarose gel and 2 liters is required as running buffer.

Table 1
Standard Conditions for Producing a PFGE Agarose Gel

Agarose concentration	0.8 %	1.0 %
Buffer	0.5X TBE	0.5X TBE
Gel volume	100 ml	100 ml
Initial pulse time	0.1 sec	3.0 sec
Final pulse time	1.0 sec	30.0 sec
Duration	6–16 h	16–20 h
Voltage	6.0 V/cm	6.0 V/cm

2. Calculate the concentration of agarose that is needed to analyze the range of DNA fragments. Increasing the agarose concentration decreases the DNA mobility within the gel, requiring a longer run time or a higher voltage. However, a higher voltage can increase DNA degradation and reduce resolution. A 0.8% agarose gel is suitable for separating DNA ranging between 50–95 kb. A 1% agarose gel is suitable for separating DNA ranging between 20 and 300 kb.

3. Place the required amount of agarose in 100 ml of 0.5X TBE, microwave until the agarose is dissolved, and cool to approximately 50°–60°C.

4. Clean the gel frame and comb with MQ water followed by ethanol. Place the gel frame on a level surface, assemble the comb, and pour the cooled agarose into the gel frame. Remove any bubbles using a pipette tip, allow the agarose to set (approx. 30–40 min), and store in the fridge until required.

5. Place the remaining 0.5X TBE buffer into the CHEF-DR® II PFGE electrophoresis tank and switch the cooling unit on. Leave the buffer circulating to cool.

6. Add the sample loading dye to the digested vaccinia virus DNA samples (Section 3.7, step 10) and incubate at 65°C for 10 min.

7. Place the agarose gel in the electrophoresis chamber; load the samples using tips with cut ends (widened bore) and appropriate DNA markers (*see* **Note 5**).

8. The DNA samples are analyzed by PFGE at 14°C in gels run with a 0.1–1.0 sec switch time for 16 h at 6 V/cm at an angle of 120° or with a switch time of 3.0–30.0 sec for 16 h at 6 V/cm, depending on the concentration of agarose used. **Table 1** summarizes the standard conditions for 0.8% and 1.0% agarose gels for PFGE.

9. Following PFGE place the agarose gel in a sealable container with 400 ml of 0.1 μg/ml ethidium bromide and gently shake for 30 min at room temperature.

10. Wash the ethidium-stained agarose gel in 400 ml of MQ water by gently shaking for 30 min.

11. Visualize DNA bands using a suitable UV system for analyzing agarose gels. An example of recombinant vaccinia virus DNA digested with the restriction enzyme *Sal*I and analyzed by PFGE is shown in **Fig. 5**.

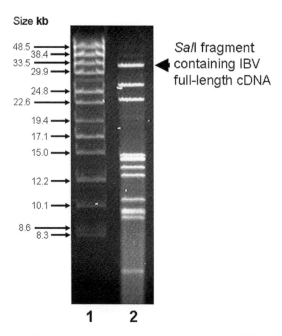

Fig. 5. Analysis of *Sal*I digested vaccinia virus DNA by PFGE. Lane 1 shows DNA markers and Lane 2 the digested vaccinia virus DNA. The IBV cDNA used does not contain a SalI restriction site; therefore the largest DNA fragment (~31 kb) generated from the recombinant vaccinia virus DNA represents the IBV cDNA with some vaccinia virus-derived DNA at both ends.

3.9. Preparation of rFP-T7 Stock Virus

Infectious recombinant IBVs are generated in situ by co-transfection of vaccinia virus DNA, containing the modified IBV cDNA and pCi-Nuc into CK cells previously infected with a recombinant fowlpox virus expressing T7 DNA-dependent RNA polymerase. This protocol covers the procedure for infecting primary avian chicken embryo fibroblasts (CEF) cells with a recombinant fowlpox virus (rFPV/T7) expressing the bacteriophage T7 RNA polymerase under the direction of the vaccinia virus $P_{7.5}$ early-late promoter *(8)*. Preparation of a 200-ml stock of rFPV/T7 uses ten T75 flasks containing confluent monolayers of CEF cells.

1. Remove the culture growth medium from the cells and infect with 2 ml of rFPV/T7 at an MOI of 0.1, previously diluted in CEF maintenance medium.
2. Incubate the infected cells for 2 h at 37°C 5% CO_2; then without removing the inoculum add 20 ml of CEF maintenance medium.

3. After 4 days postinfection check for CPE (90% of the cells should show CPE). Tap the flasks to detach the cells from the plastic and disperse the cells into the medium by pipetting them up and down.
4. Freeze-thaw the cells, as described in Section 3.1, step 1, and centrifuge at $750 \times g$, 4°C for 5 min to remove the cell debris. Store the supernatant containing the virus stock at –80°C until required.
5. Determine the titer of the virus stock using CEF cells. The titer should be on the order of 10^7 PFU/ml.

3.10. Infection CK Cells with rFPV-T7

1. Seed CK cells in 13 × 60-mm dishes to give a 50% confluent monolayers on the day after seeding. Normally, for each recovery we prepare twelve dishes, ten replicates for the recovery experiment and two controls, rFPV-T7-infected and mock-infected CK cells.
2. Remove the medium and wash the cells once with PBSa.
3. Infect the cells with rFPV-T7 at an MOI of 10. Add the virus into a final volume of 1 ml of CK cell maintenance medium per dish.
4. Incubate for 1 h at 37°C, 5% CO_2, whilst incubating the cells, prepare the solutions as outlined in Section 3.11.

3.11. Transfection of Vaccinia Virus DNA into CK Cells

During the infection of CK cells with rFPV-T7 prepare the transfection reaction reagents: rVV DNA, pCi-Nuc, and Lipofectin (Invitrogen). The reagents are added as follows:

1. Prepare the two master solutions: (A) 15 ml OPTIMEM add 100 μg of rVV DNA and 50 μg of pCi-Nuc. (B) 15 ml OPTIMEM add 300 μl of Lipofectin.
2. Incubate solutions A and B at room temperature for 30 min.
3. Mix A and B together and incubate for a further 15 min at room temperature.
4. Wash the rFPV-T7 infected CK cells twice with OPTIMEM and carefully add 3 ml of the A + B transfection mixture to ten of the dishes of the rFPV-T7-infected CK cells. The other two dishes are for the controls described above.
5. Incubate the transfected cells at 37°C, 5% CO_2 for 16 h.
6. Next morning, remove the transfection medium and add 5 ml of fresh BES medium and incubate at 37°C, 5% CO_2.
7. Two days after changing the transfection media, when FPV/IBV-induced CPE is obvious (approx. 50%), harvest the cell supernatant, place in Eppendorf tubes, centrifuge for 3 min at 13,000 rpm ($16,000 \times g$), place it into 5-ml Bijoux tubes, and filter it through a 0.22-μm (pore size) filter to remove any rFPV-T7 virus present.

3.12. Serial Passage of rIBVs

To check for the presence of any recovered rIBVs, the medium from the P_0 CK cells (Section 3.11, step 7) is passaged three times, P_1 to P_3, on CK cells (**Fig. 4B**), checking for any IBV-associated CPE. Total RNA is extracted from the P_1 to P_3 CK cells and analyzed for the presence of IBV RNA by specific RT-PCR reactions (*see* **Note 8**). For passage 1 (P_1):

1. Seed CK cells in T25 flasks to be confluent monolayers on the day required.
2. Remove the growth medium and wash once with PBSa.
3. Add 1 ml of the filtered medium from the P_0 CK cells (Section 3.11, step 7) to the confluent CK cells and incubate at 37°C, 5% CO_2 for 1 h. Then, without removing the inoculum add 4 ml of BES medium.
4. Check the cells for IBV-associated CPE over the next 2–3 days. When about 50–75% of the CK cells show a CPE, infect new cells with some of the cell medium as described in steps 1 to 3. Repeat for serial passages P_2 and P_3 CK cells. Filtration of the infected cell medium is not required after P_1.
5. After P_3 any recovered virus is used to prepare a large stock for analysis of the virus genotype and phenotype.

4. Notes

1. Vaccinia virus is classified as a category 2 human pathogen and its use is therefore subject to local regulations and rules that have to be followed.
2. Always discard any medium of solution containing vaccinia virus into a 1% solution of Virkon; leave at least 12 h before discarding.
3. Flasks of cells infected with vaccinia virus should be kept in large Tupperware boxes, which should be labeled with the word vaccinia and biohazard tape. A paper towel should be put on the bottom of the boxes to absorb any possible spillage.
4. During centrifugation of vaccinia virus infected cells use sealed buckets for the centrifugation to avoid possible spillage.
5. Vaccinia virus DNA is a very large molecule that is very easy to shear; therefore when working with the DNA be gentle and use wide bore tips, cutting the ends off ordinary pipette tips.
6. Always store vaccinia virus DNA at 4°C; do not freeze the DNA as this leads to degradation of the DNA.
7. 1% low-melting agarose can be substituted with 1% agar.
8. There is always the possibility that the recovered rIBV is not cytopathic. In this case, check for the presence of viral RNA by RT-PCR at each passage, starting at P_1.

Acknowledgments

The authors thank Dr. M. Skinner for providing pGPTNEB193. We also thank many colleagues, both past and present, who have been involved in the development of our IBV reverse genetics system. This work was supported by the Department of Environment, Food and Rural Affairs (DEFRA) project codes OD1905, OD0712, and OD0717; European Communities specific RTD program Quality of Life and Management of Living Resources QLK2-CT-1999-00002; Intervet UK; the British Egg Marketing Board (BEMB); and the Biotechnology and Biological Sciences Research Council (BBSRC) grant No. 201/15836.

References

1. Casais, R., Thiel, V., Siddell, S. G., Cavanagh, D., and Britton, P. (2001) Reverse genetics system for the avian coronavirus infectious bronchitis virus. *J. Virol.* **75**, 12359–12369.
2. Falkner, F. G., and Moss, B. (1990) Transient Dominant Selection of Recombinant Vaccinia Viruses. *J. Virol.* **64**, 3108–3111.
3. Britton, P., Evans, S., Dove, B., Davies, M., Casais, R., and Cavanagh, D. (2005) Generation of a recombinant avian coronavirus infectious bronchitis virus using transient dominant selection. *J. Virol. Meth.* **123**, 203–211.
4. Casais, R., Davies, M., Cavanagh, D., and Britton, P. (2005) Gene 5 of the avian coronavirus infectious bronchitis virus is not essential for replication. *J. Virol.* **79**, 8065–8078.
5. Hodgson, T., Britton, P., and Cavanagh, D. (2006) Neither the RNA nor the proteins of open reading frames 3a and 3b of the coronavirus infectious bronchitis virus are essential for replication. *J. Virol.* **80**, 296–305.
6. Boulanger, D., Green, P., Smith, T., Czerny, C-P., and Skinner, M. A. (1998) The 131-amino-acid repeat region of the essential 39-kilodalton core protein of fowlpox virus FP9, equivalent to vaccinia virus A4L protein, is nonessential and highly immunogenic. *J. Virol.* **72**, 170–179.
7. Mulligan, R., and Berg, P. (1981) Selection for animal cells that express the *E. coli* gene coding for xanthine-guanine phosphoribosyl transferase. *Proc. Natl. Acad. Sci. USA* **78**, 2072–2076.
8. Britton, P., Green, P., Kottier, S., Mawditt, K. L., Pénzes, Z., Cavanagh, D., and Skinner, M. A. (1996) Expression of bacteriophage T7 RNA polymerase in avian and mammalian cells by a recombinant fowlpox virus. *J. Gen. Virol.* **77**, 963–967.
9. Hiscox, J. A., Wurm, T., Wilson, L., Britton, P., Cavanagh, D., and Brooks, G. (2001) The coronavirus infectious bronchitis virus nucleoprotein localizes to the nucleolus. *J. Virol.* **75**, 506–512.
10. Ausubel, F. M., Brent, R., Kingston, R. E., et al. (1987) *Current Protocols in Molecular Biology.* Wiley, New York.
11. Sambrook, J., Fritsch, E. F., and Maniatis, T. (1989) *Molecular Cloning: A Laboratory Manual*, 2 nd Ed. Cold Spring Harbor Laboratory, New York.
12. Mackett, M., Smith, G. L., and Moss, B. (1985) The construction and characterisation of vaccinia virus recombinants expressing foreign genes. In: Glover, D. M. (ed.) *DNA Cloning: A practical Approach.* IRL Press, Oxford, pp. 191–211.
13. Smith, G. L. (1993) Expression of genes by vaccinia virus vectors In: Davison, M. J., and Elliot, R. M. (eds.) *Molecular Virology: A practical Approach.* IRL Press, Oxford, pp. 257–83.

20

Engineering Infectious cDNAs of Coronavirus as Bacterial Artificial Chromosomes

Fernando Almazán, Carmen Galán, and Luis Enjuanes

Abstract

The construction of coronavirus (CoV) infectious clones had been hampered by the large size of the viral genome (around 30 kb) and the instability of plasmids carrying CoV replicase sequences in *Escherichia coli*. Several approaches have been developed to overcome these problems. Here we describe the engineering of CoV full-length cDNA clones using bacterial artificial chromosomes (BACs). In this system the viral RNA is expressed in the cell nucleus under the control of the cytomegalovirus promoter and further amplified in the cytoplasm by the viral replicase. The BAC-based strategy is an efficient system that allows easy manipulation of CoV genomes to study fundamental viral processes and also to develop genetically defined vaccines. The procedure is illustrated by the cloning of the genome of SARS coronavirus, Urbani strain.

Key words: coronavirus; SARS; reverse genetics; infectious clone; bacterial artificial chromosome.

1. Introduction

Coronaviruses (CoV) are enveloped, single-stranded, positive-sense RNA viruses relevant to animal and human health, including the etiologic agent of the severe acute respiratory syndrome (SARS) *(1,2)*. Owing to the huge size of the CoV genomes and the presence of unstable viral sequences in bacteria, infectious clones have been engineered using nontraditional approaches that are based on the use of bacterial artificial chromosomes (BACs) *(3– 5)*, *in vitro* ligation of cDNA fragments *(6– 8)* and vaccinia virus as a vector for the propagation of CoV full-length cDNAs *(9,10)*.

From: *Methods in Molecular Biology, vol. 454: SARS- and Other Coronaviruses,*
Edited by: D. Cavanagh, DOI: 10.1007/978-1-59745-181-9_20, © Humana Press, New York, NY

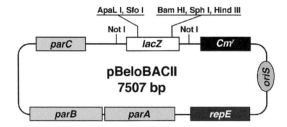

Fig. 1. Scheme of plasmid pBeloBAC11. The regulatory genes *parA*, *parB*, *parC*, and *repE*, the F-factor replication origin (*OriS*), the chloramphenicol resistance gene (*Cm'*), the *lacZ* gene, and the restriction sites that can be used to clone foreign DNAs are indicated.

In this chapter we describe the protocol for assembling CoV full-length cDNAs in the BAC plasmid pBeloBAC11 *(11)* (**Fig. 1**), using the SARS-CoV Urbani strain as an example. BACs are synthetic low-copy-number plasmids based on the well-characterized *E. coli* F-factor *(12)* that presents a strictly controlled replication leading to one or two plasmid copies per cell. These plasmids allow the stable maintenance of large DNA fragments in bacteria, minimize the toxicity problem usually observed with several CoV sequences when amplified in high-copy-number plasmids, and their manipulation is similar to that of conventional plasmids. The cDNA of the CoV genome is assembled in the BAC under the control of the cytomegalovirus (CMV) immediate-early promoter to allow the expression of the viral RNA in the nucleus by the cellular RNA polymerase II *(13)*. At the 3′-end, this cDNA is flanked by a 25-bp synthetic poly(A) and the sequences of the hepatitis delta virus (HDV) ribozyme and the bovine growth hormone (BGH) termination and polyadenylation signals to produce synthetic RNAs bearing authentic 3′-ends. This system allows the recovery of infectious virus from the cDNA clone without the need of *in vitro* ligation and transcription steps.

The BAC approach, originally applied to the transmissible gastroenteritis coronavirus (TGEV) *(3)*, has been successfully used to engineer the infectious clones of the human coronaviruses HCoV-OC43 *(5)* and SARS-CoV *(4)*, and it is potentially applicable to the cloning of other CoV cDNAs, other viral genomes, and large-size RNAs of biological relevance.

2. Materials

To reach optimal results, all solutions should be prepared in pure Milli-Q grade water that has a resistivity of 18.2 MΩ/cm.

2.1. Assembly and Manipulation of BAC Clones

2.1.1. Plasmids and Bacterial Strains

1. Plasmid pBeloBAC11 (New England Biolabs). This plasmid contains genes *parA*, *parB*, and *parC* derived from the F-factor of *E. coli* to ensure the accurate partitioning of plasmids to daughter cells, avoiding the possibility of multiple BAC coexistence in a single cell. In addition, the plasmid carries gene *repE* and the element *oriS* involved in initiation and orientation of DNA replication, a chloramphenicol resistance gene (*Cm^r*), the *lacZ* gene to allow color-based identification of recombinants by α-complementation, and the restriction sites *ApaL* I, *Sfo* I, *Bam* HI, *Sph* I, and *Hind* III to clone large fragments of DNA (**Fig. 1**).

2. *E. coli* DH10B strain (GibcoBRL. Invitrogen) [F$^-$ *mcrA* Δ (*mrr-hsd*RMS-*mcr*BC) Ø80d*lacZ*ΔM15 Δ*lac*X74 *deo*R *rec*A1 *end*A1 *ara*D139 (*ara, leu*)7697 *gal*U *gal*K λ$^-$ *rps*L *nup*G] (*see* **Note 1**).

3. DH10B electrocompetent cells. These bacterial cells can be purchased from Invitrogen (ElectroMAX DH10B cells) or prepared following the procedure described in Section 3.2.2.

2.1.2. Culture Media for E. coli

1. LB medium: 1% (w/v) tryptone, 0.5% (w/v) yeast extract, 1% (w/v) NaCl. Adjust the pH to 7.0 with 5 N NaOH. Sterilize by autoclaving on liquid cycle.

2. LB agar plates: LB medium containing 15 g/liter of Bacto Agar. Prepare LB medium and just before autoclaving add 15 g/liter of Bacto Agar. Sterilize by autoclaving on liquid cycle, and dispense in 90-mm Petri plates.

3. LB agar plates containing 12.5 μg/ml chloramphenicol. After autoclaving the LB agar medium, allow the medium to cool to 45°C, add the chloramphenicol to a final concentration of 12.5 μg/ml from a stock solution of 34 mg/ml, and dispense in 90-mm Petri plates.

4. SOB medium: 2% (w/v) tryptone, 0.5% (w/v) yeast extract, 0.05% (w/v) NaCl, 2.5 mM KCl. Adjust the pH to 7.0 with 5 N NaOH and sterilize by autoclaving on liquid cycle (*see* **Note 2**).

5. SOC medium: SOB medium containing 10 mM MgCl$_2$, 10 mM MgSO$_4$, and 20 mM glucose. After autoclaving the SOB medium, cool to 45°C and add the MgCl$_2$, MgSO$_4$, and glucose from filter sterilized 1 M stock solutions.

2.1.3. Enzymes and Buffers

1. Restriction endonucleases, shrimp alkaline phosphatase, T4 DNA ligase, Taq DNA polymerase, high-fidelity thermostable DNA polymerase, and reverse transcriptase. These enzymes can be purchased from different commercial sources.

2. Enzyme reaction buffers. Use the buffer supplied with the enzyme by the manufacturer.

2.1.4. Special Buffers and Solutions

1. LB freezing buffer: 40% (v/v) glycerol in LB medium. Sterilize by passing it through a 0.45-μm disposable filter.
2. Chloramphenicol stock (34 mg/ml). Dissolve solid chloramphenicol in ethanol to a final concentration of 34 mg/ml and store the solution in a light-tight container at –20°C. This solution does not have to be sterilized.

2.1.5. Reagents

1. Qiagen QIAprep Miniprep Kit.
2. Qiagen Large-Construct Kit.
3. Qiagen QIAEX II Kit.

2.1.6. Special Equipment

1. Equipment for electroporation and cuvettes fitted with electrodes spaced 0.2 cm.

2.2. Rescue of Recombinant Viruses

2.2.1. Cells

1. Baby hamster kidney cells (BHK-21) (ATCC, CCL-10).
2. Vero E6 cells (ATCC, CRL-1586) (*see* **Note 3**).

2.2.2. Cell Culture Medium, Solutions, and Reagents

1. Cell growth medium: Dulbecco's Minimum Essential Medium (DMEM) (Gibco-BRL. Invitrogen) supplemented with antibiotics and 10% fetal calf serum (FCS).
2. Opti-MEM I Reduced Serum Medium (GibcoBRL. Invitrogen).
3. Trypsin-EDTA solution: 0.25% (w/v) trypsin, 0.02% (w/v) EDTA.
4. Lipofectamine 2000 (Invitrogen).

3. Methods

3.1. Assembly of Full-Length CoV cDNAs in BACs

The basic strategy for the generation of CoV infectious clones using BACs is described using the SARS-CoV Urbani strain (GenBank accession number AY278741) as a model.

3.1.1. Selection of Restriction Endonuclease Sites in the Viral Genome

1. The first step for the assembly of the full-length cDNA clone is the selection of appropriate restriction endonuclease sites in the viral genome. These restriction sites must be absent in the BAC plasmid. In the case of SARS-CoV the restriction sites selected were *Cla* I, *Mlu* I, *Pme* I, *Bam* HI, and *Nhe* I (**Fig. 2A**).

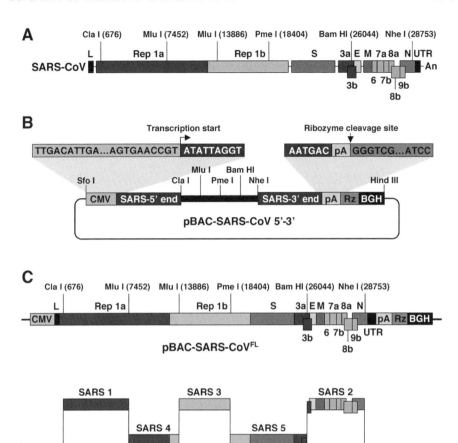

Fig. 2. Strategy to assemble a SARS-CoV infectious cDNA clone as a BAC: (A) Genetic structure of the SARS-CoV Urbani strain genome. Relevant restriction sites used for the assembly of the infectious clone are indicated. Numbers within brackets indicate the genomic positions of the first nucleotide of the restriction endonuclease recognition sequence. Letters and numbers indicate the viral genes. L, leader sequence; UTR, untranslated region; An, poly(A) tail. (B) Construction of pBAC-SARS-CoV 5′–3′. This plasmid includes the first 681 nt of the genome under the control of the CMV promoter, a multicloning site containing the restriction sites selected for the final assembly of the infectious clone, and the last 975 nt of the genome followed by a synthetic poly(A) tail (pA), the hepatitis delta virus ribozyme (Rz), and the bovine growth hormone termination and polyadenylation sequences (BGH). The CMV promoter transcription start and the ribozyme cleavage site are shown. (C) Schematic diagram showing the five-step cloning strategy used for the assembly of the SARS-CoV infectious clone. The five overlapping cDNA fragments, named SARS 1 to SARS 5, were sequentially cloned into the plasmid pBAC-SARS-CoV 5′–3′ to generate the plasmid pBAC-SARS-CoV^{FL}. Relevant restriction sites are indicated. (Reproduced from (*4*) with permission from the American Society for Microbiology.)

2. If natural preexisting restriction sites are not available in the viral genome, silent mutations have to be introduced to generate new restriction sites appropriately spaced in the viral genome, which will be used to assemble the cDNA clone (*see* **Note 4**).

3.1.2. Construction of an Intermediate BAC Plasmid as the Backbone to Assemble the Full-Length cDNA Clone

The assembly of the infectious clone in a BAC is facilitated by the construction of an intermediate BAC plasmid containing the 5′-end of the genome under the control of the CMV promoter, a multicloning site containing the restriction sites selected in the first step, and the 3′-end of the genome followed by a 25-nt synthetic poly(A), the HDV ribozyme, and the BGH termination and polyadenylation sequences. All these elements have to be precisely assembled to produce synthetic RNAs bearing authentic 5′- and 3′-ends of the viral genome (**Fig. 2B**). A detailed protocol for the generation of the SARS-CoV intermediate plasmid, pBAC-SARS-CoV 5′-3′, is described next.

1. Generate a PCR fragment containing the CMV promoter (*13*) precisely joined to the viral 5′-end by PCR using two overlapping PCR fragments as a template (*see* **Note 5**). One of these fragments should contain the CMV promoter flanked at the 5′-end by the restriction site *Sfo* I, to allow the cloning in the pBeloBAC11, and at the 3′-end by the first 20 nucleotides of the genome as overlapping sequence. The second overlapping PCR fragment should contain the 5′-end of the genome, from the first nucleotide to the restriction site *Cla* I at genomic position 676, followed by the restriction sites *Mlu* I, *Pme* I, and *Bam* HI, which will be used to assemble the infectious clone.
2. Clone the PCR fragment digested with *Sfo* I and *Bam* HI in pBeloBAC11 digested with the same restriction enzymes to generate the plasmid pBAC-SARS-CoV 5′.
3. Join the viral 3′-end with the poly(A) and the HDV-BGH sequences in a precise way by PCR using two overlapping PCR fragments as a template. One PCR fragment should contain the 3′-end of the genome, from the restriction site *Nhe* I at genomic position 28753, flanked at the 5′-end by the restriction site *Bam* HI. The other PCR fragment including the poly(A), the HDV ribozyme, and the BGH termination and polyadenylation sequences should be flanked at the 5′-end by the last 20 nucleotides of the viral genome as an overlapping sequence and at the 3′-end by the restriction site *Hind* III.
4. To generate the intermediate plasmid pBAC-SARS-CoV 5′–3′, digest with *Bam* HI and *Hind* III the PCR fragment containing the viral 3′-end followed by the poly(A) and the HDV-BGH sequences and clone it into the plasmid pBAC-SARS-CoV 5′ digested with the same restriction enzymes (*see* **Note 6**).
5. After each cloning step, the PCR-amplified fragments and cloning junctions have to be sequenced to determine that no undesired mutations were introduced.

3.1.3. Assembly of the Full-Length cDNA Clone

1. The full-length cDNA clone (pBAC-SARS-CoV^{FL}) is assembled by sequential cloning of overlapping cDNA fragments, covering the entire viral genome, into the intermediate plasmid pBAC-SARS-CoV 5′–3′ using the restriction sites selected in the first step of the cloning strategy (**Fig. 2C**) (*see* **Note 7**).
2. The overlapping cDNAs flanked by the appropriated restriction sites are generated by standard reverse transcriptase PCR (RT-PCR) (*see* **Note 5**) with specific oligonucleotides, using total RNA from infected cells as a template.
3. The genetic integrity of the cloned cDNAs is verified throughout the subcloning and assembly process by extensive restriction analysis and sequencing.

3.2. Generation and Manipulation of BAC Clones

One of the major advantages of using BAC vectors to generate infectious clones is that the manipulation of BACs is essentially the same as for a conventional plasmid with slight modifications owing to the large size of the BAC clones and the presence of this plasmid in only one or two copies per cell.

3.2.1. Amplification and Isolation of BAC Plasmids

The amplification and isolation of BAC plasmids is performed using standard procedures described for conventional plasmids but using large volumes of bacterial cultures.

3.2.1.1. ISOLATION OF BACS FROM SMALL-SCALE CULTURES

1. Small amounts of BAC DNAs are prepared from 5-ml cultures of BAC transformed DH10B cells by the alkaline lysis method. Any commercial kit can be used, but we suggest the QIAprep Miniprep Kit (Qiagen) following the recommendations for purification of large low-copy plasmids.
2. Streak the bacterial stock containing the BAC plasmid onto a LB agar plate containing 12.5 μg/ml chloramphenicol and incubate for 16 h at 37°C (*see* **Note 8**).
3. Inoculate a single colony in 5 ml of LB medium plus 12.5 μg/ml chloramphenicol in a flask with a volume of at least four times the volume of the culture and incubate for 16 h at 37°C with vigorous shaking (250 rpm) (*see* **Note 9**).
4. Harvest the bacterial cells in 15-ml centrifuge tubes by centrifugation at 6000 × *g* for 10 min at 4°C and pour off the supernatant fluid.
5. Purify the BAC plasmid following the manufacturer's instructions. Owing to the size of BAC DNAs and the need to use large culture volumes, we recommended duplicating the volume of buffers P_1, P_2, and N_3, performing the optional wash step with buffer PB, and eluting the DNA from the QIAprep membrane using buffer EB preheated at 70°C (*see* **Note 10**).

6. Depending of the BAC size, yields of 0.1–0.4 µg can be obtained. Although the BAC DNA prepared by this method is contaminated with up to 30% of bacterial genomic DNA, it is suitable for analysis by restriction enzyme digestion or PCR.

3.2.1.2. ISOLATION OF ULTRAPURE BAC PLASMIDS FROM LARGE-SCALE CULTURES

1. Large-scale preparation of ultrapure BAC DNA suitable for all critical applications, including subcloning, DNA sequencing or transfection experiments is performed by alkaline lysis with the Qiagen Large-Construct Kit, which has been specifically developed and adapted for BAC purification. This kit integrates an ATP-dependent exonuclease digestion step that enables efficient removal of bacterial genomic DNA contamination to yield ultrapure BAC DNA.
2. Inoculate a single colony from a freshly streaked plate (LB agar plate containing 12.5 µg/ml chloramphenicol) (*see* **Note 8**) in 5 ml of LB medium containing 12.5 µg/ml chloramphenicol and incubate for 8 h at 37°C with vigorous shaking (250 rpm).
3. Dilute 1 ml of the culture into 500 ml of selective LB medium (*see* **Note 9**) prewarmed to 37°C and grow the cells with vigorous shaking (250 rpm) in a 2-liter flask at 37°C for 12–16 h, to an OD at 550 nm between 1.2 and 1.5. This cell density typically corresponds to the transition from a logarithmic to a stationary growth phase (*see* **Note 11**).
4. Harvest the bacterial cells by centrifugation at 6000 × *g* for 15 min at 4°C and purify the BAC DNA with the Qiagen Large-Construct Kit according to the manufacturer's specifications (*see* **Note 12**). Depending of the size of the BAC, yields of 20–35 µg of ultrapure BAC DNA can be obtained.

3.2.2. Preparation of DH10B Competent Cells for Electroporation

Owing to the large size of BAC plasmids, the cloning of DNA fragments in BACs requires the use of DH10B competent cells with transformation efficiencies higher than 1×10^8 transformant colonies per µg of DNA. These efficiencies are easily obtained by the electroporation method, which is more reproducible and efficient than the chemical methods. Here we describe the protocol for preparing electrocompetent DH10B cells from 1 liter of bacterial culture. All the steps of this protocol should be carried out under sterile conditions.

1. Inoculate a single colony of DH10B cells from a freshly streaked LB agar plate into a flask containing 10 ml of SOB medium. Incubate the culture overnight at 37°C with vigorous shaking (250 rpm).
2. Inoculate two aliquots of 500 ml of prewarmed SOB medium with 0.5 ml of the overnight culture in separate 2-liter flasks. Incubate the flasks at 37°C with vigorous shaking (250 rpm) until the OD at 550 nm reaches 0.7 (this can take 4–5 h) (*see* **Note 13**).

3. Transfer the flasks to an ice-water bath for about 20 min. Swirl the culture occasionally to ensure that cooling occurs evenly. From this point on, it is crucial that the temperature of the bacteria not rise above 4°C.

4. Transfer the cultures to two ice-cold 500-ml centrifuge bottles and harvest the cells by centrifugation at $6000 \times g$ (6000 rpm in a Sorvall GS3 rotor) for 10 min at 4°C. Discard the supernatant and resuspend each cell pellet in 500 ml of ice-cold 10% glycerol in sterile water.

5. Harvest the cells by centrifugation at $6000 \times g$ (6000 rpm in a Sorvall GS3 rotor) for 15 min at 4°C. Carefully pour off the supernatant and resuspend each cell pellet in 250 ml of ice-cold 10% glycerol (*see* **Note 14**).

6. Repeat step 5 reducing the resuspension volume to 125 ml for each cell pellet.

7. Harvest the cells by centrifugation at $6000 \times g$ (6000 rpm in a Sorvall GS3 rotor) for 15 min at 4°C. Carefully pour off the supernatant (*see* **Note 14**) and remove any remaining drops of buffer using a Pasteur pipette attached to a vacuum line.

8. Resuspend the cells in a final volume of 3 ml of ice-cold 10% glycerol, avoiding the generation of bubbles. This volume has been calculated to reach an optimal cell concentration of $2–4 \times 10^{10}$ cells/ml.

9. Transfer 50 µl of the suspension to an ice-cold electroporation cuvette (0.2-cm gap) and test whether arcing occurs when an electrical discharge is applied with the electroporation apparatus using the conditions described in Section 3.2.3.3, step 4. Arcing is usually manifested by the generation of a popping sound in the cuvette during the electrical pulse. If arcing occurs, wash the cell suspension once more with 100 ml of 10% glycerol and repeat steps 7 and 8.

10. Dispense 100-µl aliquots of the final cell suspension into sterile, ice-cold 1.5-ml microfuge tubes, freeze quickly in a dry-ice methanol bath, and transfer to a −70°C freezer. Electrocompetent DH10B cells can be stored at −70°C for up to 6 months without loss of transforming efficiency.

3.2.3. Cloning of DNA Fragments in BACs

The same standard techniques used for the cloning of DNA in conventional plasmids are applied to BACs with special considerations owing to the large size of BAC plasmids.

3.2.3.1. PREPARATION OF BAC VECTORS AND DNA INSERTS

1. Digest the BAC vector and foreign DNA with a two- to threefold excess of the desired restriction enzymes for 3 h using the buffers supplied with the enzymes. Use an amount of target DNA sufficient to yield 3 µg of the BAC vector and 0.25–0.5 µg of the desired DNA insert. Check a small aliquot of the digestion by agarose gel electrophoresis to ensure that the entire DNA has been cleaved.

2. When two enzymes requiring different buffers are used to digest the DNA, carry out the digestion sequentially with both enzymes. Clean the DNA after the first digestion by extraction with phenol:chloroform and standard ethanol precipitation

or by using the Qiagen QIAEX II Gel Extraction Kit following the manufacturer's instructions for purifying DNA fragments from aqueous solutions (*see* **Note 15**).

3. Purify the digested BAC vector and the DNA insert by agarose gel electrophoresis using the Qiagen QIAEX II Gel Extraction Kit following the manufacturer's instructions (*see* **Notes 15** and **16**). Determine the concentration of the BAC vector and the insert by UV spectrophotometry or by quantitative analysis on an agarose gel.

4. If the BAC vector was digested with only one restriction enzyme or with restriction enzymes leaving compatible or blunt ends, the digested BAC vector has to be dephosphorylated prior to its purification by agarose gel electrophoresis to suppress self-ligation of the BAC vector. We recommend cleaning the DNA before the dephosphorylation reaction as described in step 2 and using shrimp alkaline phosphatase following the manufacturer's specifications.

3.2.3.2. Ligation Reaction

1. For protruding-ended DNA ligation, in a sterile microfuge tube mix 150 ng of purified digested BAC vector, an amount of the purified insert equivalent to a molar ratio of insert to vector of 3:1, 1.5 μl of 10X T4 DNA ligase buffer containing 10 mM ATP, 3 Weiss units of T4 DNA ligase, and water to a final volume of 15 μl. In separate tubes, set up two additional ligations as controls, one containing only the vector and the other containing only the insert. Incubate the reaction mixtures for 16 h at 16°C (*see* **Note 17**).

2. In the case of blunt-ended DNAs, to improve the ligation efficiency use 225 ng of vector, the corresponding amount of insert, and 6 Weiss units of T4 DNA ligase, and incubate the reaction mixtures for 20 h at 14°C.

3.2.3.3. Transformation of DH10B Competent Cells by Electroporation

1. Thaw the electrocompetent DH10B cells at room temperature and transfer them to an ice bath.

2. For each transformation, pipette 50 μl of electrocompetent cells into an ice-cold sterile 1.5-ml microfuge tube and place it on ice together with the electroporation cuvettes.

3. Add 2 μl of the ligation reaction (about 20 ng of DNA) and incubate the mixture of DNA and competent cells on ice for 1 min. For routine transformation with supercoiled BACs, add 0.1 ng of DNA in a final volume of 2 μl. Include all the appropriate positive and negative controls.

4. Set the electroporation machine to deliver an electrical pulse of 25 μF capacitance, 2.5 kV, and 100 Ω resistance (*see* **Note 18**).

5. Add the DNA/cells mixture into the cold electroporation cuvette avoiding bubble formation and ensuring that the DNA/cells mixture sits at the bottom of the cuvette. Dry the outside of the cuvette with filter paper and place the cuvette in the electroporation device.

6. Deliver an electrical pulse at the settings indicated above. A time constant of 4–5 msec should be registered on the machine (*see* **Note 19**).
7. Immediately after the electrical pulse, remove the cuvette and add 1 ml of SOC medium prewarmed at room temperature.
8. Transfer the cells to a 17 × 100-mm polypropylene tube and incubate the electroporated cells for 50 min at 37°C with gentle shaking (250 rpm).
9. Plate different volumes of the electroporated cells (2.5, 20, and 200 μl) onto LB agar plates containing 12.5 μg/ml chloramphenicol and incubate them at 37°C for 16–24 h (*see* **Note 20**).

3.2.3.4. SCREENING OF BACTERIAL COLONIES BY PCR

1. The recombinant colonies containing the insert in the correct orientation are identified by direct PCR analysis of the bacterial colonies using specific oligonucleotides and conventional Taq DNA polymerase (*see* **Note 21**).
2. For each bacterial colony prepare a PCR tube with 25 μl of sterile water.
3. Using sterile yellow tips, pick the bacterial colonies, make small streaks (2–3 mm) on a fresh LB agar plate containing 12.5 μg/ml chloramphenicol to make a replica, and transfer the tips to the PCR tubes containing the water (*see* **Note 22**). In separate tubes, set up positive and negative controls. Leave the tips inside the PCR tubes for 5 min at room temperature.
4. During this incubation time, prepare a 2X master mix containing 2X PCR buffer, 3 mM MgCl$_2$ (which has to be added only in the case that the PCR buffer does not contain MgCl$_2$), 0.4 mM dNTPs, 2 μM of each primer, and 2.5 U of Taq DNA polymerase per each 25 μl of master mix. Prepare the appropriate amount of 2X master mix taking into consideration that the analysis of each colony requires 25 μl of this master mix.
5. Remove the yellow tip and add 25 μl of 2X master mix to each PCR tube.
6. Transfer the PCR tubes to the thermocycler and run a standard PCR, including an initial denaturation step at 95°C for 5 min to liberate and denature the DNA templates and to inactivate proteases and nucleases.
7. Analyze the PCR products by electrophoresis through an agarose gel.
8. Pick the positive colonies from the replica plate and isolate the BAC DNA as described in Section 3.2.1 for further analysis.

3.2.3.5. STORAGE OF BACTERIAL CULTURES

1. Mix 0.5 ml of LB freezing medium with 0.5 ml of an overnight bacterial culture in a cryotube with a screw cap.
2. Vortex the culture to ensure that the glycerol is evenly dispersed, freeze in ethanol-dry ice, and transfer to –70°C for long-term storage.
3. Alternatively, a bacterial colony can be stored directly from the agar plate without being grown in a liquid medium. Using a sterile yellow tip, scrape the bacteria from the agar plate, and resuspend the cells into 200 μl of LB medium in a cry-

otube with a screw cap. Add an equal volume of LB freezing medium, vortex the mixture, and freeze the bacteria as described in step 2 (*see* **Note 23**).

3.2.4. Modification of BAC Clones

The modification of BAC clones is performed using the same techniques as for conventional plasmids with the modifications described in this chapter. We recommend introducing the desired modifications into intermediate BAC plasmid containing the different viral cDNA fragments used during the assembly of the full-length cDNA clone, and then inserting the modified cDNA into the infectious clone by restriction fragment exchange.

3.3. Rescue of Recombinant Viruses

Infectious virus is recovered by transfection of susceptible cells with the full-length cDNA clone. When the transfection efficiency of the susceptible cells is very low, we recommend first transfecting BHK-21 cells and then plating these cells over a monolayer of susceptible cells to allow virus propagation. BHK-21 cells are selected because they present good transfection efficiencies and support the replication of most known CoVs after transfection of the viral genome. The transfection of BACs containing large inserts into mammalian cells has been optimized in our laboratory. The best results were provided by the cationic lipid Lipofectamine 2000 (Invitrogen). The following protocol is indicated for a 35-mm-diameter dish and can be up- or down-scaled if desired (*see* **Note 24**).

1. One day before transfection, plate 4×10^5 BHK cells in 2 ml of growth medium without antibiotics to obtain 90–95% confluent cell monolayers by the time of transfection (*see* **Note 25**). Also plate susceptible cells (Vero E6 in the case of SARS-CoV) at the required confluence for the amplification of the recombinant virus after transfection.
2. Before transfection, equilibrate the Opti-MEM I Reduced Serum Medium (GibcoBRL. Invitrogen) at room temperature and put the DNA (*see* **Note 26**) and the Lipofectamine 2000 reagent on ice. For each transfection sample, prepare transfection mixtures in sterile microfuge tubes as follows:

   a. Dilute 5 µg of the BAC clone in 250 µl of Opti-MEM medium. Mix carefully, avoiding prolonged vortexing or pipetting to prevent plasmid shearing.
   b. Mix Lipofectamine 2000 gently before use. Dilute 12 µl of Lipofectamine 2000 in 250 µl of Opti-MEM medium, mix by vortexing, and incubate the diluted Lipofectamine 2000 at room temperature for 5 min.
   c. Combine the diluted DNA with diluted Lipofectamine 2000, mix carefully, and incubate for 20 min at room temperature.

3. During this incubation period, wash the BHK-21 cells once with growth medium without antibiotics and leave the cells in 1 ml of the same medium per dish.

4. Add the 500 µl of the DNA/Lipofectamine 2000 mixture onto the washed cells and mix by rocking the plate back and forth. Incubate the cells at 37°C for 6 h (*see* **Note 27**).
5. Remove the transfection medium, wash the cells with trypsin-EDTA solution, and detach the cells using 300 µl of trypsin-EDTA solution. Add 700 µl of growth media to collect the cells and reseed them over a confluent monolayer of susceptible cells containing 1 ml of normal growth medium.
6. Incubate at 37°C until a clear cytopathic effect is observed.
7. Analyze the presence of virus in the supernatant by titration.
8. Clone the virus and analyze the genotypic and phenotypic properties of the recovered virus.

4. Notes

1. *E. coli* DH10B strain is a recombination-defective strain used for the propagation of BACs to avoid unwanted rearrangements.
2. SOB medium should be Mg^{2+}-free to avoid arcing during the electroporation step.
3. Vero E6 cells can yield up to 10^8 PFU/ml of SARS-CoV.
4. The silent mutations introduced in the viral genome to generate new restriction sites can be used as genetic markers to identify the virus recovered from the infectious clone.
5. To reduce the number of undesired mutations, perform all PCR reactions with a high-fidelity polymerase, according to the manufacturer's instructions.
6. The CMV promoter and the BGH termination and polyadenylation sequences can be amplified from pcDNA3.1 (Invitrogen). Alternatively, these sequences together with the HDV ribozyme can be amplified from plasmid pBAC-TGEV $5'-3'$ that is available from the authors upon request.
7. In general, the cloning of CoV full-length cDNAs in BACs allows the stable propagation of the infectious clone in *E. coli* DH10B cells. If a residual toxicity, characterized by a small colony phenotype and a delay in the bacterial growth, is observed during the assembly of the infectious clone, we recommend inserting the cDNA fragment responsible for this toxicity in the last cloning step to minimize the toxicity problem. Additionally, the infectious clone can be stabilized by the insertion of a synthetic intron to disrupt the toxic region identified in the viral genome *(14)*. The intron insertion site has to be precisely designed to generate $5'$ and $3'$ intron splice sites matching the consensus sequences of mammalian introns *(15)* and to restore the viral sequence after intron splicing during the translocation of the viral RNA expressed in the cell nucleus to the cytoplasm. The probability of splicing for every insertion site is estimated using the HSPL program, designed to predict splice sites in human DNA sequences *(16)*.
8. Cultures of BAC transformed bacteria should be grown from a single colony isolated from a freshly streaked selective plate. Subculturing directly from glycerol

stocks or plates that have been stored for a long time may lead to loss of the construct.

9. LB broth is the recommended culture medium, since richer broths such as TB (Terrific Broth) lead to extremely high cell densities, which can overload the purification system, resulting in lower yield and less purity of the BAC DNA.

10. When other kits are used instead of the QIAprep Miniprep Kit (Qiagen), equivalent modifications have to be included to optimize the recovery of BAC DNA.

11. To avoid DNA degradation and unwanted rearrangements owing to culture over-aging, it is important to prevent the culture from growing up to the late stationary growth phase.

12. The use of a swinging bucket rotor is recommended for the last isopropanol precipitation step to facilitate the further resuspension of the BAC DNA. After washing with 70% ethanol, air-dry the pellet for only 5 min. Never use vacuum, as overdrying the pellet will make the BAC DNA difficult to dissolve. Carefully remove any additional liquid drops, add 250 µl of 10 mM Tris-HCl (pH 8.5) (DNA dissolves better under slightly alkaline conditions), and resuspend the DNA overnight at 4°C. To prevent plasmid shearing, avoid vortexing or pipetting to promote resuspension of the BAC DNA. Transfer the DNA to a clean 1.5-ml microfuge tube, remove any possible resin traces by centrifugation for 1 min in a table-top microfuge, and keep the supernatant in a clean tube at 4°C. If the purified BAC DNA is not going to be used for a long period of time we recommend storage at –20°C. Avoid repeated freeze-thaw cycles to prevent plasmid shearing.

13. For efficient cell transformation, bacterial culture OD at 550 nm should not exceed 0.8. To ensure that the culture does not grow to a higher density, OD measurement every 20 min after 3 h of growth is highly recommended.

14. Take care when decanting the supernatant as the bacterial pellets lose adherence in 10% glycerol.

15. The Qiagen QIAEX II resin can be used to efficiently purify DNA fragments from 40 bp to 50 kb from aqueous solutions and from standard or low-melt agarose gels in TAE or TBE buffers. Other commercial kits are available, but check whether they have been optimized for purification of DNA fragments larger than 10 kb, as most BAC constructs used during the assembly of the infectious clone are larger than 10 kb.

16. Ethidium bromide-DNA complex excitation by UV light may cause photo-bleaching of the dye and single-strand breaks. To minimize both effects, use a long-wavelength UV illumination (302 nm instead of 254 nm) to cut the desired DNA bands from the agarose gel.

17. The large size of the BAC vectors reduces the ligation efficiency. To increase this efficiency, it is essential to use larger amounts of vector, insert, and T4 DNA ligase than when using conventional plasmids.

18. Most electroporation machines contain programs with defined parameters for transforming specific cell types. In this case, choose the program containing the conditions closest to those described in this protocol.

19. The presence of salt increases the conductivity of the solution and could cause arcing during the electrical pulse, drastically reducing the transformation

efficiency. If arcing occurs, use a smaller amount of the ligation reaction in the electroporation or remove salt from the DNA using any commercial kit or by extraction with phenol:chloroform followed by precipitation with ethanol and 2 M ammonium acetate.

20. Plating volumes higher than 200 µl of electroporated cells on a single plate may inhibit the growth of transformants owing to the large number of dead cells resulting from electroporation. If only small numbers of transformant colonies are expected, the recommendation is to spread 200 µl-aliquots of the electroporated cells on different plates.

21. A mix of small and large colonies indicates that the cloned DNA fragment presents some toxicity when amplified in *E. coli*. Choose the small colonies, which may contain the correct insert, and always grow the bacteria containing this recombinant BAC plasmid at 30°C to minimize the toxicity problem. In this case, we strongly recommend inserting this toxic DNA fragment into the infectious clone in the last cloning step, in order to reduce the manipulation and minimize the possibility of introducing unwanted mutations. Infectious BAC cDNA clones presenting a residual toxicity should be grown at 30°C.

22. It is important to avoid overloading the reaction by adding too many bacteria, which may alter the ionic balance of the reaction and inhibit the amplification by the Taq polymerase.

23. We recommend using this storage method for BAC clones that present a residual toxicity and are not fully stable when amplified in *E. coli*.

24. All work involving SARS-CoV has to be performed in a Biosafety Level 3 (BSL3) laboratory, following the guidelines of the European Commission and the National Institutes of Health.

25. Do not add antibiotics to media during transfection as this causes cell death. A healthy cell culture is critical for an efficient transfection. The use of low-passage-number cells is recommended.

26. Use a BAC DNA isolated with the Qiagen Large-Construct Kit since a DNA preparation of high purity is required in the transfection step.

27. If susceptible cells are directly transfected, incubate them at 37°C until the cytopathic effect is observed and proceed to clone and characterize the recovered virus. In this case, optimization of the transfection of the desired cells with the BAC clone using Lipofectamine 2000 should be required. For transfection optimization, use a similar size plasmid expressing GFP. This plasmid is available from the authors upon request.

References

1. Enjuanes, L., Brian, D., Cavanagh, D., Holmes, K., Lai, M. M. C., Laude, H., Masters, P., Rottier, P. J. M., Siddell, S. G., Spaan W. J. M., Taguchi, F., and Talbot, P. (2000) Coronaviridae. In: van Regenmortel, M. H. V., Fauquet, C. M., Bishop, D. H. L., Carstens, E. B., Estes, M. K., Lemon, S. M., Maniloff, J., Mayo, M. A., McGeoch, D. J., Pringle, C. R., and Wickner R. B. (eds.) *Virus Taxonomy: Seventh Report of the International Committee on Taxonomy of Viruses*, Academic Press, New York, pp. 835–849.

2. Masters, P. S. (2006) The molecular biology of coronaviruses. *Adv. Virus Res.* **66**, 193–292.

3. Almazán, F., González, J. M., Pénzes, Z., Izeta, A., Calvo, E., Plana-Durán, J., and Enjuanes, L. (2000) Engineering the largest RNA virus genome as an infectious bacterial artificial chromosome. *Proc. Natl. Acad. Sci. USA* **97**, 5516–5521.

4. Almazán, F., DeDiego, M. L., Galán, C., Escors, D., Álvarez, E., Ortego, J., Sola, I., Zúñiga, S., Alonso, S., Moreno, J. L., Nogales, A., Capiscol, C., and Enjuanes, L. (2006) Construction of a severe acute respiratory syndrome coronavirus infectious cDNA clone and a replicon to study coronavirus RNA synthesis. *J. Virol.* **80**, 10900–10906.

5. St-Jean, J. R., Desforges, M., Almazán, F., Jacomy, H., Enjuanes, L., and Talbot, P. J. (2006) Recovery of a neurovirulent human coronavirus OC43 from an infectious cDNA clone. *J. Virol.* **80**, 3670–3674.

6. Yount, B., Curtis, K. M., and Baric, R. S. (2000) Strategy for systematic assembly of large RNA and DNA genomes: transmissible gastroenteritis virus model. *J. Virol.* **74**, 10600–10611.

7. Yount, B., Denison, M. R., Weiss, S. R., and Baric, R. S. (2002) Systematic assembly of a full-length infectious cDNA of mouse hepatitis virus strain A59. *J. Virol.* **76**, 11065–11078.

8. Yount, B., Curtis, K. M., Fritz, E. A., Hensley, L. E., Jahrling, P. B., Prentice, E., Denison, M. R., Geisbert, T. W., and Baric, R. S. (2003) Reverse genetics with a full-length infectious cDNA of severe acute respiratory syndrome coronavirus. *Proc. Natl. Acad. Sci. USA* **100**, 12995–13000.

9. Thiel, V., Herold, J., Schelle, B., and Siddell, S. G. (2001) Infectious RNA transcribed *in vitro* from a cDNA copy of the human coronavirus genome cloned in vaccinia virus. *J. Gen. Virol.* **82**, 1273–1281.

10. Casais, R., Thiel, V., Siddell, S. G., Cavanagh, D., and Britton, P. (2001) Reverse genetics system for the avian coronavirus infectious bronchitis virus. *J. Virol.* **75**, 12359–12369.

11. Wang, K., Boysen, C., Shizuya, H., Simon, M. I., and Hood, L. (1997) Complete nucleotide sequence of two generations of a bacterial artificial chromosome cloning vector. *Biotechniques* **23**, 992–994.

12. Shizuya, H., Birren, B., Kim, U. J., Mancino, V., Slepak, T., Tachiiri, Y., and Simon, M. (1992) Cloning and stable maintenance of 300-kilobase-pair fragments of human DNA in *Escherichia coli* using an F-factor-based vector. *Proc. Natl. Acad. Sci. USA* **89**, 8794–8797.

13. Dubensky, T. W., Driver, D. A., Polo, J. M., Belli, B. A., Latham, E. M., Ibanez, C. E., Chada, S., Brumm, D., Banks, T. A., Mento, S. J., Jolly, D. J., and Chang, S. M (1996) Sindbis virus DNA-based expression vectors: utility for *in vitro* and *in vivo* gene transfer. *J. Virol.* **70**, 508–519.

14. González, J. M., Pénzes, Z., Almazán, F., Calvo, E., and Enjuanes, L. (2002) Stabilization of a full-length infectious cDNA clone of transmissible gastroenteritis coronavirus by insertion of an intron. *J. Virol.* **76**, 4655–4661.

15. Senapathy, P., Shapiro, M. B., and Harris, N. L. (1990) Splice junctions, branch point sites, and exons: sequence statistics, identification, and applications to genome project. *Methods Enzymol.* **183**, 252–278.

16. Solovyev, V. V., Salamov, A. A., and Lawrence, C. B. (1994) Predicting internal exons by oligonucleotide composition and discriminant analysis of spliceable open reading frames. *Nucleic Acids Res.* **22**, 5156–5163.

21

Systematic Assembly and Genetic Manipulation of the Mouse Hepatitis Virus A59 Genome

Eric F. Donaldson, Amy C. Sims, and Ralph S. Baric

Abstract

We have developed a DNA assembly platform that utilizes the nonspecific, highly variable sequence signatures of type IIs restriction enzymes to assemble a full-length molecular clone of murine hepatitis coronavirus (MHV) strain A59. The approach also allows changes to be engineered into a DNA fragment by designing primers that incorporate the restriction site and the mutations of interest. By adding the type IIs restriction site in the proper orientation, subsequent digestion removes the restriction site and leaves a sticky end comprising the mutation of interest ready to ligate to a second fragment generated in parallel as its complement. In this chapter, we discuss the details of the method to assemble a full-length infectious clone of MHV and then engineer a specific mutation into the clone to demonstrate the power of this unique site-directed "No See'm" mutagenesis approach.

Key words: MHV; coronavirus; murine; infectious clone; No See'm technology; site-directed mutagenesis; type IIs restriction enzymes; reverse genetics.

1. Introduction

Mouse hepatitis virus (MHV) strain A59 is an extensively studied group 2 coronavirus (CoV). Its genome is an approximately 31.5-kb single-stranded positive-sense RNA, which contains a $5'$ cap and poly(A) tail *(1,2)*. The first two-thirds of the genome encodes the nonstructural proteins required for viral RNA replication and transcription, and the final third comprises six additional open reading frames (ORFs) from which a nested set of $3'$ co-terminal subgenomic mRNAs is transcribed (**Figs. 1A,B,3C**). Upon entry into the cell, the

From: *Methods in Molecular Biology, vol. 454: SARS- and Other Coronaviruses,*
Edited by: D. Cavanagh, DOI: 10.1007/978-1-59745-181-9_21, © Humana Press, New York, NY

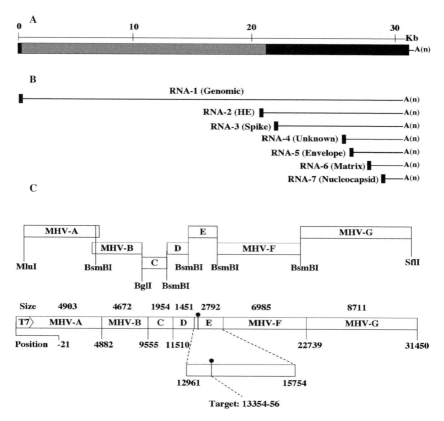

Fig. 1. The MHV genome, subgenomic transcription, and strategy for infectious clone assembly: (A) The MHV genome is ~31.5 kb, with the first two-thirds comprising the nonstructural replicase genes required for viral RNA synthesis (gray rectangle), and the final third encoding the structural genes important for assembly of the viral particle (large black rectangle). The 5′-end is capped and contains a leader sequence (small black box) and the 3′-end has a poly(A) tail. (B) The structural and accessory genes are encoded as a nested set of co-terminal mRNAs, each of which contains the leader sequence derived from the 5′-end of the genome and a poly(A) tail. (C) For assembly of an infectious clone of MHV, the genome was divided into seven stable contiguous cDNA fragments known as MHV-A, MHV-B, MHV-C, MHV-D, MHV-E, MHV-F, and MHV-G. Unique type IIs restriction sites were used to generate unique junctions, which when ligated together produce a full-length cDNA. This allows manipulation of any region of the genome by targeting the fragment of interest and engineering No See'm sites to incorporate the mutation. In this example we target genomic position 13354-56 (indicated by ●), which falls within the MHV-E fragment. Numbers on top of the assembled genome indicate fragment sizes, while bottom numbers represent genomic positions.

N-encapsidated viral RNA is released from the virion and immediately translated by host machinery, resulting in large polyprotein precursors, which are autocleaved to generate intermediates and eventually 16 mature proteins, most of which are thought to function in viral RNA synthesis *(3–6)*. The subgenomic mRNAs are generated from a similar-sized subgenomic negative strand, which is synthesized only after the viral replication complex is functional. The negative subgenomic RNA serves as a template for transcription of similar-sized subgenomic mRNAs, which contain the leader sequence as well as the ORF that encodes the protein directly downstream of the leader sequence *(7–10)*. All subgenomic mRNAs are co-terminal, with each subsequent mRNA containing the sequence for all downstream ORFs, although only the 5′-most ORF is translated. This allows for rapid detection of viral replication by RT-PCR using primers that anneal to the leader and the N-gene, which results in a ladder of different sized products (**Figs. 1B,3B,C**) *(7–10)*.

In cell culture, murine delayed brain tumor (DBT) cells infected with wild-type MHV experience cytopathology characterized by the formation of syncytia, caused by interaction of spike glycoproteins, which anchor in the cell membrane and interact with other spikes in neighboring cells. Wild-type infections generate uniform circular plaques ~5 mm in diameter (**Fig. 4A**).

Prior to the development of an infectious clone of MHV, reverse genetics of the viral coding sequence was restricted to the ORFs downstream of the replicase. Targeted recombination was the primary method for manipulation of the structural and accessory ORFs of MHV, and while extremely powerful, this methodology did not provide an approach to engineer changes into the viral replicase genes *(2,11)*. Moreover, RNA recombination can result in aberrant recombination events resulting in second-site changes *(12)*, necessitating an approach to manipulate the entire genome as a recombinant molecule.

Full-length cDNA constructs revolutionized reverse genetic applications in the entire MHV genome. Our laboratory developed a full-length molecular clone of MHV-A59 (icMHV) by implementing a No See'm approach (**Figs. 1C,2A,B**) *(13)*. Our strategy was to divide the genome into pieces that could be stably subcloned into *E. coli*-based plasmids to facilitate targeted mutagenesis studies and would be easy DNA to store and maintain, using the same approach that was successful with the transmissible gastroenteritis virus (TGEV) infectious clone *(13,14)* (**Figs. 1C,2A,B**). However, the MHV molecular clone was difficult to establish as several toxic regions existed in the genome. Consequently, several of the MHV cDNA fragments required subdivision into smaller subclones and cloning into transcription and translation negative vectors (**Fig. 2B**).

To ensure unidirectional ligation of all fragments, native or engineered type IIs restrictions sites were used to form junctions at the ends of each fragment, which allowed the restriction site to be removed by restriction digestion *(13)*.

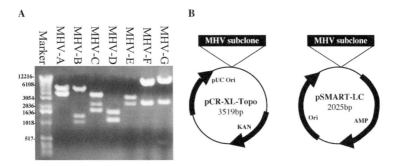

Fig. 2. Engineering the clones of MHV: (A) The MHV genome was divided into seven contiguous fragments, named MHV-A through MHV-G, each flanked by unique type IIs restriction endonuclease sites, which when digested leave unique sticky ends that facilitate unidirectional assembly. Top bands represent MHV fragments, while all other bands represent digested vector. (B) The cloning vectors used to stably clone the contiguous fragments of MHV.

A T7 promoter site was added at the 5′-end of the genome to facilitate *in vitro* transcription of the full-length cDNA fragment after ligation, and a poly(A) tail was included at the 3′-end *(13)* (**Fig. 1C**). The purpose of this review is to provide a detailed protocol that allows for efficient, systematic assembly and mutagenesis of coronavirus full-length cDNAs using class IIs restriction enzymes and the No See'm based mutagenesis approach. The pros and cons of this approach and an appraisal of what the future holds for this technology are discussed at the end (*see* **Note 8**).

2. Materials

2.1. Cell Culture and Lysis

1. Two cell lines are used for MHV full-length RNA transfection experiments: DBT cells and baby hamster kidney cells expressing the MHV receptor (BHK-MHVr) *(13)*.
2. Minimum essential medium (Gibco -Invitrogen, Carlsbad, CA) supplemented with 10% tryptose phosphate broth, 10% Fetal Clone II (HyClone, Logan, UT), and 1% gentamicin/kanamycin. In addition, to select for BHK-MHVr cells expressing MHVr, geneticin (0.8 mg/ml) is added to the medium. The resistant clones were selected by 3X cell sorting for CEACAM1a expression and are maintained in geneticin as previously described by our group *(13)*.
3. Trypsin (0.25%) for removing cells from bottom of flask and versene (0.5 mM) for washes.

2.2. Transformation and Amplification of Plasmids

1. Seven plasmids containing MHV fragments A-G cloned into pCR-XL-Topo (Invitrogen, Carlsbad, CA) or pSMART (Lucigen, Middleton, WI) vectors, as follows: pCR-XL-Topo-MHV-A, pSMART-MHV-B, pSMART-MHV-C, pSMART-MHV-D, pSMART-MHV-E, pSMART-MHV-F, and pSMART-MHV-G.
2. Chemically competent Top10 cells (Invitrogen) for transformations and maintenance of plasmid DNA.
3. Super Optimal Catabolite-repression (SOC) medium (Invitrogen, Carlsbad, CA) is used to stabilize the transformed chemically competent cells prior to plating them.
4. Luria-Bertani (LB) plates supplemented with ampicillin (75 μg/ml) for pSMART vectors or kanamycin (50 μg/ml) for pCR-XL-Topo for growing colonies of bacteria containing the transformed plasmids.
5. LB broth supplemented with ampicillin (75 μg/ml) for pSMART or kanamycin (50 μg/ml) for pCR-XL-Topo, 5 ml per colony for growing up individual colonies.

2.3. Plasmid Purification, Restriction Digestion Screen, and Digestion

1. Qiaprep Spin Miniprep Kit (Qiagen, Valencia, CA.; cat. no. 27106) is used to purify plasmid DNA according to the manufacturer's directions.
2. Restriction enzymes and reagents as follows (New England BioLabs, Ipswich, MA): *Mlu*I (0.5–1 U/μl final volume), *Bsm*BI (0.5–1 U/μl final volume), *Bgl*I (0.5—1 U/μl final volume), *Ahd*I (0.5–1 U/μl final volume), *Sfi*I (0.5–1 U/μl final volume), 10X bovine serum albumin (10 mg/ml), calf intestinal alkaline phosphatase (CIP) (1 U/μl); NEB 10X buffers 1–4 are used to digest the cDNA from the vector.
3. Agarose gel for electrophoresis (0.8–1.0% w/v) of restriction digestions.
4. LB broth supplemented with ampicillin (75 μg/ml) for pSMART or kanamycin (50 μg/ml) for pCR-XL-Topo vectors, 20 ml per culture for growing up larger stocks of colonies that look right by a restriction screen.

2.4. Fragment Purification and Ligation

1. Qiaex II Gel Extraction Kit (Qiagen, Valencia, CA; Cat. no. 20010), 3 M sodium acetate pH 5.2, elution buffer from Qiagen miniprep kit for purifying bands cut out of the agarose gel.
2. Reagents for chloroform extraction/isopropanol precipitation of fragments: chloroform, isopropanol, 70% ethanol, 95% ethanol, and elution buffer.
3. A DNA spectrophotometer for quantifying the concentration of individual purified cDNA fragments.
4. T4 DNA ligase plus 10X ligase buffer (New England Biolabs) for ligating the full-length cDNA.

2.5. In Vitro *Transcription and Electroporation*

1. MHV N-gene amplified with SP6 promoter on the 5'-end is used as a template for generating N-gene transcripts. Primers for Sp6-N have been published *(13)*. Transfecting in parallel with N-gene increases viral replication by roughly 15-fold.
2. mMessage mMachine T7 Transcription Kit (Ambion, Austin, TX; cat. no. 1344) and mMessage mMachine SP6 Transcription Kit (Ambion, Austin, TX; cat. no. 1340) for generating full-length MHV RNA or SP6-N-gene RNA.
3. One 0.4-cm Gene Pulser Cuvette (Bio Rad, Hercules, CA) for each transfection.
4. BHK-MHVr cells are seeded at low density and grown in 150-cm^2 flasks to 70% confluence, washed three times in nuclease-free PBS, and resuspended in cold PBS at a concentration of 1.0×10^7 cells/ml, and $\sim 5 \times 10^6$ DBT cells are infected by plating the electroporated BHK-MHVrs on top. We see greatly reduced transfection efficiencies if cells are allowed to approach confluence prior to harvest.
5. Bio Rad (Hercules, CA) Gene Pulser Excel electroporator for doing the transfection.

2.6. *Plaque Purification, Harvesting Viral RNA, and RT-PCR*

1. 2X DMEM medium (Gibco)
2. Fetal Clone II (Hyclone) or fetal bovine serum (Gibco)
3. Low-melting-temperature cell culture grade agarose (Cambrex, Rockland, ME; cat. no. 50000)
4. Phosphate buffered saline (Gibco).
5. Trizol reagents (Invitrogen, Carlsbad, CA) for harvesting total RNA from infected cells.
6. 75% ethanol prepared in DepC water (Invitrogen, Carlsbad, CA)
7. PCR/sequencing primers flanking the region of the mutation.
8. SuperScript III Reverse Transcription kit (Invitrogen; cat. no. 18080-044) for reverse transcription of viral RNA to cDNA.
9. pCR-XL-Topo cloning kit (Invitrogen, Carlsbad, CA; cat. no. K4500) for cloning the amplicon for sequencing.

2.7. *Designing and Implementing Mutations*

1. MHV genomic sequence (Accession no. NC_001846) for designing primers.
2. NebCutter 2.0 restriction digestion tool (New England BioLabs http://tools.neb.com/NEBcutter2/index.php) for analyzing sequences to ensure that the type IIs restriction site selected does not occur naturally in the wild-type sequence and is not introduced by the mutation.
3. Web Primer on-line primer design tool (http://seq.yeastgenome.org/cgi-bin/web-primer) for development of primers for engineering mutations and for sequencing to verify that the correct change was incorporated into the virus.

4. Expand Long PCR Kit (Roche, Basel, Switzerland; Cat. no. 11 681 834 001) for amplification of the mutant fragments.
5. dNTPs (10 mM) prepared in nuclease-free water.
6. Appropriate type IIs restriction enzyme for digestion of the two fragments.
7. T4 DNA ligase for assembling the full-length mutant fragment.
8. pCR-XL-Topo cloning kit for cloning the fragment into the pCR-XL-Topo vector, and for transforming the new plasmid for growth in LB medium.

3. Methods

Assembling the full-length clone of MHV involves a series of steps, each of which is important for amplifying and purifying the reagents necessary for the next step. Briefly, plasmids are transformed into chemically competent *E. coli* and plated in the presence of appropriate antibiotic selection. Colonies are isolated and grown overnight at room temperature (*see* **Note 6**); the DNA is harvested and inserts verified by a screening restriction digestion. It is important to screen the DNA prior to setting up the large-scale preps to ensure plasmid integrity. We typically harvest 50–75 μg of each plasmid DNA, restrict the DNA with the appropriate restriction endonucleases, and purify the viral cDNAs from agarose gels. It is equally important to keep the digested fragments free of contaminants, including residual carbohydrates from the gel extraction procedure or contamination with other fragments, particularly a wild-type version of a mutated fragment (*see* **Note 2**). Following gel extraction, chloroform extractions are performed to ensure that all carbohydrate contaminants are removed prior to preparing the ligation reaction (*see* **Note 4**).

It is also important to maintain an RNAse-free environment (as much as possible) after ligation of the full-length clone. Always wear gloves, keep the bench top RNAse-free by spraying with RNAse Zap (Ambion, Austin TX; cat. no. 9780), and always use DepC water (*see* **Note 2**) and barrier tips (*see* **Note 3**). In addition, washing the BHK-MHVr cells with PBS made with DepC water may help prevent degradation of the electroporated full-length viral RNA (*see* **Note 2**).

3.1. Transformation and Restriction Screening of the Seven Clones of MHV

1. The seven fragments of MHV are maintained in either pCR-XL-Topo or pSMART vectors, and each vector encodes a different antibiotic resistance cassette. For pCR-XL-Topo cloning vector encodes kanamycin resistance, whereas the pSMART vector encodes ampicillin resistance (pSMART vectors are available with kanamycin or ampicillin resistance; in this case it is ampicillin). LB broth and LB agar plates with each antibiotic are required for growing the newly

transformed bacteria. Although other vector-fragment combinations exist, we focus on the following in this report: pCR-XL-Topo-MHV-A, pSMART-MHV-B, pSMART-MHV-C, pSMART-MHV-D, pSMART-MHV-E, pSMART-MHV-F, and pSMART-MHV-G *(13)*. Plasmids are stored at approximately 20 ng/µl at –80°C.

2. Thaw each plasmid and the chemically competent Top10 cells (one vial per plasmid) on ice. Add 100 ng of each plasmid to an appropriately labeled vial of cells (50 µl) and incubate on ice for 30 min.

3. Transform the cells by heat shock in a water bath maintained at 42°C for 2 min, followed by 2 min on ice.

4. Add 200 µl SOC medium (without antibiotic) and rock the vials at room temperature for 2 h.

5. Inoculate two LB plates (with appropriate antibiotic) for each plasmid, using 25 µl and 100 µl of culture and incubate the plates at room temperature for 48 h.

6. Pick at least five colonies of each plasmid (from one or both plates) and inoculate each into a 15-ml conical containing 5 ml of LB broth with appropriate antibiotic. Grow the colonies for 16–24 h at 28.5°C in an incubator shaker set at 250 rpm. We have found that plasmid stability is enhanced at the lower temperature.

7. Prepare a library plate by inoculating an LB agar plate (with appropriate selection) with 25 µl of the supernatant of each colony. The library plate should contain all of the colonies for one plasmid, and is grown at room temperature for 16–24 h and then stored at 4°C. Then spin the cultures at 3000 × g for 10 min at 4°C, aspirate, and discard the supernatant.

8. Plasmids are isolated from *E. coli* using the Qiagen Miniprep Kit, according to the manufacturer's instructions. Follow the plasmid DNA purification scheme using the Qiaprep Spin Miniprep Kit and a microcentrifuge protocol, and elute the plasmid DNA in 50 µl of elution buffer preheated to 70°C.

9. Screen each colony by restriction digestion to verify the insert size and restriction map.

   a. pCR-XL-Topo-MHV-A: 10 µl plasmid, 1 µl *Mlu*I, 1 µl *Bsm*BI, 2 µl NEB Buffer 3, 6 µl water. Incubate at 55°C for 1 h.

   b. pSMART-MHV-B: 10 µl plasmid, 1 µl *Bgl*I, 2 µl NEB Buffer 3, 6 µl water. Incubate at 37°C for 1 h. Add 1 µl *Bsm*BI and incubate at 55°C for an additional 1 h.

   c. pSMART-MHV-C: 10 µl plasmid, 1 µl *Bgl*I, 2 µl NEB Buffer 3, 6 µl water. Incubate at 37°C for 1 h. Add 1 µl *Bsm*BI and incubate at 55°C for an additional 1 h.

   d. pSMART-MHV-D: 10 µl plasmid, 1 µl *Bsm*BI, 1 µl BSA, 2 µl NEB Buffer 4, 5 µl water. Incubate at 55°C for 1 h. Add 1 µl *Ahd*I and incubate at 37°C for an additional 1 h.

   e. pSMART-MHV-E: 10 µl plasmid, 1 µl *Bsm*BI, 2 µl NEB Buffer 3, 7 µl water. Incubate at 55°C for 1 h.

   f. pSMART-MHV-F: 10 µl plasmid, 1 µl *Bsm*BI, 2 µl NEB Buffer 3, 7 µl water. Incubate at 55°C for 1 h.

   g. pSMART-MHV-G: 10 µl plasmid, 1 µl *Sfi*I, 1 µl *Bsm*BI, 1 µl BSA, 2 µl NEB Buffer 2, 5 µl water. Incubate at 55°C for 1 h.

10. Run the restriction digestions out on a 0.8–1.0% agarose gel in TAE buffer (4.84 g/liter Tris base, 2 ml/liter 0.5 M EDTA pH 8.0, 1.14 ml/liter glacial acetic acid; pH to 8.5) at 100 mA, and determine that the plasmid DNAs have the appropriate banding pattern (**Fig. 2A**). The digests were designed to ensure that the desired MHV insert is the slowest migrating band in the gel (**Fig. 2A**). MHV-A is 5000 bp, MHV-B is 4672 bp, MHV-C is 1954 bp, MHV-D is 1451 bp, MHV-E is 2792 bp, MHV-F is 6985 bp, and MHV-G is 8711 bp (**Figs. 1C,2A**).

3.2. Amplification and Digestion of the Seven Fragments of MHV

1. Identify replicates on the library plate for each fragment that looks right by restriction screen.
2. Prepare a 50-ml conical containing 15 ml of LB broth with appropriate antibiotic and inoculate it with a streak from the library plate for each plasmid to be amplified. In general, one 15-ml culture of each plasmid DNA insert is sufficient for assembling at least two full-length MHV clones. Incubate for 12–16 h at 28.5°C in an incubator shaker set at 250 rpm. Then add 5 ml of LB plus antibiotic and allow growth to continue for an additional 4–6 h under the same conditions.
3. Spin the cultures at 3000 × *g* for 10 min at 4°C and discard the supernatants. Resuspend the pellet in 750 μl of Qiagen Miniprep P1 Buffer and transfer equally to three 1.5-ml microfuge tubes. Purify plasmids using the Qiaprep Miniprep Kit, and elute to a final volume of 50 μl in elution buffer heated to 70°C. Combine all three 50-μl aliquots of each fragment in a single tube.
4. Digest the plasmids as follows:

   a. pCR-XL-Topo-MHV-A: 150 μl plasmid, 10 μl MluI, 20 μl NEB Buffer 3, 20 μl water. Incubate at 55°C for 1.5 h. Add 5 μl of CIP and incubate at 37°C for 1 h. CIP catalyzes the removal of 5′ phosphate groups from DNA, which prevents concatamers of MHV-A from occurring during ligation *(15)*. CIP cannot be completely heat inactivated, so the reaction is chloroform extracted and isopropanol precipitated at this point to remove the CIP.

      i. For the chloroform extraction all steps are performed at room temperature: Add 25 μl of water, 50 μl 3 M sodium acetate, and 275 μl of chloroform to the reaction. Shake by hand for 2 min, and then spin at full speed in a microcentrifuge for 2 min. Remove the aqueous phase to a fresh tube and add an equal volume of isopropanol. Incubate at room temperature for 10 min, followed by centrifugation at full speed in a microcentrifuge for 10 min. Remove the supernatant and resuspend the pellet in 1 ml of 70% ethanol. Spin for 5 min at full speed, remove the supernatant, and resuspend in 95% ethanol. Spin for 5 min, remove supernatant, and air-dry the pellet for no more than 5 min. Resuspend the pellet in 170 μl elution buffer heated to 70°C.

      ii. Add 10 μl *Bsm*BI and 20 μl NEB Buffer 3. Incubate for 2 h at 55°C.

b. pSMART-MHV-B: 150 μl plasmid, 10 μl *BgI*I, 20 μl NEB Buffer 3, 10 μl water. Incubate at 37°C for 1.5 h. Add 10 μl *Bsm*BI and incubate at 55°C for an additional 2 h.

c. pSMART-MHV-C: 150 μl plasmid, 10 μl BglI, 20 μl NEB Buffer 3, 10 μl water. Incubate at 37°C for 1.5 h. Add 10 μl *Bsm*BI and incubate at 55°C for an additional 2 h.

d. pSMART-MHV-D: 150 μl plasmid, 10 μl *Bsm*BI, 2 μl BSA, 20 μl NEB Buffer 4, 8 μl water. Incubate at 55°C for 2 h. Add 10 μl *Ahd*I and incubate at 37°C for an additional 1.5 h. *Ahd*I cuts the vector into two fragments, which allows the MHV-D fragment to be the largest band.

e. pSMART-MHV-E: 150 μl plasmid, 10 μl BsmBI, 20 μl NEB Buffer 3, 20 μl water. Incubate at 55°C for 2 h.

f. pSMART-MHV-F: 150 μl plasmid, 10 μl BsmBI, 20 μl NEB Buffer 3, 20 μl water. Incubate at 55°C for 2 h.

g. pSMART-MHV-G: 150 μl plasmid, 10 μl SfiI, 2 μl BSA, 20 μl NEB Buffer 2, 18 μl water. Incubate at 55°C for 1.5 h. Add 5 μl of CIP and incubate at 37°C for 1 h. Do the chloroform extraction as described above (Section 3.2.4.a.i) and resuspend the pellet in 170 μl of elution buffer. Add 10 μl of *Bsm*BI and 20 μl NEB buffer 3 and incubate at 55°C for 2 h.

5. Isolate the digested fragments by electrophoresis on a 0.8–1.0% gel in wells capable of holding 75 μl each (three wells per reaction) (**Fig. 2A**). The restriction products should be run for enough time to allow for high resolution of individual fragments.

6. Excise the top band of the correct size for each fragment from the gel, making sure that it corresponds to the size of the MHV fragment (**Fig. 2A**) and is not an undigested vector-MHV fragment. Place the excised band in a 1.5-ml microfuge tube.

7. Use the Qiaex II Gel Extraction Kit for extracting the DNA from the agarose as follows:

   a. Use the Qiaex II Agarose gel extraction protocol with the following modifications:

      i. Resuspend the excised band in 620 μl of QX1 buffer, 12.5 μl 3 M sodium acetate, and 11 μl QIAEX II resin. Incubate in a water bath at 55°C for a total of 12–16 min, agitating every 2 min. For fragments larger than 4 kb (MHV-A, MHV-B, MHV-F, and MHV-G), add 360 μl of water after 6 min of incubation. Follow the protocol for the remaining steps and elute to a final volume of 35 μl. Add the three replicates for each extraction into one tube (~100 μl).

8. A final chloroform only extraction is performed to remove any residual impurities that remain after the DNA extraction (*see* **Note 1**). Use a 1:1 volume of chloroform, shake for 2 min, and spin at top speed in a microcentrifuge at room temperature for 2 min. Remove the aqueous phase to a fresh tube.

3.3. Assembling the Viral Genome

1. Quantitate each purified MHV fragment with a UV spectrophotometer (optional).
2. Set up a shotgun ligation by adding relatively equivalent amounts of each fragment to a fresh 1.5-ml microfuge tube. For example, add 25 μl of each of the fragments with the lowest concentrations, 20 μl for fragments of intermediate concentrations, and only 15 μl of fragments with the highest concentrations, achieving a goal of relatively equivalent numbers of molecules of each fragment in the ligation mix (1 μg of 1000 bp DNA = 1.52 pmol = 9.1 × 10^{11} molecules). The reaction size will vary each time but a typical reaction is similar to this:

   a. 20 μl of MHV-A.
   b. 25 μl of MHV-B.
   c. 25 μl of MHV-C.
   d. 25 μl of MHV-D.
   e. 25 μl of MHV-E.
   f. 15 μl of MHV-F.
   g. 20 μl of MHV-G.
   h. 15 μl of water.
   i. 20 μl of 10X ligation buffer.
   j. 10 μl of ligase.
   k. Incubate the ligase reaction overnight at 4°C.

3. Chloroform extract and isopropanol precipitate the reaction as described above (Section 3.2.4.a.i), and resuspend the cDNA pellet in 10 μl of DepC water heated to 70°C.
4. *In vitro* transcription of the full-length cDNA construct is conducted using the mMessage mMachine T7 Transcription Kit using the following recipe per reaction:

   a. 7.5 μl of GTP (30 mM).
   b. 25 μl 2X NTP/CAP (15 mM ATP, 15 mM CTP, 15 mM UTP, 3 mM GTP, and 12 mM cap analog).
   c. 5 μl 10X Buffer.
   d. 7.5 μl of full-length cDNA template.
   e. 5 μl enzyme (RNA polymerase).
   f. Incubate the reaction at 40.5°C for 25 min, then at 37°C for 50 min, and then at 40.5°C for 25 min additional. Use a PCR-cycler for most consistent results.
   g. The remaining 2.5 μl of the ligation reaction and 5 μl of the RNA transcription reaction can be run on a gel to determine if full-length ligation occurred and whether *in vitro* transcription was successful (**Fig. 3A**). Use LE Agar (0.5%w/v) for the gel with 10% SDS added to the gel buffer and the running buffer. Treat 5 μl of RNA with DNAse and then add 2 μl of 10% SDS and 4 μl 5 mM EDTA prior to running the gel. For optimal resolution run the gel overnight at 30 mA.

5. Preparation of SP6-N-gene. Primers for adding the SP6 promoter onto the N-gene have been published previously *(13)*. Use these primers (10 mM) along with the

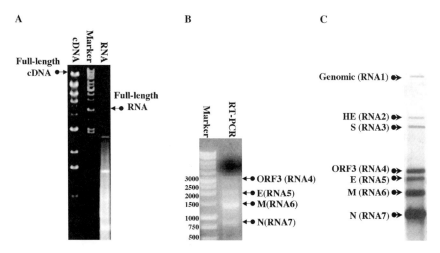

Fig. 3. *In vitro* assembly of full-length cDNAs and transcripts and the recovery of infectious virus: (A) Running the cDNA from an assembly ligation results in a ladder of intermediates, as well as a full-length product ~31.5 kb. After *in vitro* transcription, running RNA on a gel results in a smear of RNA including a full-length viral RNA, which runs at 16–17 kb. (B) Detecting viral replication can be accomplished using primers that anneal to the leader sequence and the N-gene. Using the antisense primer to perform reverse transcription and then both primers to amplify the target via RT-PCR generates a series of bands corresponding to several subgenomic mRNAs. (C) A northern blot showing the seven mRNAs of icMHV, using a probe that anneals to the 300 5' nucleotides of the N-gene allows detection of genomic and all subgenomic mRNAs.

1 µl of purified MHV-G fragment (Section 3.2, step 8) per reaction and reagents from the Expand Long Kit as follows:

a. 1 µl cDNA from MHV-G fragment.
b. 1 µl sense primer (10 mM).
c. 1 µl antisense primer (10 mM).
d. 1.75 µl dNTP (10 mM).
e. 5 µl Buffer #1.
f. 0.75 µl Expand Taq.
g. 39.5 µl water.
h. Then run PCR through the following cycles: 1 cycle of 94°C for 5 min, followed by 30 cycles of 94°C for 1 min, 58°C for 1 min, 68°C for 3 min, and then 1 cycle of 68°C for 10 min followed by 4°C until ready for gel electrophoresis.
i. The PCR product is then purified as described above (Section 3.2 steps 5–8).

6. *In vitro* transcription of the N-gene is conducted using the mMessage mMachine SP6 Transcription Kit with the following reaction recipe per reaction:

    a. 12.5 μl 2X NTP/CAP (10 mM ATP, 10 mM CTP, 10 mM UTP, 2 mM GTP, and 8 mM cap analog).

    b. 2.5 μl 10X Buffer..

    c. 3.5 μl SP6-N-gene template

    d. 2.5 μl enzyme (RNA polymerase).

    e. 4 μl nuclease-free water.

    f. Incubate the reaction at 40.5°C for 25 min, then at 37°C for 50 min, and then at 40.5°C for 25 min more. Use a PCR-cycler for the most consistent results.

    g. Do one reaction for each transfection and one additional reaction for the negative control.

3.4. Preparation of Cells

1. Cell culture must be planned a day or two in advance, as three T150 flasks of BHK-MHVr cells at ~70% confluency and one T150 flask of DBTs at 80–90% confluency are required to do one transfection and one negative control (*see* **Note 5**).

2. Cells are maintained in the medium described above (Section 2.1, steps 1–3). For passage, the medium is aspirated and the attached cells are washed. Cells are removed from the flask with 0.25% trypsin, and then resuspended and passaged in the medium. Splitting a confluent T150 flask of BHK-MHVr cells (e.g., ~2–3 × 10^7 cells) the night before at 1:5 will generally produce a T150 at ~70% confluency the next day.

3. Preparation of BHK-MHVr cells for transfection (per T150 flask):

    a. Remove the medium and wash the cells. Remove the cells from the flask with trypsin. Add 2 ml of trypsin to the flask and incubate for 5 min at 37°C. Remove cells from flask by tapping gently.

    b. Resuspend the trypsinized cells in 8 ml of medium and transfer cells from all three flasks to a sterile 50-ml conical.

    c. Spin the cells at ~2000 × *g* for 5 min and remove the medium, leaving the pellet in the conical.

    d. Resuspend the pellet in 10 ml of ice-cold PBS and spin at 2000 × *g* for 5 min.

    e. Remove the PBS and resuspend the pellet in an additional 10 ml of ice-cold PBS.

    f. Transfer 10 μl of the cell suspension to a hemocytometer for counting, and spin the remaining suspension at ~2000 × *g* for 5–10 min, while counting the cells.

    g. Remove the PBS from the cells and resuspend the pellet in the volume of PBS that will give a concentration of 10^7 cells/ml. Counting 10 μl of cells from a 10-ml preparation translates into 10^4 cells per 1 mm^2 (grid of 16 squares arranged 4 × 4) on the hemocytometer. Count all the cells in three of the 16 square grids and average them. Then, multiply this number by $1 × 10^4$, which gives the concentration in cells/ml. In general, to get to a final concentration of $1 × 10^7$ cells/ml resuspend the final pellet in 1/100 of the final count. For example, if the final average count is 200 cells, this would calculate to $2.0 × 10^6$ cells/ml in a 10-ml volume for a total of 2.0×10^7 cells. Resuspend in 2 ml of PBS to get to a final concentration of 1×10^7 cells/ml. Keep cells on ice until ready for electroporation. A total volume of 1.6 ml of 1×10^7 cells/ml is required for one transfection and one negative control.

4. Preparation of DBT cells for infection. Only one confluent T150 flask of DBT cells is required.

   a. Remove the medium and wash the cells. Remove the cells from the flask with trypsin.
   b. Resuspend the cells in 10 ml of medium and transfer 1 ml of cells to a fresh T75 flask labeled for infection ($\sim 5.0 \times 10^6$ cells). Bring up to 9.5 ml total volume with medium. Make one flask per transfection and one for a negative control.

3.5. Transfection

1. Add 800 µl of BHK-MHVr cells resuspended in PBS (free of magnesium or calcium) to a concentration of 10^7 cells to each 0.4-cm Gene Pulser Cuvette labeled for the appropriate transfection or control. Do one for each transfection and one for a negative control.
2. Add 22.5 µl of N-transcript from the SP6-N-gene transcription reaction to each cuvette, including the negative control.
3. Add 45 µl of the full-length RNA from the transcription reaction to the appropriately labeled cuvette.

   a. The remaining 5 µl is saved to run on an RNA gel (**Fig. 3A**).

4. Set the Bio Rad Gene Pulser Excel electroporator at 25 µF and 850 V.
5. Place each cuvette into the Shock Pod shocking chamber and pulse three times.
6. Allow each cuvette to incubate at room temperature for 10 min.
7. Remove the electroporated cells from the cuvette into the appropriately labeled T75 flask of DBT cells (optional). Alternatively, allow the electroporated MHVr cells to settle onto a T75 flask and incubate at 37°C for 1–2 days to allow for the development of cytopathology.
8. Incubate the flasks at 37°C in an incubator with 5% CO_2.

3.6. Detection of Cytopathic Effect

1. Wild-type MHV from the clone will generally produce cytopathic effects (CPE) within 8–12 h, and the syncytium across the monolayer is usually complete by 24 h. We typically recover between 1000 and 10,000 plaques (infectious center assay) following electroporation of transcripts. CPE is very obvious for DBT cells infected with MHV. Excess spike glycoproteins accumulate at and protrude from the cell surface, which facilitates fusion with the neighboring cells. This creates large multinucleated syncytia, which are usually observed by 8–12 h posttransfection. As the cells die, they lift off the flask, and so an efficient transfection

of full-length viral RNA will result in a rapid clearing of the monolayer, usually within 16–24 h.

2. Mutants generally come up more slowly and may take up to 72 h for CPE to be obvious. Highly debilitated mutants may not display obvious CPE. Passage usually selects for second-site revertant viruses that replicate efficiently, so extensive passage usually results in recombinant viruses containing multiple mutant alleles scattered across the genome. Reversions at the site of mutation can be reduced by fixing each engineered mutation with two or more nucleotide changes in the codon.

3. Harvest supernatants by spinning the at ∼3000 × g for 10 min at room temperature and aliquoting to fresh 2-ml screw cap tubes. Store at –80°C. These are labeled as passage zero (P0).

3.7. Plaque Purification

1. Plaque purification allows isolation of single colonies of virus for propagation of stocks and verification of genotype (**Fig. 4**). This is done by plating a serial dilution of supernatants from P_0 onto 60-mm dishes of DBT cells and then adding an agar overlay to prevent diffusion of the virus into the medium.

2. One confluent T150 flask of DBT cells is sufficient to seed twenty 60-mm dishes labeled by dilution and virus. Resuspend washed cells in a total of 61 or 81 ml of medium and aliquot 3 or 4 ml of cells onto each of twenty 60-mm plates. Repeat as necessary for the correct number of plates. The cells are then incubated at 37°C overnight. They are ready to use at ∼95–100% confluency.

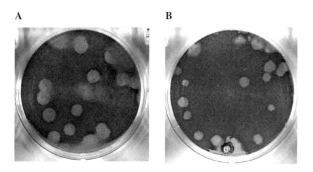

A   B

Fig. 4. Plaques produced by infection with recombinant MHV and a mutant: (A) DBT cells were infected with recombinant icMHV and overlaid with agar. Plaques formed at 24–36 h, and cells were stained with neutral red. MHV produces clear, uniform plaques of approximately 5 mm in diameter. (B) In contrast, some mutant forms of MHV produce differential plaques sizes. Here we show a mutant with a double amino acid change in Nsp10 (the lysines at positions 24 and 25 in Nsp10 were changed to alanines), which results in smaller overall plaque sizes, as well as a variety of different sizes, some with irregular borders.

3. When the cells are confluent, remove the medium. Dilute the supernatant from P_0 using dilutions of one part P_0 supernatant to nine parts PBS, ranging from 10^{-1} to 10^{-7} dilutions.

4. Add 200 µl of the appropriate dilution onto a 60-mm plate of DBT cells and spread evenly by mixing back and forth every 15 min, while incubating at room temperature for 1 h. This is passage one (P_1).

5. Prepare the overlay ahead of time. Make enough to aliquot 5 ml of overlay onto each plate. For 100 ml:

    a. Autoclave 49 ml sterile water and 0.8 g of LE agar (low-melt) and incubate in a water bath at 55°C for 15 min.
    b. Add 40 ml 2X medium heated to 37°C.
    c. Add 10 ml FCII or FBS heated to 37°C.
    d. Add 1 ml of Gen/Kan heated to 37°C.

6. Carefully add 5 ml of overlay onto each plate. This must be accomplished at close to 40°–42°C as the medium rapidly solidifies below these temperatures and higher temperatures will adversely affect cell viability. Allow the plates to incubate 24–48 h at 37°C.

7. Plaques for ic MHV resemble wt MHV (**Fig. 4A**), whereas mutant plaques can vary greatly (**Fig. 4B**). Select five plaques for each virus. For each plaque to be picked: use sterile technique to cut $\sim^1/_3$ in. off the end of a P1000 barrier tip, place the tip over the plaque, and push it through the agar, making sure that the surface of the plate and the agar are both manipulated so as to ensure that the entire plaque is extracted without any neighboring plaques. Place the picked medium into 300 µl PBS and incubate at 4°C for 30 min.

8. The 300 µl containing the plaque and PBS is then used to infect a 60-mm plate of DBT cells, brought up as described above. To infect cells, remove the medium and pipette all 300 µl of the plaque and PBS onto a 60-mm plate of DBT cells, spread evenly by mixing back and forth every 10–15 min, and incubate at room temperature for 1 h. Then add 4 ml of medium to each plate. This is passage 2 (P_2). Allow the infections to proceed at 37°C until CPE is obvious (usually 24–36 h). We typically only plaque purifiy recombinant viruses 1X as additional passages increase the probability of second-site mutations or reversions (*see* **Note 7**).

9. Remove the medium from the 60-mm dishes, centrifuge at $\sim$3000 × g for 10 min at room temperature and aliquot the supernatant into 1.5-ml microfuge tubes labeled P_2.

10. Resuspend the cells in 1 ml of Trizol reagents and dispense them into a 1.5-ml microfuge tube for isolation of total RNA.

3.8. Harvesting RNA, Reverse Transcription, and Verification of Viral Genotype

1. Total RNA is harvested using Trizol reagents following the standard protocol. Briefly, approximately 4×10^6 cells are harvested in 1 ml of Trizol and the RNA is frozen at −80°C until further use:

    a. All steps are performed at room temperature. Start by thawing the frozen cells in Trizol and allowing them to sit at room temperature for 5 min.

    b. Add 200 µl chloroform and shake by hand for 15 sec.

    c. Spin at 12,000 × g for 15 min.

    d. Remove aqueous phase to a fresh tube.

    e. Add 500 µl of isopropanol and incubate for 10 min.

    f. Spin at 12,000 × g for 10 min.

    g. Remove supernatants and resuspend the pellet in 1 ml of 75% ethanol diluted in DepC water.

    h. Spin at 7500 × g for 5 min.

    i. Remove ethanol and air-dry the pellet for 5–10 min.

    j. Dissolve the pellet 100 µl of DepC water heated to 70°C and incubate at 55°C for 10 min.

2. Design PCR/sequencing primers flanking the region of the mutation. Use the antisense primer as the gene-specific primer for SuperScript III reverse transcription with modifications to the protocol:

    a. Add 2 µl of RNA and 2 µl of antisense primer to a fresh tube and incubate at 70°C for 10 min.

    b. Add 2 µl of 0.1 M DTT.

    c. Add 2 µl of 10 mM dNTPs.

    d. Add 4 µl of 5X first-strand buffer.

    e. Add 7 µl of nuclease-free water.

    f. Add 1 µl of SuperScript III.

    g. Incubate at 55°C for 1 h.

    h. Incubate at 70°C for 20 min to inactivate the RT.

3. Amplify the region of interest with a 50 µl PCR reaction appropriate for the size and composition of the cDNA target.

4. Isolate the DNA from PCR following the same procedures as described in Section 3.2, steps 5–7.i modifying the recipes for the 50-µl reaction size.

5. Once isolated, clone the target into a vector of choice. We use pCR-XL-Topo, following the manufacturer's recommended protocol.

6. Transform, grow-up, and screen as described in Section 3.1. Screen by restriction digestion with *Eco*RI, and sequence clones that are of the appropriate size to verify genotype.

3.9. Designing Mutants Using the No See'm Technology

1. Identify a target of interest. This can be a single/double/triple codon change, a deletion, etc.

2. Determine which type IIs restriction enzymes *do not* cut the vector, the target fragment, and the fragment with the newly designed mutations incorporated into the sequence. This can be accomplished with any restriction analysis tool; however, NEBcutter2.0 works well enough and is user friendly.

A   **Map mutation and select restriction site**

```
Wt       AACCACTAATCAGGATTCTT    BbsI    GAAGACNN
Target   AACCACTAATGTCGATTCTT            CTTCTGNNNNNNN
```

```
MHV(+)13335-GCCGGAGGCAACCACTAATCAGGATTCTTATGGTGGTGCTTCC-13377
MHV(-)13335-CGGCCACCGTTGGTGATTAGTCCTAAGAATACCACCACGAAGG-13377
```

B   **Design primers to engineer mutation and restriction site**

```
MutS: 5-GAAGACGTCGATTCTTATGGTGGTGCTTCC-3
MutA: 5-GAAGACGAATCAGCATTAGTGGTTGCCTCCGGC-3
```

```
Wt(+)GCCGGAGGCAACCACTAATCAGGATTCTTATGGTGGTGCTTCC
     ||||||||||||||||||   |||||
   3-CGGCCTCCGTTGGTGATTACAGCTAAGCAGAAG-5 MutA
```

```
      MutS 5-GAAGACGTCGATTCTTATGGTGGTGCTTCC-3
             |    |||||||||||||||||||||||||
Wt(-) CGGCCACCGTTGGTGATTAGTCCTAAGAATACCACCACGAAGG
```

C   **PCR, digestion, and ligation**

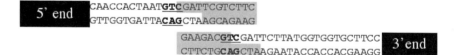

```
5' end  CAACCACTAATGTCGATTCGTCTTC
        GTTGGTGATTACAGCTAAGCAGAAG
                  GAAGACGTCGATTCTTATGGTGGTGCTTCC
                  CTTCTGCAGCTAAGAATACCACCACGAAGG  3' end
```

```
5' end  CAACCACTAATGTCGATTCTTATGGTGGTGCTTCC  3' end
        GTTGGTGATTACAGCTAAGAATACCACCACGAAGG
```

Fig. 5. Engineering mutations with the No See'm approach: (A) The position of interest, a glutamate (CAG) to alanine (GTC) mutation at position 13354-56, is targeted and mapped to the MHV-E fragment. The vector sequence, the wild-type fragment, and the mutated fragment are analyzed to determine which type IIs restriction enzymes do not cut any of them, and one of these is selected as the restriction site to add. In this case BbsI is selected. (B) Primers are designed that add the mutation and the restriction site in proper orientation so that upon digestion the restriction site is eliminated and complementing sticky ends are produced. The BbsI cut sites are highlighted in gray. The mutated codon is bold and underlined. (C) PCR is conducted using vector specific

3. Select one of the noncutters to add onto the end of each fragment. In **Figs. 1C** and **5**, we selected a glutamate codon at position 13354-56 (CAG) and changed it to an alanine (GTC). To stabilize the mutation against reversion to the wt genotype, we generally alter the codon at multiple positions. This mutation maps to MHV-E

◄A   **Map mutation and select restriction site**   —

```
Wt        AACCACTAATCAGGATTCTT      BbsI      GAAGACNN
Target    AACCACTAATGTCGATTCTT                CTTCTGNNNNNN
```

```
MHV(+)13335-GCCGGAGGCAACCACTAATCAGGATTCTTATGGTGGTGCTTCC-13377
MHV(-)13335-CGGCCACCGTTGGTGATTAGTCCTAAGAATACCACCACGAAGG-13377
```

B   **Design primers to engineer mutation and restriction site**

```
MutS:  5-GAAGACGTCGATTCTTATGGTGGTGCTTCC-3
MutA:  5-GAAGACGAATCAGCATTAGTGGTTGCCTCCGGC-3
```

```
Wt(+)GCCGGAGGCAACCACTAATCAGGATTCTTATGGTGGTGCTTCC
     ||||||||||||||||||||   |||||
      3-CGGCCTCCGTTGGTGATTACAGCTAAGCAGAAG-5 MutA
```

```
         MutS 5-GAAGACGTCGATTCTTATGGTGGTGCTTCC-3
                     |   ||||||||||||||||||||||||
Wt(-) CGGCCACCGTTGGTGATTAGTCCTAAGAATACCACCACGAAGG
```

C   **PCR, digestion, and ligation**

Fig. 5. (Continued) primers along with the newly designed primers that incorporate the mutation of interest to produce two amplicons. These are cloned and purified, and then digested with BbsI and BsmBI to produce a mutated MHV-E fragment that can then be incorporated into the full-length clone.

as the fragment target, and enzyme *Bbs*I does not cut the sequence or the pCR-XL-Topo vector.

4. Design primers that incorporate the mutation into the target fragment utilizing the nonspecific site of the restriction cut site to engineer a junction oriented such that upon cleavage the restriction site is removed and complementing sticky ends remain (**Fig. 5**).

5. Amplify both fragments by PCR using the newly designed primers along with vector-specific sequencing primers and reaction conditions appropriate for the size of each target. For MHV-E fragment in pCR-XL-Topo M13F and M13R are used. Run PCR reactions on a gel, extract and purify amplicons of the correct size, clone into pCR-XL-Topo (or another vector of choice), transform, pick colonies and grow larger cultures, screen by restriction digestion, and sequence verify each amplicon (as described above).

6. Digest each amplicon using the appropriate reaction conditions and enzymes. In the case of MHV-E in **Fig. 5**, we amplified with vector primers M13R and M13F, which ensured that the *Bsm*BI sites were preserved at both ends of the fragment, so each amplicon has a *Bsm*BI site and a newly engineered *Bbs*I site. These amplicons are digested first with BbsI and NEB buffer 2 for 1.5 h at 37°C, followed by digestion with *Bsm*BI at 55°C for 2 h. In addition, we digested wild-type MHV-E with just *Bsm*BI, and purified the vector band (the smaller of the two). All of these digests are purified as described above.

7. With this strategy, both amplicons and the vector are ligated in a single ligation reaction:

   a. 11 μl of each purified band (amplicon 1, amplicon 2, vector).
   b. 4 μl ligation buffer.
   c. 3 μl ligase.
   d. Incubate at 4°C overnight.

8. Transform 5 μl of the ligation, plate, pick colonies and grow larger cultures, screen by restriction digestion, and sequence verify each amplicon.

9. Build MHV virus using the newly mutated cDNA.

4. Notes

1. Always do a chloroform-only extraction of the viral fragments after digestion as impurities left behind by the extraction kit can interfere with the DNA ligase.
2. Always use PBS prepared in DepC water for washing cells. This reduces the opportunity for RNAses in the medium to degrade the full-length viral transcript prior to electroporation.
3. Always use barrier tips for pipetting, particularly when working with wild-type MHV along side a mutant.
4. Run digestions of wild-type fragment and mutant fragment on different gels with new running buffer each time. Wild-type MHV is so efficient at replication that

even very low levels of cross-contamination of fragments will result in wild-type MHV out-competing any mutant.

5. For best results use BHK-MHVr cells at or below 80% confluency.
6. Clone stability is enhanced by growing everything at room temperature.
7. Avoid passage of highly attenuated mutant viruses because reversions to wild-type and second-site reversions occur rapidly upon passage.
8. There are several strengths and a few weaknesses to the No See'm infectious clone approach. Strengths over a single full-length clone include: (a) rapid mutagenesis of independent subclones in parallel; (b) mixing and matching of existing mutants from different subclones; (c) increased stability of smaller cDNAs; (d) safety of molecular clone of highly pathogenic viruses; (e) requirement for high individual expertise, which minimizes the chances of harmful use of the coronavirus molecular clone; (f) high recovery of recombinant viruses; (g) speed of recovery of recombinant virus; and (h) reduced sequencing costs (sequence subclone only). Weaknesses include: (a) technical expertise required for assembly; (b) the difficulty of *in vitro* transcription of full-length coronavirus RNAs; (c) troubleshooting the lengthy assembly process; and (d) obtaining reliable sequence information for the wild-type genome.

Well over a hundred mutants have been engineered into the clone of MHV using this approach, which far exceeds the number for all other coronavirus reverse genetic systems reported. Moreover, this technology has been used to rapidly develop reagents in response to newly identified emerging coronaviruses. For example, SARS-CoV was identified in the fall of 2002, and a full-length infectious clone was available the summer of 2003 *(16)*. Piecemeal assembly of smaller subclones is within range of commercial DNA synthesis companies, allowing for synthetic reconstruction of coronavirus genomes. However, attempts to reconstruct other coronavirus genomes using this method have proven more difficult. Current problem areas that must be solved for the advancement of this technology include: (a) unreliable sequence information in the public databases; (b) mistake-prone amplification and cloning of AT-rich genomes (such as in human coronavirus NL63); (c) clone stability and sequence toxicity issues with different viruses; (d) size issues of the different fragments for different viruses; and (e) cell tropism issues with newly identified animal viruses (Bat CoV).

Once an infectious clone is established for a virus, rapid production of candidate vaccine strains *(17– 19)* and vectors for therapeutic gene delivery in animals and humans *(20)* is possible. Attenuating mutations can be identified, combinations can be engineered into the infectious clone, and recombinant viruses can be generated and tested. In SARS-CoV, this approach has identified zoonotic candidate vaccine strains *(17,18)* and a novel strategy for preventing recombination between wild-type and recombinant viruses by rewiring the viral communication network *(19)*. This strategy provides a recombination safe vaccine platform that will likely redefine coronavirus vaccinology.

A tremendous amount of creativity in the coronavirus field has resulted in four independent and unique reverse genetic systems for a single virus family *(11,13,14,21,22)*, all of which have helped pave the way for the emergence of the golden age of coronavirus genetics. We anticipate the discovery and genetic analysis of a large number of unique genetic functions in the coronavirus genome, and expect that these discoveries will translate to a better understanding of coronavirus replication, transcription, assembly, and mechanisms of pathogenesis of these important human and animal pathogens.

Acknowledgments

We are grateful to Boyd Yount, who developed this protocol and has taught many of us how to use it.

References

1. Lai, M. M., and D. Cavanagh, (1997) The molecular biology of coronaviruses *Adv. Virus. Res.* **48**, 1–100.
2. Masters, P. (2006) The molecular biology of coronaviruses *Adv. Virus Res.* **66**, 193–292.
3. Bost, A. G., et al. (2000) Four proteins processed from the replicase gene polyprotein of mouse hepatitis virus colocalize in the cell periphery and adjacent to sites of virion assembly *J. Virol.* **74**(7), 3379–3387.
4. Brockway, S. M., et al. (2003) Characterization of the expression, intracellular localization, and replication complex association of the putative mouse hepatitis virus RNA-dependent RNA polymerase *J. Virol.* **77**(19), 10515–10527.
5. Denison, M. R., et al. (1998) Processing of the MHV-A59 gene 1 polyprotein by the 3C-like proteinase *Adv. Exp. Med. Biol.* **440**, 121–127.
6. Schiller, J. J., Kanjanahaluethai, A., and Baker, S. C. (1998) Processing of the coronavirus MHV-JHM polymerase polyprotein: identification of precursors and proteolytic products spanning 400 kilodaltons of ORF1a *Virology* **242**(2), 288–302.
7. Baric, R. S., Curtis, K. M., and Yount, B. (2001) MHV subgenomic negative strand function *Adv. Exp. Med. Biol.* **494**, 459–465.
8. Sawicki, S. G., and Sawicki, D. L. (1986) Coronavirus minus-strand RNA synthesis and effect of cycloheximide on coronavirus RNA synthesis *J. Virol..* **57**(1), 328–334.
9. Sawicki, S. G., and Sawicki, D. L. (1990) Coronavirus transcription: subgenomic mouse hepatitis virus replicative intermediates function in RNA synthesis *J. Virol.* **64**(3), 1050–1056.
10. Sethna, P. B., Hung, S. L., and Brian, D. A. (1989) Coronavirus subgenomic minus-strand RNAs and the potential for mRNA replicons *Proc. Natl. Acad. Sci. USA* **86**(14), 5626–5630.

11. Kuo, L., et al. (2000) Retargeting of coronavirus by substitution of the spike glycoprotein ectodomain: crossing the host cell species barrier *J. Virol.*. **74**(3), 1393–1406.

12. Schickli, J. H., et al. (1997). The murine coronavirus mouse hepatitis virus strain A59 from persistently infected murine cells exhibits an extended host range *J. Virol.* **71**(12), 9499–9507.

13. Yount, B., et al. (2002) Systematic assembly of a full-length infectious cDNA of mouse hepatitis virus strain A59 *J. Virol.* **76**(21), 11065–11078.

14. Yount, B., Curtis, K. M., and Baric, R. S. (2000) Strategy for systematic assembly of large RNA and DNA genomes: transmissible gastroenteritis virus model *J. Virol.* **74**(22), 10600–10611.

15. Sambrook, J., Fritsch, E. F., and Maniatis, T. (1989) *Molecular Cloning: A Laboratory Manual*. 2nd Ed. Cold Spring Harbor Laboratory, New York, Section 5.72.

16. Yount, B., et al. (2003) Reverse genetics with a full-length infectious cDNA of severe acute respiratory syndrome coronavirus *Proc. Natl. Acad. Sci. USA* **100**(22), 12995–13000.

17. Baric, R. S., et al. (2006) SARS coronavirus vaccine development. *Adv. Exp. Med. Biol.* **581**, 553–560.

18. Deming, D., et al. (2006) Vaccine efficacy in senescent mice challenged with recombinant SARS-CoV bearing epidemic and zoonotic spike variants. *PLoS Med.* **3**(12), e525.

19. Yount, B., et al. (2006) Rewiring the severe acute respiratory syndrome coronavirus (SARS-CoV) transcription circuit: engineering a recombination-resistant genome *Proc. Natl. Acad. Sci. USA* **103**(33), 12546–12551.

20. Yount, B., et al. (2005) Severe acute respiratory syndrome coronavirus group-specific open reading frames encode nonessential functions for replication in cell cultures and mice. *J. Virol.*, **79**(23), 14909–14922.

21. Almazan, F., et al. (2000) Engineering the largest RNA virus genome as an infectious bacterial artificial chromosome *Proc. Natl. Acad. Sci. USA.* **97**(10), 5516–5521.

22. Thiel, V., et al. (2001) Infectious RNA transcribed *in vitro* from a cDNA copy of the human coronavirus genome cloned in vaccinia virus *J. Gen. Virol.* **82**(Pt 6), 1273–1281.

VII

IDENTIFYING CORONAVIRUS RECEPTORS

22

Identification of Sugar Residues Involved in the Binding of TGEV to Porcine Brush Border Membranes

Christel Schwegmann-Wessels and Georg Herrler

Abstract

Coronaviruses most often infect the respiratory or intestinal tract. Transmissible gastroenteritis virus (TGEV), a group 1 coronavirus, infects the porcine small intestine. Piglets up to the age of 3 weeks die from diarrhea caused by the viral gastroenteritis unless they are protected by antibodies. In addition to the cellular receptor, porcine aminopeptidase N, the TGEV spike protein binds to sialic acid residues. We have shown that the sialic acid binding activity mediates the binding of TGEV to a mucin-like glycoprotein present in porcine brush border membranes. This was shown by performing a virus overlay binding assay with proteins obtained from brush border membranes by lectin precipitation. Because of the reactivity with specific lectins we assume that the recognized glycoprotein has the characteristics of a mucin.

Key words: transmissible gastroenteritis virus; sialic acid binding activity; brush border membranes; piglets; lectin precipitation; virus overlay binding assay

1. Introduction

Transmissible gastroenteritis virus (TGEV) infects susceptible cells by binding to the cellular receptor porcine aminopeptidase N (pAPN) *(1)*. It has been shown that the spike (S) protein of TGEV has, in addition to its pAPN binding site, a sialic acid binding activity, which is located at the N-terminal region of the protein *(2)*. We have shown that binding to sialoglycoproteins in cell culture and in porcine brush border membrane preparations may be a pathogenicity factor affecting the enteropathogenicity of TGEV *(3,4)*. Virus mutants that were not

From: *Methods in Molecular Biology, vol. 454: SARS- and Other Coronaviruses,*
Edited by: D. Cavanagh, DOI: 10.1007/978-1-59745-181-9_22, © Humana Press, New York, NY

able to bind to sialic acids did not infect piglets *(5,6)*. Thus, binding to sialogly-coproteins or mucins in the gastrointestinal tract may help the virus to reach the cellular receptor pAPN. The sialoglycoproteins serve as first attachment factors.

To further characterize the sugar moieties of the sialoglycoproteins that are bound by TGEV, we isolated brush border membranes from piglets and analyzed them by performing lectin precipitation. Lectin precipitates were further used for a virus overlay binding assay. The lectins used in this study were wheat germ agglutinin (WGA), which specifically binds N-acetylglucosamine and sialic acids; peanut agglutinin (PNA), which specifically recognizes galactose-β(1-3)-N-acetylgalactosamine, a disaccharide present in O-glycosylated proteins; and jacalin, which binds to the same sugars as PNA. The advantage of jacalin is that in contrast to PNA it binds to galactose-β(1-3)-N-acetylgalactosamine without prior neuraminidase treatment. In this way the specific lectin precipitation can be well combined with the virus binding to sialic acids. A high-molecular-mass protein that is recognized by the sialic acid binding activity of TGEV was also readily recognized by WGA and jacalin, suggesting that this sialoglycoprotein is highly O-glycosylated. Therefore we designated it mucin-like glycoprotein (MGP).

2. Materials

2.1. Brush Border Membrane Preparation

1. Preparation buffer A: 100 mM mannit, 10 mM HEPES, Tris pH 7.4/4 °C.
2. Preparation buffer B: 100 mM mannit, 10 mM HEPES, 0.1 mM $MgSO_4X$ 7 H_2O, Tris pH 7.4/4 °C.
3. Vesicle buffer: 100 mM mannit, 100 mM KCl, 10 mM HEPES, Tris pH 7.4/4 °C.

2.2. Virus Purification

1. Dulbecco's modified Eagle medium (DMEM) with 4500 mg/liter glucose (Gibco/BRL) supplemented with 10% fetal calf serum (Gibco/BRL).
2. PBS Dulbecco's without calcium and magnesium (Gibco/BRL).
3. Sucrose (Roth) freshly dissolved in PBS in concentrations of 20, 50, and 60% w/w.
4. Neuraminidase from *Vibrio cholerae* (1 U/ml, DADE Behring), store at 4 °C (*see* **Note 1**).

2.3. Lectin Precipitation

1. Neuraminidase from *Vibrio cholerae* (1 U/ml, DADE Behring). Store at 4 °C.
2. Sodium acetate buffer solution (50 mM), pH 5.5 with sodium chloride (154 mM) and calcium chloride (9 mM) (*see* **Note 2**). Store at 4 °C.

3. Protease inhibitors: leupeptin (Roche), Pefa Block®SC (Roche).
4. TBS pH 7.5: Tris 0.05 M, sodium chloride 0.15 M. Adjust pH with HCl.
5. NP40 lysis buffer: Sodium deoxycholate 0.5 %,TBS pH 7.5, Nonidet® P40 (Roche). Add 1 mM $MgCl_2$, $CaCl_2$, and $MnCl_2$.
6. Wheat germ agglutinin agarose (Sigma), peanut agglutinin agarose (Vector laboratories), jacalin agarose (Sigma). Store at 4 °C.
7. SDS sample buffer (2X): 100 mM Tris-HCl, pH 6.8, 4% (w/v) sodium dodecyl sulfate (SDS), 20% glycerol, 0.02% bromophenol blue. Store in aliquots at –20 °C.

2.4. SDS-Gel Electrophoresis (SDS-PAGE)

1. Separating buffer: 1.5 M Tris-HCl, pH 8.8. Store at 4 °C.
2. Stacking buffer: 1 M Tris-HCl, pH 6.8. Store at 4 °C.
3. 30% acrylamide/bis solution. Neurotoxic! Store at 4 °C.
4. 10% (w/v) SDS prepared in water. Store at room temperature.
5. 10% (w/v) ammonium persulfate (APS) prepared in water. Freeze in aliquots at –20 °C. Aliquot in use must be stored at 4 °C.
6. N,N,N′,N′-tetramethyl-ethylendiamine (TEMED, Bio-Rad). Toxic! Store at 4 °C (*see* **Note 3**).
7. 10X running buffer: 10 g SDS, 30 g Tris, 144 g glycin filled up to 1 liter with water. Store at room temperature (*see* **Note 4**).
8. Prestained molecular weight markers: Rainbow marker (Amersham) or PageRuler Prestained Protein Ladder (Fermentas). Store at –20 °C.

2.5. Western Blotting

1. Buffer anode I: 1 M Tris 300 ml, ethanol 200 ml, H_2O 500 ml. Adjust to pH 9 with HCl (*see* **Note 5**).
2. Buffer anode II: 1 M Tris 25 ml, ethanol 200 ml, H_2O 770 ml. Adjust to pH 7.4 with HCl.
3. Buffer cathode: aminocapronic acid 5.25 mg, 1 M Tris 25 ml, ethanol 200 ml, H_2O 770 ml. Adjust to pH 9 with HCl.
4. Nitrocellulose membrane (Schleicher & Schuell/Whatman).
5. Filter paper (Schleicher & Schuell/Whatman, 2043A, ref.no.: 10381185).
6. Blocking buffer: 0.5 % blocking reagent (Roche, article number: 1096176) in PBS-0.05% T. Stir at 80 °C for 30 min. After solution store at 4 °C.

2.6. Virus Overlay Binding Assay

1. Phospate buffered saline with/without Tween (PBS, PBS-0.1% T, PBS-0.05% T): Prepare a 10X stock of PBS with NaCl 80 g, KCl 2 g, Na_2HPO_4 11.5 g, KH_2PO_4 2 g. The pH has to be 7.5. Dilute 100 ml with 900 ml water for use. Add 0.1 % or 0.05 % Tween, depending on the buffer needed, and stir well (*see* **Note 6**).

2. Virus and antibody dilution buffer: PBS-0.05% T.
3. Purified virus (TGEV-NA).
4. First antibody: anti-TGEV-S protein monoclonal antibody 6A.C3 from L. Enjuanes *(7)*.
5. Second antibody: Anti-mouse Ig, biotinylated species-specific whole antibody from sheep (Amersham/GE Healthcare).
6. Streptavidin biotinylated horseradish peroxidase complex (Amersham/GE Healthcare).
7. Enhanced chemoluminescent substrates: BM Chemoluminescence Blotting Substrate (Roche) or Super Signal® (Pierce).

3. Methods

3.1. Brush Border Membrane Preparation

The brush border membrane preparation is performed according to the protocol of Schröder et al. *(8)*.

1. The mucosa of fresh jejunum from sacrificed suckling piglets is abraded from the serosa and frozen in liquid nitrogen. The samples are stored at $-80\,^{\circ}\text{C}$ until use.
2. All preparation steps are performed on ice at $4\,^{\circ}\text{C}$.
3. The mucosa is weighed and 150 ml of preparation buffer A is added.
4. After thawing on ice the mucosa is mixed in a regular household mixer on level 4 twice for 1 min each. After each mixing step, the suspension is filled in a tumbler and left on ice for 2 min. Then the foam is sucked off.
5. The volume of the homogenate is determined in a measuring cylinder.
6. The basolateral brush border membranes are precipitated by adding 1 M MgCl_2 in an amount of 1/100 of the homogenate volume. The MgCl_2 solution is added drop by drop under continuous stirring on ice followed by further stirring on ice for 15 min.
7. The complete homogenate is then centrifuged for 18 min, $3687 \times g$ at $4\,^{\circ}\text{C}$ (*see* **Note 7**). After centrifugation the supernatant is poured into fresh centrifuge beakers.
8. In a second centrifugation step for 40 min, $16,000 \times g$ at $4\,^{\circ}\text{C}$, the apical brush border membrane fraction is pelleted (*see* **Note 8**). The pellet is floated in 35 ml of preparation buffer B; i.e., the buffer is rinsed over the pellet until the sediment is detached from the bottom. The floating pieces of the pellet are then homogenized ten times using a potter.
9. To further purify the preparation, the MgCl_2 precipitation is repeated followed by a centrifugation for 10 min, $3293 \times g$ at $4\,^{\circ}\text{C}$ (*see* **Note 9**).
10. The supernatant is then centrifuged for 30 min, $26,430 \times g$ at $4\,^{\circ}\text{C}$ (*see* **Note 10**). The resulting pellet is resuspended in 35 ml vesicle buffer and homogenized ten times with the potter.

11. After a final centrifugation step for 40 min, 31,300 × g at 4 °C, the brush border membranes are homogenized in 1 to 3 ml vesicle buffer by using a 1-ml syringe and a 0.45 × 23-gauge needle (*see* **Note 11**).

12. The protein content of the brush border membrane preparation is estimated by using the Bradford assay (BIORAD). To make all protein binding sites available, surface-active saponine (1%) is added in a ratio of 1:2 to all samples. After incubation for 20 min at room temperature the Bradford staining reagent is added.

3.2. Virus Purification

1. ST (swine testicular) cell monolayers in 145-mm cell culture dishes are used for TGEV infection at a multiplicity of infection of 1. For infection, the cells are washed three times with sterile PBS (with Ca^{2+} and Mg^{2+}) and inoculated with a volume of 300 µl virus suspension per dish. With this small virus volume the plates are incubated at 37 °C for 1 h with careful shaking (*see* **Note 12**). Then the inoculum is removed and 20 ml DMEM is added to each plate followed by an incubation at 37 °C for about 24 h (*see* **Note 13**).

2. As soon as the CPE is observed, the cell culture supernatant is harvested and centrifuged at 1900 × g for 10 min, 4 °C, to remove cell debris. The resulting supernatant is subjected to ultracentrifugation at 120,000 × g for 1 h, 4 °C, to pellet the virus particles (*see* **Note 14**). The resulting pellets are resuspended in PBS (total volume 400 to 500 µl).

3. To make all sialic acid binding sites on the S protein available, a neuraminidase treatment of the concentrated virus particles is performed. Neuraminidase from *Vibrio cholerae* (1 U/ml, DADE Behring) is used in a final concentration of 50 mU/ml. The sample is then incubated at 37 °C for 30 min and cooled on ice.

4. For purification of the virions, a continuous sucrose gradient (20–50% w/w in PBS) with a 60% sucrose cushion is prepared. The sucrose solutions are stirred at 50°–60 °C and stored at room temperature until use. For the gradient centrifugation, SW 50.1 tubes, ultraclear (Beckman), are used (*see* **Note 15**). Per tube, 0.5 ml of the 60% sucrose cushion is applied overlaid by a gradient of 20 and 50% sucrose (using a gradient mixer and a pump). 2.5 ml of the 20% and 2 ml of the 50% solution are put into the gradient mixer. The gradient has a height corresponding to a 4-ml volume. The virus suspension is then carefully put on the top of the gradient using a 100-µl pipette. For purification, the gradient is ultracentrifuged at 150,000 × g for 3 h, 4 °C (*see* **Note 16**).

5. After centrifugation, the visible virus band is carefully harvested using a 1-ml syringe and a 0.9 × 40-gauge needle (*see* **Note 17**). After dilution with PBS to a total volume of 4.5 ml, the virions are pelleted by ultracentrifugation (SW 50.1) at 150,000 × g for 1 h, 4 °C (*see* **Note 16**). The viral particles are resuspended in a small volume of PBS (depending on the pellet size in 100 to 300 µl). The sample is stored in aliquots at –20 °C. The purity of the preparation is tested in an SDS-PAGE with Coomassie staining of the proteins in the gel. The protein concentration is determined photometrically at 280 nm (*see* **Note 18**).

3.3. Lectin Precipitation

1. Brush border membrane vesicles (1 mg/ml) are incubated at a ratio of 1:2 with neuraminidase from *Vibrio cholerae* (1 U/ml, DADE Behring) for 1 h at 37 °C (*see* **Note 1**). Control samples are treated with sodium acetate buffer in the same manner (in all samples: addition of protease inhibitors: leupeptin and pefa-block).
2. Each sample is diluted in 1 ml NP40 lysis buffer containing 1 mM MgCl$_2$, 1 mM CaCl$_2$, and 1 mM MnCl$_2$ (without protease inhibitors). For membrane lysis, the samples are incubated on ice for 15 min. Membrane debris is centrifuged at 16,000 × g, 30 min, 4 °C (*see* **Note 19**).
3. 330 μl of each supernatant is put on 50 μl lectin agarose (wheat germ agglutinin, jacalin, peanut agglutinin) and the tubes are shaken overnight at 4 °C.
4. The next day the agarose samples are washed three times with NP40 lysis buffer (after centrifugation at 16,000 × g, 3 min, 4 °C). Bound glycoproteins are eluted with 50 μl twofold SDS sample buffer for 10 min at 96 °C. After a final centrifugation step at 16,000 × g, 5 min, 4 °C, the supernatants containing the lectin-specific glycoproteins are used for SDS-gel electrophoresis.

3.4. SDS-Gel Electrophoresis (SDS-PAGE)

1. These instructions are intended for the use of minigels (70 × 80 × 0.75 mm). Glass plates are cleaned with a rinsable detergent and rinsed with distilled water. After drying they are rinsed with 70% ethanol.
2. Prepare an 8% gel, 0.75 mm thick by mixing 4.6 ml distilled water with 2.5 ml Tris-HCl (1.5 M, pH 8.8), 2.7 ml acrylamide/bis solution (30%), 100 μl SDS (10% in water), 100 μl ammonium persulfate solution (10% in water), and 10 μl TEMED. Pour the gel (~4 ml), leaving about 1 cm of space for the stacking gel and overlay with isopropanol. The gel should polymerize within 30 min.
3. Pour off the isopropanol. Prepare the stacking gel by mixing 3.4 ml distilled water with 0.63 ml Tris-HCl (1 M, pH 6.8), 0.83 ml acrylamide/bis solution (30%), 50 μl SDS (10% in water), 50 μl ammonium persulfate solution (10% in water), and 10 μl TEMED. Pour about 0.5–1 ml of this mixture over the polymerized separating gel and insert the comb. The stacking gel should polymerize within 30 min.
4. Prepare the running buffer by diluting 100 ml of the 10X running buffer with 900 ml of water. Use a measuring cylinder. Invert the bottle to mix.
5. Once the stacking gel has solidified, assemble the gel running unit. Carefully remove the comb and wash the wells with running buffer by using, e.g., a Hamilton Microlitre syringe.
6. Add the running buffer to the lower and upper chambers of the gel unit and load 20 μl of each sample in a well. Load one well with 2.5–5 μl prestained molecular weight marker filled up to 20 μl with 2X SDS sample buffer.
7. Connect the gel unit to a power supply. Run the gel at 80 V until the dye front of the sample buffer is a sharp line and leaves the stacking gel. It is then possible to run the gel at 150 V until the dye front runs off the gel. The whole running time of the gel is about 1.5 h.

3.5. Western Blotting

1. The samples separated by SDS-PAGE are transferred to a nitrocellulose membrane electrophoretically. For this purpose a semidry blotting system with two graphite electrodes is used. Six sheets of filter paper have to be soaked in anode I buffer, three filters in anode II buffer, and nine sheets of paper have to be soaked in the cathode buffer. The nitrocellulose membrane is wetted with pure water before use. Prior to use, all sheets and the membrane are cut to a size of 6 × 8 cm (just a bit larger than the size of the separating gel).
2. The power supply is switched off and the gel unit is disassembled. The stacking gel is cut and discarded. One edge of the separating gel is cut to have a better orientation for blotting.
3. The anode I papers are taken with a pinzette and excess buffer and air bubbles are eliminated from the papers by stripping with clean gloves. The papers are then placed onto the anode and the anode II papers placed onto the anode I papers like a sandwich.
4. The nitrocellulose membrane is carefully taken out of the pure water and put on the papers. A little water is put on the upper side of the membrane, the better to get the gel from the glass slides by creating capillary forces. The free side of the gel is laid on the membrane, and the remaining glass slide on the other side is lifted carefully. A spacer is used to loosen the gel from the slide. After removing possible air bubbles, the cathode papers are put on the gel as described above for the anode papers.
5. With a glass pipette residual air bubbles are stripped out of the sandwich and the cathode is put on the top. About 2 kg are put on the blotting apparatus (e.g., two bottles with 1 liter each) and the apparatus is connected to the power supply. The running conditions are 0.8 mA/cm^2. For one membrane with 6 × 8 cm use 38 mA and let the apparatus run for 1 h.
6. Once the transfer is completed, switch off the power supply and carefully lift the upper electrode. Normally some of the papers stick to this upper side. Search for the position of the gel and control if the prestained marker is visible on the nitrocellulose membrane. Discard the gel and mark the bands of the marker with a ballpoint pen because they sometimes get lost during later incubation steps.
7. The nitrocellulose is than incubated in 20 ml blocking buffer overnight at 4 °C on a rocking platform (*see* **Note 20**).

3.6. Virus Overlay Binding Assay

1. For the virus overlay binding assay the blocking buffer is discarded and after a short rinse with PBS-0.1% T, the membrane is washed with PBS-0.1% T three times for 10 min each on a rocking platform at room temperature (*see* **Note 21**).
2. The membrane is incubated for 1 h at 4 °C with purified TGEV (around 8 µg protein/blot) diluted in 500 µl of PBS-0.05% T covered by Parafilm (*see* **Note 22**).

3. After another washing period (three times for 10 min each, at room temperature with PBS-0.1% T) the first antibody (mab antiS 6A.C3 from L. Enjuanes) is diluted 1:200 in PBS-0.05% T. Again 500 μl of the dilution are dropped onto the membrane covered by Parafilm (*see* **Note 23**). Incubation was at 4 °C for 1 h.
4. The membrane is then again washed three times for 10 min each at room temperature with PBS-0.1% T.
5. The secondary antibody (anti-mouse Ig, biotinylated species-specific whole antibody from sheep, Amersham/GE Healthcare) is prepared 1:1000 in 10 ml PBS-0.05% T and the membrane is incubated in this solution for 1 h at 4 °C under shaking.

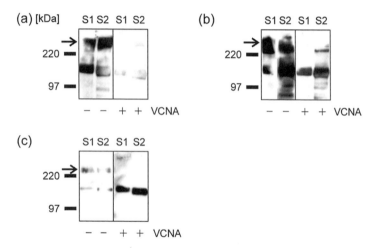

Fig. 1. Binding of TGEV to lectin-precipitated brush border membrane proteins from two different suckling piglets (S1, S2). Brush border membranes were treated either with neuraminidase from *Vibrio cholerae* (+VCNA) or mock-treated (–VCNA) and then precipitated with: (a) wheat germ agglutinin (WGA), (b) jacalin, or (c) peanut agglutinin (PNA) agarose. A virus overlay binding assay was performed after blotting of the brush border membrane proteins. Two main bands were recognized by the virus after lectin precipitation. The lower band of about 150 kDa presents the cellular receptor porcine aminopeptidase N (pAPN). The PNA-precipitation (c, right side) shows clearly that binding to pAPN is not sialic-acid-dependent. TGEV binding to the high-molecular-mass band (arrow) is clearly sialic-acid-dependent, as after VCNA treatment binding is eliminated or at least strongly reduced. The strong precipitation of MGP by WGA gives a hint that it is highly sialylated or has a strong content of N-acetylglucosamine. Jacalin binds to galactose-β(1-3)-N-acetylgalactosamine, a disaccharide present in O-glycosylated proteins. The strong MGP band (b) demonstrates that this protein is highly O-glycosylated. Therefore we designated it mucin-like glycoprotein (MGP). For the same reason it is precipitated by PNA (c), but PNA precipitation is not as good without neuraminidase treatment. After neuraminidase treatment (+VCNA) the virus is not able to bind to MGP because of the missing sialic acids.

6. After the next washing period (three times for 10 min each, at room temperature with PBS-0.1% T), the membrane is incubated with streptavidin biotinylated horseradish peroxidase complex (Amersham/GE Healthcare) diluted 1:5000 in 10 ml PBS-0.05% T for 1 h at 4 °C on a rocking platform.

7. The peroxidase complex dilution is discarded followed by three washings for 10 min each with PBS-0.1% T at room temperature.

8. The bound peroxidase is detected by chemoluminescence using the BM Chemoluminescence Blotting Substrate (Roche) or Super Signal® (Pierce) for the detection of weaker signals (*see* **Note 24**). The chemoluminescence signal can be visualized with an X-ray film or by using the Gel Documentation System ChemiDoc from BioRad (**Fig. 1**).

4. Notes

1. Neuraminidase from *Vibrio cholerae* is no longer available from DADE Behring. Alternatively, one can use, e.g., type V neuraminidase from *Clostridium perfringens* (Sigma).

2. The pH of sodium acetate buffer solution has to be adjusted with acetic acid.

3. A small aliquot of the TEMED bottle is stored at 4 °C for current use. In this way the original bottle is not opened too frequently, retaining the quality of the TEMED.

4. Never adjust the pH of the running buffer, e.g. with HCl. This will disturb the running conditions during electrophoresis. The buffer has to be used without pH adjustment.

5. It is better for the environment to use ethanol for buffers instead of methanol, which has to be discarded separately under specific conditions.

6. The pH of the PBS solution has to be at 7.5 without any adjustment. It just has to be checked.

7. For example, $3687 \times g$ corresponds to 6000 rpm in a GSA rotor.

8. $16,000 \times g$ corresponds to 12,500 rpm in a GSA rotor.

9. $3293 \times g$ corresponds to 6000 rpm in an SS34 rotor.

10. $26,430 \times g$ corresponds to 17,000 rpm in an SS34 rotor.

11. $31,300 \times g$ corresponds to 18,500 rpm in an SS34 rotor. For the homogenization of the pellet, first carefully rinse the buffer over the pellet to loosen parts of it. Homogenization has to be performed on ice because it takes some time until the whole pellet is suspended.

12. A small virus volume is important to get an efficient infection. In this way the particles have the best contact with the cell surface.

13. For virus production at least five or, better, ten dishes with a diameter of 145 mm (Greiner) have to be used.

14. $120,000 \times g$ corresponds to 28,000 rpm in an SW32 Beckman rotor. The centrifuge tubes have to be filled with 35 ml supernatant each. Do not use a smaller volume because then the tubes will be damaged by the centrifugation forces. Use ultraclear SW28 centrifuge tubes to see the white virus pellets better. Yellow pellets could be a hint that cellular proteins are still present in the supernatant.

For this reason the first centrifugation step for getting rid of the cell debris is very important.

15. It is important to use ultraclear tubes. In this way the "white" band formed by purified virus particles can be seen at the respective density after centrifugation.

16. 150,000 × *g* corresponds to 35,000 rpm in an SW55 Beckman rotor.

17. The "white" virus particle band can best be seen with the help of scattering light produced by an external light source. The needle has to be carefully directed vertically into the tube from the top up to the height of the band. The harvested volume should not exceed 1 ml per tube.

18. Different dilutions of BSA or transferrin are used as a standard. For the first measurement, the virus preparation is diluted 1:100. Sometimes higher concentrations have to be measured to get an extinction reading in a representative range. With the standard staining methods that are used by most of the assays for the determination of protein concentration (e.g., BCA assay) it is not possible to detect the TGEV proteins, so the photometric measurement at 280 nm is used.

19. 16,000 × *g* corresponds to 14,000 rpm in an Eppendorf microcentrifuge.

20. For blocking, different buffers as well as nonfat dry milk were tried. A lot of them, especially the nonfat dry milk, lead to high backgrounds in the virus overlay binding assay. With blocking reagent from Roche, high backgrounds owing to the wrong blocking substance can be avoided.

21. Vigorous washing of the membrane is necessary as in a virus overlay binding assay high backgrounds could be a problem.

22. For reduction of a possible background, virus and antibodies are diluted in PBS-0.05% T instead of just PBS.

23. By using Parafilm one needs just 500 µl virus or antibody dilution for one membrane. In this way the virus preparation lasts for a longer time.

24. Usually Super Signal® (Pierce) is used, as the signals after the virus overlay assay may be quite weak. If the signals are stronger, it is better to use BM Chemoluminescence Blotting Substrate (Roche) as the background is lower with this substrate.

Acknowledgments

The authors would like to thank Professor Gerhard Breves and Professor Bernd Schröder for the brush border membrane preparation protocol and Marion Burmester for technical assistance. This work was supported by Deutsche Forschungsgemeinschaft (SFB 280).

References

1. Delmas, B., Gelfi, J., L' Haridon, R., Vogel, L. K., Sjostrom, H., Noren, O., and Laude, H. (1992) *Nature* **357**, 417–420.
2. Schultze, B., Krempl, C., Ballesteros, M. L., Shaw, L., Schauer, R., Enjuanes, L., and Herrler, G. (1996) *J. Virol.* **70**, 5634–5637.

3. Schwegmann-Wessels, C., Zimmer, G., Schroder, B., Breves, G., and Herrler, G. (2003) *J. Virol.* **77**, 11846–11848.
4. Schwegmann-Wessels, C., Zimmer, G., Laude, H., Enjuanes, L., and Herrler, G. (2002) *J. Virol.* **76**, 6037–6043.
5. Bernard, S., and Laude, H. (1995) *J. Gen. Virol.* **76**(Pt 9), 2235–2241.
6. Krempl, C., Schultze, B., Laude, H., and Herrler, G. (1997) *J. Virol.* **71**, 3285–3287.
7. Gebauer, F., Posthumus, W. P., Correa, I., Sune, C., Smerdou, C., Sanchez, C. M., Lenstra, J. A., Meloen, R. H., and Enjuanes, L. (1991) *Virology* **183**, 225–238.
8. Schröder, B., Hattenhauer, O., and Breves, G. (1998) *Endocrinology* **139**, 1500–1507.

23

Pseudotyped Vesicular Stomatitis Virus for Analysis of Virus Entry Mediated by SARS Coronavirus Spike Proteins

Shuetsu Fukushi, Rie Watanabe, and Fumihiro Taguchi

Abstract

Severe acute respiratory syndrome (SARS) coronavirus (CoV) contains a spike (S) protein that binds to a receptor molecule (angiotensin-converting enzyme 2; ACE2), induces membrane fusion, and serves as a neutralizing epitope. To study the functions of the S protein, we describe here the generation of SARS-CoV S protein-bearing vesicular stomatitis virus (VSV) pseudotype using a VSVΔG*/GFP system in which the G gene is replaced by the green fluorescent protein (GFP) gene (VSV-SARS-CoV-St19/GFP). Partial deletion of the cytoplasmic domain of SARS-CoV S protein (SARS-CoV-St19) allowed efficient incorporation into the VSV particle that enabled the generation of a high titer of pseudotype virus. Neutralization assay with anti-SARS-CoV antibody revealed that VSV-SARS-St19/GFP pseudotype infection is mediated by SARS-CoV S protein. The VSVΔG*/SEAP system, which secretes alkaline phosphatase instead of GFP, was also generated as a VSV pseudotype having SARS-CoV S protein (VSV-SARS-CoV-St19/SEAP). This system enabled high-throughput analysis of SARS-CoV S protein-mediated cell entry by measuring alkaline phosphatase activity. Thus, VSV pseudotyped with SARS-CoV S protein is useful for developing a rapid detection system for neutralizing antibody specific for SARS-CoV infection as well as studying the S-mediated cell entry of SARS-CoV.

Key words: SARS-CoV; coronavirus; pseudotype; vesicular stomatitis virus; spike protein; attachment; fusion; cell entry.

From: *Methods in Molecular Biology, vol. 454: SARS- and Other Coronaviruses,*
Edited by: D. Cavanagh, DOI: 10.1007/978-1-59745-181-9_23, © Humana Press, New York, NY

1. Introduction

Entry of SARS coronavirus (SARS-CoV) into target cells is mediated by binding of the viral spike (S) protein to the receptor molecules, angiotensin-converting-enzyme 2 (ACE2) *(1)*. Studies on SARS using infectious SARS-CoV require care because of the highly pathogenic nature of this virus, so an alternative methodology is needed. Recently, pseudotyped retrovirus particles bearing SARS-CoV S protein have been generated by several laboratories *(2–4)*. These pseudotyped viruses have been shown to have a cell tropism identical to authentic SARS-CoV and their infectivity is dependent on ACE2, indicating that the infection is mediated solely by SARS-CoV S protein. Pseudotyped viruses have proved to be a safe viral entry model because of an inability to produce infectious progeny virus. A quantitative assay of pseudovirus infection could facilitate the research on SARS-CoV entry, cell tropism, and neutralization antibody.

Another pseudotyping system with a vesicular stomatitis virus (VSV) particle was previously reported to produce pseudotypes of envelope glycoprotein of several RNA viruses (i.e., measles virus, hantavirus, Ebola virus, and hepatitis C virus) *(5–8)*. This system (VSVΔG*/GFP system) may be useful for research on envelope glycoprotein owing to its ability to grow high titers in a variety of cell lines. The pseudotype virus titer obtained from the VSVΔG*/GFP system ($>10^5$ infectious units (IU)/ml) is generally higher than that of the pseudotyped retrovirus system *(6)*. Furthermore, infection of pseudotyped VSV in target cells can be detected as GFP-positive cells within 16 h postinfection (hpi) because of the powerful GFP-expression in the VSVΔG*/GFP system *(6)*. In contrast, the time required for the pseudotyped retrovirus system to detect infection is 48 hpi *(9,10)*, which is similar to that for the SARS-CoV to replicate to the level of producing plaques or cytopathic effects on infected cells. Thus, pseudotyping of SARS-CoV S protein using the VSVΔG*/GFP system may have greater advantages than retrovirus pseudotypes for studying the function of SARS-CoV S protein as well as for developing a rapid system for detection of neutralizing antibody specific for SARS-CoV.

Here we describe protocols for introducing SARS-CoV-S protein into VSV particles using the VSVΔG*/GFP system. The infection of VSV pseudotype bearing SARS-CoV S protein (VSV-SARS-St19/GFP) was easily detected in target cells as an expression of the GFP protein. The following methods were originally designed to produce VSV-SARS-St19/GFP and to measure the infection efficiency. In addition to a significant advantage of the VSV-SARS-St19/GFP for safe and rapid analyses of infection, the VSVΔG*/SEAP system, in which the G gene is replaced with the secreted alkaline phosphatase (SEAP) gene, may be superior to high-throughput quantitative analysis of S-mediated cell entry. The protocol for analyzing pseudotypes using the VSVΔG*/SEAP system is also described briefly.

2. Materials

1. Cell lines: The human embryonic kidney 293T (ATCC CRL-11268) is used to produce VSV pseudotypes bearing SARS-CoV-S proteins. The Vero E6 (ATCC Vero clone CRL 1586) is used for target cells of VSV pseudotype infection.
2. Dulbecco's Modified Eagle's Medium with high glucose concentration supplemented with 5% fetal calf serum (DMEM-5%FCS) is used for growing the 293T and Vero E6 cells.
3. Phosphate-buffered saline (PBS): 0.14 M NaCl, 2 mM KCl, 3 mM Na_2HPO4, 1.5 mM KH_2PO4, pH7.2. Autoclave to sterilize it.
4. Polyfect transfection reagent (QIAGEN, Hilden, Germany)
5. Mammalian expression plasmid encoding SARS-CoV-S protein with 19 amino acid truncation in the C terminus (*see* **Note 1**).
6. The VSVΔG*/GFP-G or VSVΔG*/SEAP-G (*see* **Note 2**).
7. 0.22-μm pore size sterile filter.
8. Digital fluorescence microscope for detecting GFP expression.
9. Reporter Assay Kit-SEAP (Toyobo, Osaka, Japan) and luminescence microplate reader for SEAP activity analyses.

3. Methods

A schematic description of the production of VSV pseudotype is shown in **Fig. 1**.

3.1. Expression of SARS-CoV S Protein and Production of VSV Pseudotype

1. Plate 293T cells onto a type I collagen-coated T-75 flask in DMEM-5%FCS at a 20–30 % confluent (generally 2.5×10^6 cells per T-75 flask).
2. After 3 h, transfect the 293T cells with 18.5 μg of expression plasmid encoding SARS-CoV-S protein (pKS/SARS-St19) using Polyfect transfection reagent according to the procedure recommended by the manufacturer. Incubate the cells in a CO_2 incubator for 48 h in DMEM-5% FCS. During this incubation, 293T cells grow to reach confluence.
3. Remove the supernatant from the cells. Wash the cells with PBS and add 10 ml of fresh DMEM-5%FBS. Inoculate the 1×10^6 IU of VSVΔG*/GFP-G or VSVΔG*/SEAP-G to produce pseudotypes to express GFP or SEAP, respectively. Incubate the cells in CO_2 incubator for absorption.
4. After 1 h, wash the cells with PBS three times. Then add the 15 ml of fresh DMEM-5%FCS.
5. After 24 h, collect the culture supernatants that contain VSV pseudotype-bearing SARS-CoV S protein (VSV-SARS-St19/GFP or VSV-SARS-St19/SEAP).
6. Centrifuge the supernatants at $3000 \times$ g for 5 min to remove cell debris, and then filter through a 0.22-μm pore size filter. Store at –80°C.

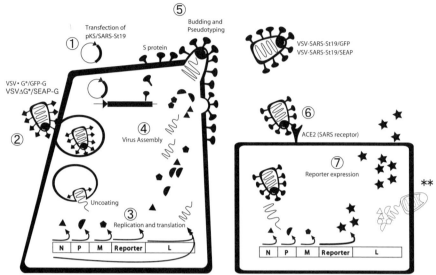

Fukushi et al. Pseudotyped vesicular stomatitis virus for analysis of virus entry mediated by SARS coronavirus spike proteins

Fig. 1 Schematic illustration of the production and infection of VSV pseudotype bearing SARS-CoV S protein: (1) Transfection of pKS/SARS-St19. It provides the S protein of SARS-CoV and cells express the S protein on the cell surface. (2) Infection of VSVΔG*/GFP-G or VSVΔG*/SEAP-G. These viruses possess the genome containing the reporter gene instead of the VSV-G gene. (3) Virus replication and translation. All viral components except the G protein will be supplied by these viruses. (4) Virus assembly, budding, and pseudotyping. Translated viral proteins are assembled and viral particles bud from the plasma membrane. Since the S protein, which was provided by the expression plasmid, is expressed on the cell surface, a virus can incorporate it into the virus particle. (5) VSV-SARS-St19/GFP and VSV-SARS-St19/SEAP. Pseudotyped VSV possessing SARS-S protein is released into culture supernatant. (6) Infection of the pseudotyped VSV. These viral particles can infect cells expressing the receptor for SARS-CoV, ACE2. (7) Estimation of the infectivity of the viruses. The infection of the pseudotyped viruses is estimated by the expression level of the reporter protein. **Progeny viruses are produced by the infection of VSV-SARS-St19/GFP or VSV-SARS-St19/SEAP. However, these viruses do not have infectivity because they have no glycoprotein (virus with the thin lines).

3.2. Determining Infectivity of VSV-SARS-St19/GFP

The infectivity of VSV-SARS-St19/GFP, harboring the VSVΔG*/GFP genome, can be determined as the number of GFP-positive cells.

1. Mix serially diluted VSV-SARS-St19/GFP with DMEM-5%FCS and inoculate the mixture into Vero E6 cells seeded in 96-well culture plates.

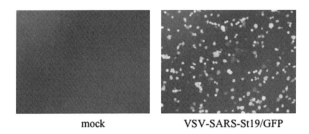

mock   **VSV-SARS-St19/GFP**

Fig. 2. Detection of VSV-SARS-St19/GFP infection. The VSV-SARS-St19/GFP was inoculated to Vero E6 cells. GFP expression was examined under a fluorescent microscope.

2. Incubate the cells in a CO_2 incubator for 7 h. Longer incubation (overnight to 24 h) may be beneficial for clearer fluorescent intensities on GFP-positive cells.
3. Detect GFP-positive cells using fluorescent microscopy (**Fig. 2**).
4. Determine the IU of the pseudotype. The IU is defined as a virus titer endpoint determined by limiting dilution.

3.3. Neutralization Assay of VSV-SARS-St19/GFP Infection Using Antibodies Specific to SARS-CoV

1. Mix serum samples serially diluted with DMEM-5%FCS (50 μl by volume) with the same volume of DMEM-5%FCS containing 3000 IU of VSV-SARS-St19/GFP.
2. Incubate for 1 h at 37°C for neutralization.
3. Inoculate the mixture to Vero E6 cells seeded on 96-well culture plates.
4. Follow steps 2 and 3 of Section 3.2.
5. Take photographs of cells expressing GFP under fluorescent microscope.
6. Count the GFP-positive cells on the photographs using the ImageJ software (http://rsb.info.nih.gov/ij/). The results of the neutralization assay using rabbit anti-SARS-CoV antibodies are shown in **Fig. 3**.

4. Analyzing Infectivity of VSV Pseudotype as a SEAP Expression

The infectivity of VSV-SARS-St19/SEAP, for which VSVΔG*/SEAP-G is used to produce pseudotype bearing S protein, can be determined as SEAP activities in the culture supernatants.

1. Mix serially diluted VSV-SARS-St19/SEAP with DMEM-5%FCS and inoculate the mixture into Vero E6 cells seeded on 96-well culture plates (*see* **Note 3**).
2. Incubate the cells in a CO_2 incubator for 1 h.
3. Remove the inoculum, wash the cells with PBS three times, and then add 100 μl of fresh DMEM-5%FCS (*see* **Note 4**).

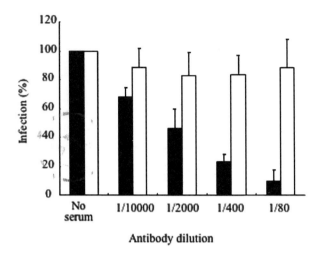

Fig. 3. Neutralization of infection of VSV-SARS-St19/GFP. The VSV-SARS-St19/GFP mixed with serially diluted rabbit anti-SARS-CoV was inoculated into Vero E6 cells. Rabbit anti-SARS-CoV-N peptide was used as negative control serum. The number of GFP-positive cells in the absence of serum was set as 100%.

4. Incubate the cells in a CO_2 incubator overnight.
5. Determination of SEAP activity is performed by a specific SEAP assay kit. Several kits from different manufacturers are now available, and since manufacturers' protocol details vary, only a general description follows based on the Toyobo kit we have used.

The technique uses 96-well plates. Twenty µl of supernatants from cell cultures are added to the wells with an equal volume of endogenous alkaline phosphatase inhibitor. Incubate the mixture at 37°C for 30 min. Next, add 160 µl of chemiluminescent substrate to the mixture. An incubation of 30 min at 37°C allows the chemiluminescence reaction. Use a luminescence microplate reader to detect the chemiluminescence signal.

5. Notes

1. To generate high virus titer of VSV pseudotype, prepare the plasmid encoding a C-terminal truncated version of the S protein, because it has been shown that truncation of C-terminal 19 amino acids leads to efficient incorporation of S protein into the VSV particles, and then the VSV pseudotype shows an efficient infection to target cells *(11)*.
2. The VSVΔG*/GFP-G is a VSV-G protein-bearing VSV pseudotype in which VSV-G gene is replaced with GFP gene. The VSVΔG*/SEAP-G is same as VSVΔG*/GFP-G except for having SEAP gene instead of the GFP gene.

Pseudotypes having VSV-G protein are used as "seed" viruses for generating S protein bearing VSV pseudotypes. Since the VSVΔG*/GFP or VSVΔG*/SEAP system was developed by Prof. M. A. Whitt (University of Tennessee Health Science Center, TN), ask him for sharing and using the system when starting experiments.

3. As the SEAP gene is obtained by modification of human placental alkaline phosphatase gene, some cell lines derived from placenta that express alkaline phosphatase similar to SEAP should not be used as target cells. For the analysis of VSV-SARS-St19/SEAP infection, we suggest using Vero E6 cells that show a very low level of alkaline phosphatase activity in the culture supernatant.

4. The medium containing VSV-SARS-St19/SEAP may have strong SEAP activity since it is derived from culture medium of 293T cells inoculated VSVΔG*/SEAP-G (see steps 3–6 in Section 3.1). In order to remove carryover SEAP activities derived from VSVΔG*/SEAP-G the Vero E6 cells have to be washed with PBS at least three times.

Acknowledgments

We thank Dr. M. A. Whitt for providing the VSVΔG*/GFP-G or VSVΔG*/SEAP-G. This work was supported in part by a grant from the Ministry of Health, Labor, and Welfare of Japan.

References

1. Li, W, Moore, M. J., Vasilieva, N., et al. (2003) Angiotensin-converting enzyme 2 is a functional receptor for the SARS coronavirus. *Nature* **426**, 450–454.
2. Hofmann, H., Geier, M., Marzi, A., et al. (2004) Susceptibility to SARS coronavirus S protein-driven infection correlates with expression of angiotensin converting enzyme 2 and infection can be blocked by soluble receptor. *Biochem. Biophys. Res. Commun.* **319**, 1216–1221.
3. Nie, Y., Wang, P., Shi, X., et al. (2004) Highly infectious SARS-CoV pseudotyped virus reveals the cell tropism and its correlation with receptor expression. *Biochem. Biophys. Res. Commun.* **321**, 994–1000.
4. Simmons, G., Reeves, J.D., Rennekamp, A.J., Amberg, S.M., Piefer, A.J., Bates, P. (2004) Characterization of severe acute respiratory syndrome-associated coronavirus (SARS-CoV) spike glycoprotein-mediated viral entry. *Proc. Natl. Acad. Sci. USA* **101**, 4240–4245.
5. Matsuura, Y., Tani, H., Suzuki, K., et al. (2001) Characterization of pseudotype VSV possessing HCV envelope proteins. *Virology* **286**, 263–275.
6. Ogino, M., Ebihara, H., Lee, B. H., et al. (2003) Use of vesicular stomatitis virus pseudotypes bearing Hantaan or Seoul virus envelope proteins in a rapid and safe neutralization test. *Clin. Diagn. Lab. Immunol.* **10**, 154–160.
7. Takada, A., Robison, C., Goto, H., et al. (1997) A system for functional analysis of Ebola virus glycoprotein. *Proc. Natl. Acad. Sci. USA* **94**, 14764–14769.

8. Tatsuo, H., Okuma, K., Tanaka, K., et al. (2000) Virus entry is a major determinant of cell tropism of Edmonston and wild-type strains of measles virus as revealed by vesicular stomatitis virus pseudotypes bearing their envelope proteins. *J. Virol.* **74**, 4139–4145.

9. Moore, M. J., Dorfman, T., Li, W., et al. (2004) Retroviruses pseudotyped with the severe acute respiratory syndrome coronavirus spike protein efficiently infect cells expressing angiotensin-converting enzyme 2. *J. Virol.* **78**, 10628–10635.

10. Nie, Y., Wang, G., Shi, X., et al. (2004) Neutralizing antibodies in patients with severe acute respiratory syndrome-associated coronavirus infection. *J. Infect. Dis.* **190**, 1119–1126.

11. Fukushi, S., Mizutani, T., Saijo, M., et al. (2005) Vesicular stomatitis virus pseudotyped with severe acute respiratory syndrome coronavirus spike protein. *J. Gen. Virol.* **86**, 2269–2274.

Index

Printed in the United States of America